A Textbook on
GEONOMY

J. A. Jacobs

Killam Memorial Professor of Science
and Director of
Institute of Earth and Planetary Physics
University of Alberta, Canada

ADAM HILGER
LONDON

ISBN 0 85274 212 6 (Library)
ISBN 0 85274 289 4 (Paperback)

Published by Adam Hilger Ltd
Rank Precision Industries
29 King Street, London WC2E 8JH
Printed in Great Britain by
William Clowes & Sons, Limited, London, Beccles and Colchester

Contents

Preface

The past decade has witnessed a spectacular increase in our knowledge of the earth, the solar system (particularly the moon, Mars and Venus), and of the outermost reaches of space itself. This book is concerned primarily with the physics (and chemistry) of the earth. This subject cannot be divorced from the larger question of the origin of the earth and the solar system. Thus the first two chapters are devoted to these broader issues. The manuscript was essentially completed by December 1972 and tries to give an account of our knowledge of the earth up to that time. The subject is developing extremely rapidly, however, and inevitably parts of the text will be out of date by the time the book is published. To keep the length of the book within bounds, some selection of subject matter was necessary—if certain aspects have been discussed in more detail than others, it probably reflects the personal interest of the author rather than its own importance to the general field. It is hoped, however, that a reasonably balanced account has been given. Again because of the vast literature on the subject, many references have had to be omitted—it is hoped none of the more important papers have been overlooked.

I gave much thought to the title of this book. A number of geophysicists have proposed the name 'geonomy' for the branch of planetary science that treats geological and geophysical studies of the solid body of the earth. The book is to some extent broader in scope than the adoption of the proposal might suggest, for it deals in part with broader aspects of planetary science. On the other hand, it is to some extent narrower, for it deals but briefly with the geology of the crust. Nevertheless, I have written the book in the spirit of what its advocates intend the word 'geonomy' to signify, and so have chosen the word in the hope that its use will become more general.

J. A. Jacobs

The Solar System

1.1 Radioactive decay and the age of the earth

The phenomenon of radioactivity was discovered in 1896 by Henri Becquerel. Most naturally occurring elements have isotopes with stable nuclei, although there are several with unstable nuclei. These unstable atoms break down or 'decay' spontaneously, the isotopes which undergo decay becoming, as a result, nuclei of different elements. The fundamental equation of radioactive decay was shown by Rutherford in 1900 to be

$$\frac{dP}{dt} = -\lambda P \tag{1.1}$$

where P is the number of parent atoms at time t, and λ is the decay constant. Integration of equation (1.1) gives

$$P = P_0\, e^{-\lambda t} \tag{1.2}$$

where P_0 is the number of parent atoms at time $t=0$.

Radioactive decay rates are often expressed in terms of the half-life, $T_{1/2}$, which is the time in which one half of the radioactive nucleus will decay. It follows from equation (1.2) that

$$T_{1/2} = \frac{1}{\lambda} \ln 2 \simeq \frac{0.693}{\lambda} \tag{1.3}$$

Solving for t, equation (1.2) gives

$$t = \frac{1}{\lambda} \ln\left(\frac{P_0}{P}\right) = \frac{1}{\lambda} \ln\left(1 + \frac{D}{P}\right) \tag{1.4}$$

where D equals $P_0 - P$, which is equal to the number of daughter atoms at time t.

A radioactive nucleus may decay in a number of ways. It may emit an alpha particle which is the nucleus of an He^4 atom, and thus consists of two protons and two neutrons bound tightly together. In α-decay the daughter product forms the nucleus of a new element whose atomic number is two units less than that of the parent and whose mass number is four units lower. As an example

$$_{92}U^{238} \longrightarrow {}_{90}Th^{234} + \alpha$$

Another form of radioactive decay is by β-decay, a negative electron being emitted from the unstable parent nucleus. The charge on the nucleus would increase by one, the nuclear mass remaining essentially unchanged.

Another mode of decay is by electron capture. In this process the nucleus captures an orbital electron from the innermost shell (the K shell) and the electron combines with a proton, changing the latter into a neutron. Electron capture thus reduces the atomic number by one unit with no significant change in mass number. The nucleus produced in electron capture may be in an excited state and may then decay to a lower state of energy by emitting the excess energy in the form of a quantum of electromagnetic radiation (a γ ray). Some nuclei may decay in more than one way, one of the most important being K^{40} which decays spontaneously in two ways namely

$$K^{40} + \text{orbital electron} \longrightarrow Ar^{40} + \gamma$$

and

$$K^{40} \longrightarrow Ca^{40} + \beta^-.$$

Because of the widespread abundance of 'common' Ca^{40}, only the K^{40}–Ar^{40} branch of the decay is used to any extent in age determinations.

Equation (1.4) depends on two assumptions: that λ is constant, and that the only alteration in amount of daughter or parent in the system is due to radioactive decay. The second assumption that the system has remained closed during its history is often questionable. On the other hand, the energies involved in α- and β-decay are enormous compared with possible external influences, and the assumption that λ is constant is not likely to be in error for these transitions. In the case of electron capture, a very small change in the decay rate of Be^7 (only 0·07 per cent) has been detected in the laboratory by changing the external parameters. The Be^7 electron capture process, however, involves very low energies and it is very unlikely that the decay constant can vary in the case of K^{40}.

For the purposes of age determinations, three criteria must be met by a radioactive isotope: the half-life must be approximately of the order of the age of the earth (4500 m yr); it must be reasonably abundant in terrestrial rocks; and significant enrichments of the daughter product must occur. Table 1.1 gives the half-lives of the commonly used isotopes, and Table 1.2 the abundances of these elements in certain rock types. The dual decay of K^{40} to Ca^{40} and Ar^{40} has already been discussed. Rb^{87} decays to Sr^{87} with the emission of a low energy β-particle. U^{238}, U^{235} and Th^{232} decay through three separate series of radioactive intermediate products to the stable end-products Pb^{206}, Pb^{207} and Pb^{208}, respectively. It follows from equation (1.4) that to obtain an age we have to determine the amounts of the parent and daughter isotopes present in the sample. This is done by means of a mass spectrometer. The experimental details will not be

2

Table 1.1 Values of half-lives and decay constants

Isotope	$\lambda(10^{-10}\ \text{yr}^{-1})$	$T_{1/2}(10^9\ \text{yr})$
K^{40}	$\lambda_e = 0\cdot585$	$11\cdot8$
	$\lambda_\beta = 4\cdot72$	$1\cdot47$
	$\lambda = 5\cdot31$	$1\cdot31$
Rb^{87}	$0\cdot147$	$47\cdot0$
	or	or
	$0\cdot139$	$50\cdot0$
U^{238}	$1\cdot54$	$4\cdot51$
U^{235}	$9\cdot72$	$0\cdot713$
Th^{232}	$0\cdot499$	$13\cdot9$
Re^{187}	$0\cdot161$	$43\cdot0$

(After York and Farquhar (1971))

Table 1.2 Approximate estimates of concentrations of radioactive and 'common daughter' elements in common rocks

	U (p.p.m.)	Th (p.p.m.)	Pb (p.p.m.)	K %	Rb (p.p.m.)	Sr (p.p.m.)
Granite	4	15	20	3·5	200	300
Basalt	1	3	4	0·75	30	470
Ultramafic	0·02	0·08	0·1	0·004	0·5	50
Shale	4	12	20	2·7	140	300

(After York and Farquhar (1971))

discussed here, but a good account of the techniques and interpretation of the results has been given by York and Farquhar (1971).

A minimum age of the earth is given by the age of the oldest terrestrial rocks. Table 1.3 gives evidence for the existence of a continental crust on earth for more than 3000 m yr. Many K–Ar ages have now been reported for stony meteorites ranging from 400 to 5000 m yr. There is, however, a significant preponderance of ages between 4000–4800 m yr. Rb–Sr studies of meteorites show that a major chemical differentiation took place in these bodies some 4300–4700 m yr ago. Lead-isotope analyses also yield an age of 4500 m yr for various meteorite bodies.

Table 1.3 Mineral and whole rock ages greater than 3×10^9 yr

Location	Method	Mineral	Age (10^9 yr)
Kola Peninsula, USSR	K–Ar	Biotite	3·46
Ukraine, USSR	K–Ar	Biotite	3·05
Swaziland	Rb–Sr*	Whole rock	3·07±0·06
			3·44±0·30
Transvaal, S. Africa	Rb–Sr*	Whole rock	3·20±0·07
Congo	Rb–Sr*	Microcline	3·52±0·18
Minnesota, USA	U–Pb	Zircon	$\gtrsim 3\cdot3$
Montana, USA	U–Pb	Zircon	$\gtrsim 3\cdot1$

*$\lambda = 1\cdot39 \times 10^{-11}\ \text{y}^{-1}$.
(After York and Farquhar (1971))

Estimates of the age of the earth may also be made from theories of the evolution of lead-isotope ratios (see York and Farquhar (1971) for a summary of such methods). Table 1.4 gives recent estimates by this method using both meteoritic and terrestrial lead data, and indicates an age of the earth around 4500 m yr. All of these estimates assume that the earth's lead at one time had the same isotopic composition as that now found in iron meteorites.

Table 1.4   Estimates of the age of the earth combining meteoritic and terrestrial lead data

Author	Age (10^9 yr)
Patterson, Tilton and Inghram (1955)	4·5
Patterson (1956)	4·55 ± 0·07
Ostic, Russell and Reynolds (1963)	4·53 ± 0·03
Tilton and Steiger (1965)	4·75 ± 0·05
Ulrych (1967)	4·53 ± 0·04

(After York and Farquhar (1971))

1.2 Composition of stellar and cosmic matter; the origin of the elements

Our knowledge of the chemical composition of the universe is obtained from spectroscopic examination of the radiation from the sun, the stars and interstellar gas, and from direct chemical analysis of lunar samples and meteorites (see Fig. 1.1). Although the abundances of the different elements vary considerably, it may be seen that only ten elements (all with $Z < 27$) are quantitatively important, and of these the lightest elements, hydrogen and helium, far outweigh the other eight. With the exception of hydrogen and helium, the chemical composition of the universe is everywhere much the same. Moreover, the abundances of the elements, with certain exceptions, show a rapid exponential decrease with increasing atomic number up to about $Z = 30$, followed by an almost constant value for the heavier elements. It is possible that absolute abundances of the elements depend on nuclear rather than chemical properties, and are related to the inherent stability of the nuclei. Terrestrial matter, like meteoritic matter, differs from stellar material chiefly in the rarity of the gaseous elements. This is not surprising and is probably the result of the events which led to the formation of the solar system.

It is believed that the elements were formed in stellar interiors, through processes that involved fusion of hydrogen nuclei and the 'burning' of helium nuclei to form the elements of low mass (deuterium, lithium, beryllium and boron are possible exceptions: they seem to have been produced at lower temperatures by high-energy particle irradiation). These processes, however, are not capable of explaining the existence of elements of higher mass, which require the bombard-

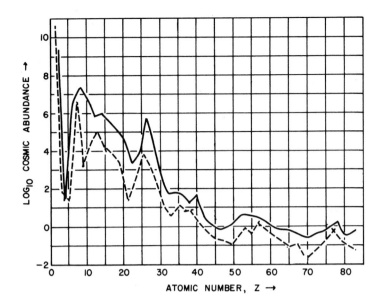

Fig. 1.1 Cosmic abundances of the elements (normalized to 10^6 for Si). The solid lines show elements of even atomic number, and the broken line elements of odd atomic number

ment of the lighter nuclei by neutrons. If the bombardment is sufficiently *slow* (the *s*-process), the elements formed are able to decay between subsequent neutron additions (through β-particle emission) to more stable nuclei. In the *r*-process all neutron additions take place much more *rapidly* before β-decays have a chance to occur. The details of these processes have been thoroughly investigated (see e.g. Fowler and Stephens (1968)). As an example, only the r-process is able to produce transuranic elements.

Equally important as the question of the formation mechanism of the elements is the question of when this formation took place. Some answers to this question have been obtained by radioactive measurements. I^{129} decays through β-emission to Xe^{129} with a half-life of 16·4 m yr. This half-life is so short that no I^{129} now exists in nature. However, this isotope should be produced at the time of formation of the other elements. It can be expected that the condensation of planetary material would incorporate any of the iodine present, but would be unlikely to incorporate much xenon. Xe^{129} formed within the condensed materials would be trapped and might be observed today—very small amounts of excess Xe^{129} have been found in a number of meteorites. Reynolds (1963), who has been foremost amongst those working in this field, has coined the term 'xenology' for the study of xenon isotope abundances in meteorites.

Absolute formation intervals calculated by the Xe^{129}–I^{129} method generally give values between 40 and 300 m yr, depending on the nucleosynthesis model assumed. Variations in relative formation intervals are much less, but are difficult to assess because of different methods of data treatment between different

5

laboratories. Podosek (1970) has subjected all of the Berkeley data to a common method of analysis and found fifteen different meteorites possessing formation intervals within a period of 15 m yr. Nevertheless, real differences existed between some of the samples. The determination of differences in meteorite formation times (or, more precisely, in cooling times) as short as a few million years now appears possible by the Xe^{129}–I^{129} method. When this information is combined with differences in meteorite crystallization times that can now be obtained more precisely by the Rb^{87}–Sr^{87} method (Papanastassiou and Wasserburg (1969)), valuable information can be obtained on the early chronology of the solar system.

Other noble-gas components in meteorites that arise from an extinct radio-nuclide and have the potential for the determination of formation intervals are the heavy xenon isotopes $Xe^{131-136}$ which arise from the fission of Pu^{244}. These fission Xe components are much more difficult to identify than the excess Xe^{129}, and were not discovered until precise isotopic data were obtained on the calcium-rich achondrites, a class of meteorites high in uranium (supposedly chemically similar to plutonium) and very low in trapped Xe (Rowe and Kuroda (1965)). The Pu^{244}/U^{238} ratio at the time meteorites began retaining xenon not only gives an independent measure of the formation interval of meteorite bodies, but is also important in theories of galactic nucleosynthesis.

Pu^{244} has now been discovered in nature in an old Precambrian rare-earth mineral called bastnaesite from the Mountain Pass deposit, California (Hoffman et al. (1971)). Although its existence as an extinct radioactivity had been postulated to explain the Xe isotope ratios observed in meteorites, this is the first indication of its present existence in nature. Pu^{244} has a half-life of about 82 m yr so that the solar system was formed about 60 half-lives ago. The difficulty of finding natural Pu^{244} is that its average concentration should be only $(\frac{1}{2})^{60}$ of its original value, and it could thus only hope to have been found in some mineral in which it was enriched several orders of magnitude. The discovery, if substantiated, has profound cosmological significance. Considered in conjunction with the primeval meteoritic abundance of I^{129}, it can be concluded that about 180 m yr elapsed from the time when a portion of interstellar gas collapsed into the gas cloud that was to form the solar system and the time when the newly formed meteorites were cool enough to retain noble Xe gas. Hoffman et al. point out that the present abundance of Pu^{244} due to the influx of Pu^{244} nuclei in cosmic rays may be comparable with, or even greater than, that surviving from primordial earth material. In this respect, Flesicher and Naeser (1972) have found that the Mountain Pass bastnaesite has an apparent Cretaceous fission track age and thus does not reveal any anomalous fission tracks due to Pu^{244}. Further work is necessary on this very important matter.

Blake and Schramm (1973) have suggested the chronological possibilities of the r-only isotope Cm^{247} (half-life $\sim$ 16 m yr). Since I^{129} was present in the early solar system material and I^{129} and Cm^{247} have essentially the same half-lives and both nuclei are produced in the r-process, Cm^{247} should also have been present. Attempts to detect its presence are now being made.

6

The only nucleo-cosmochronometers established at present are for r-process nucleosynthesis. A short-lived s-process chronometer would be extremely important, since it could provide estimates (independent of any model) of the time interval Δ between the solidification of material in the solar system and the last s-process nucleosynthetic events contributing to the elemental abundances of the solar system. Blake *et al.* (1973) have considered the s-only isotope Pb^{205}, which decays by electron capture to Tl^{205} (with a half-life ~ 15 m yr). Present indications are that Δ for the s-process is comparable to Δ for the r-process, which implies that Δ represents a time scale for the separation of all the solar system material from all nucleosynthetic sources.

The possibility of the p-process isotope Sm^{146} as an extinct natural radionuclide was suggested by Kohman in 1954. Sm^{146} decays to Nd^{142} emitting an α-particle with a half-life of 100 m yr, but the search for an isotopic anomaly in Nd has until recently met with no success. However, Notsu *et al.* (1973) have now obtained the first possible evidence for extinct Sm^{146} in a Ca-rich achondrite ('Juvinas'). It must be pointed out that the formation interval obtained from the decay of Sm^{146} is not necessarily the same as that obtained from the decay of I^{129} or Pu^{244}, since the latter times are produced mainly by the r-process and Sm^{146} by the p-process. Again, the end-point of the formation interval in the case of I^{129}–Xe^{129} or Pu^{144}–Xe is the onset of Xe retention, whereas for Sm^{146}–Nd^{142} the end-point corresponds to the time of the fixation of rare-earth elements in different phases by solidification. An additional potential advantage in studying the Sm–Nd isotope system is the presence of the long-lived radioactive isotope Sm^{147} which could possibly be used for the determination of the solidification age.

1.3 Origin of the solar system

Of all the problems in science, one of the most fascinating is that of the origin and evolution of the solar system. There can be but few scientists who have not at some time speculated on this question. It must be realized, however, that it is by no means certain *a priori* that the problem can be given a definite answer. It may well be that all memory has been lost of the circumstances under which our solar system was born. All that we can do now is to attempt to derive its present state from an assumed event which occurred in the distant past, so that in a sense the method is one of trial and error. It is far more reasonable to assume that planets are normally present in the vicinity of certain stars than to suppose that our planetary system is unique or at least very rare. There are about 200 000 million stars in our galaxy alone. The origin of the solar system is thus part of a very much larger problem–the evolution of the sun and stars in general.

The various theories which have been suggested for the origin of the solar system may be divided into two main classes: those which regard the origin to be the result of a gradual evolutionary process and those which attribute it to some cataclysmic action, usually associated with a hypothetical close encounter of the sun with a star in the distant past. No attempt will be made to trace the historical development of the different theories that have been proposed. Cataclysmic theories are associated with the names of Moulton and Chamberlin at the turn of the century and later with the work of Jeans and Jeffreys. Just as the moon by its gravitational attraction raises tides on the earth, a close encounter of the sun with a star would raise immense gaseous tides on the sun. According to Jeans and Jeffreys, matter would be pulled away from the sun to form a long gaseous filament. This filament would become unstable and soon break up into several parts, each forming a distinct aggregation of matter which would later develop by cooling and contraction into a planet. There are, however, many difficulties with such a theory. Nölcke (1930) showed that a filament of matter drawn out from the sun in such a manner would rapidly disperse, its density being well below the Roche limit.* Furthermore the temperature of the gaseous filament would have exceeded 1 million °K, at which temperature the mean velocity of hydrogen atoms is more than 150 km/s. Spitzer (1939) showed that the escape velocity (see Appendix A) would be reached within a few hours and that the filament would then dissipate. Some of the material would escape into interstellar space while the rest would form an extended gaseous nebula around one or more of the stars involved.

The early evolutionary hypotheses, dating back to Kant in 1755, proved equally unsuccessful. Typical of such theories is the nebular hypothesis of Laplace. Laplace supposed that originally the sun was a rotating, gaseous nebula. Under its general gravitational attraction it would gradually contract, its rotation becoming more rapid. When the centrifugal force in the outer layers of the nebula exceeded the gravitational attraction of the nebula as a whole, gaseous matter would be thrown off, the expelled material forming a ring revolving in the equatorial plane of the nebula. As the nebula continued to contract, the material of the ring slowly collected into a single aggregation of gaseous matter which, on further condensation and cooling, developed into a planet revolving around the central body. As a result of further contraction of the nebula and its increased rotation, additional material was thrown off, to become another planet by the method described above. It was later shown that if the mass of the present planets were spread out in Laplacian rings, such rings would never coalesce into planets. More important, however, is that one would expect on Laplace's hypothesis that the sun would rotate with its maximum possible angular velocity compatible with stability. In point of fact, although the sun has 0·999 of the total mass of the solar system, it possesses less than 0·02 of the angular momentum. Even if all the angular momentum of the solar system were concentrated in the sun, its period of rotation would be about 12 h and the centrifugal force at its equator would be only about 5 per cent of the force of gravity. Thus rings of matter could not possibly be thrown off in the manner suggested by Laplace.

Such considerations raise the question of whether the solar system just after

---

* At the Roche density the self-gravitation of the gas cloud would just balance the solar tidal force. Gravitational stability would exist, and hence planetary condensation proceed, when the critical density is well exceeded. For further details see Appendix A.

its formation was essentially the same as it is today or whether significant evolutionary changes have taken place. Any model based on the assumption that the angular momentum and total mass of the solar system have remained unchanged cannot account for present conditions; the composition of the planets is so highly selective that the original mass must have greatly exceeded its present value. The early monistic theories of the origin of the solar system, such as that of Laplace, are almost certainly incorrect but for entirely different reasons from those given historically for their rejection. Additional problems for the earth are the origin of its atmosphere, hydrosphere and continents. Have the continents grown from nuclei during geologic time or have they always been roughly the same size and merely been reworked? This question will be considered in Chapter 8.

The failure of theories of the cataclysmic type to provide a satisfactory explanation for the origin of the solar system led to a reconsideration of those based on the gradual evolution of some primordial system. In all such theories the sun is assumed to be at the centre of a diffuse, gaseous cloud which may have been an interstellar cloud into which the sun had passed, or a gaseous cloud from which the sun had condensed. These new attempts are associated with such names as von Weizsacker, Kuiper, ter Haar and the school of cosmogony that grew up around the Russian astronomer Schmidt.

In all theories of the origin of the solar system up to about 1940 only mechanical (i.e. gravitational) forces were considered. A radical departure from this point of view was made by Alfvén (1942–1945, 1954), who suggested that electromagnetic forces play a key role. In his original theory, Alfvén assumed that the sun had a general magnetic field and that the gaseous cloud which surrounded it was ionized by radiation from the sun. The degree of ionization of any particular element will depend upon temperature and thus upon distance from the sun. The gaseous cloud falls in towards the sun under its gravitational attraction, but the magnetic field around the sun impedes the fall of the ionized constituents. Thus there is a gradual diffusion of the non-ionized constituents through those that are ionized. Any given constituent as it falls towards the sun sooner or later reaches a distance at which, because of its increasing temperature, thermal ionization sets in and the magnetic field begins to act as a brake. Alfvén proposed that four main clouds, differing in composition, were produced by this process. Angular momentum is transferred from the sun to the gas clouds by electromagnetic forces, causing a concentration of gas in the equatorial planes; through condensation the gas is then transformed into small solid or liquid bodies, and the planets formed by the agglomeration of these bodies. The moon and Mars are supposed to condense out of impurities in one cloud (consisting mainly of helium), the earth, Venus and Mercury from another cloud (consisting mainly of hydrogen), the four major planets from a third cloud (consisting mainly of carbon), and Pluto (and perhaps Triton, a satellite of Neptune) from the fourth cloud (consisting mainly of iron and silicon). One of the main difficulties with this theory is that the postulated magnetic fields must be much larger than seem plausible. In particular for the sun, a surface field of 300 000 Γ (gauss) is required, but the field at the present time is less than 2 Γ. In spite of this and other difficulties, Alfvén made a very real contribution to the problem of the origin of the solar system by demonstrating the importance of electromagnetic forces.

Theories of the origin of the solar system considered so far have assumed the formation of the planets to be independent of that of the sun, the planets forming after the sun had become a normal star. Some success (e.g. Hoyle (1955), (1960); McCrae (1960)) has been obtained in recent years starting from a different premise, namely that the origin and formation of the planets are related to the formation of the sun itself. A number of authors have also shown that some of the elements and compounds found in the earth and in meteorites are not compatible with the earth having condensed out of a hot gas cloud of solar composition. In this case the separation of the elements of the earth from the surrounding hydrogen must have taken place at low temperatures.

Most theories of the origin of the planets assume that they accreted from relatively cold homogeneous material. The division of the earth (and possibly the other terrestrial planets) into a crust, mantle and core, has then been presumed to be the result of segregation processes which occurred sometime during or after the accretion of the planet. However, in recent years the possibility that the earth accreted inhomogeneously has been considered. Calculations by Larimer (1967) on the condensation sequence of elements and compounds from the solar nebula led other workers to suggest that the different meteorite classes, the moon and the various regions of the earth represent the accretion of material that has been condensed out of the solar nebula at different temperatures. Clark *et al.* (1972) later suggested that the planets accreted *during* condensation and that internal zoning is an original feature of a planet. They proposed that the iron body that is now the earth's core condensed first and became the nucleus upon which the silicate mantle was later deposited. The last accumulates would be FeS, Fe_3O_4, the volatiles and hydrated silicates. Anderson and Hanks (1972) have examined the condensation sequence in more detail and have shown that earlier difficulties with such a model in accounting for the composition of the moon and explaining how the earth's core became molten may be overcome. The accretion of the earth and formation of the core are discussed in more detail in §§7.3 and 10.5. The ultimate size of a planet will depend on the amount of condensable material available to it in its orbit and the temperature of the nebula in its vicinity before a high temperature phase of the sun, such as T-Tauri, blew the uncondensed and smaller condensed particles out of the inner solar system.

It is beyond the scope of this book to discuss the question of the origin of the solar system in any more detail: some recent reviews have been given by ter Haar (1967); Herczeg (1968); and Williams and Cremin (1968). Meteorites will be considered briefly in §1.6. Because of the great increase in our knowledge of the moon, Mars and Venus following the successful launching of space craft, these three bodies will be discussed separately in Chapter 2. The origin of the solar system remains one of the oldest unsolved problems in natural philosophy. At a meeting organized by the Royal Society of London and the Royal Astronomical

Society on March 27, 1972 it was suggested that there are as many theories on the subject as there are researchers in the field; ter Haar even went so far as to suggest that a solution is no nearer now than it was in Descartes' time 340 years ago!

1.4 Variation of the gravitational constant G with time

In 1938, Dirac suggested that the gravitational 'constant' G might vary inversely as the age of the universe. Later Brans and Dicke (1961) modified the theory of general relativity to make it compatible with the requirements of Mach's principle—in their cosmology G would be expected to decrease with time, presently at a rate of 1–3 parts in 10^{11} per yr (Dicke (1962)). Revisions of such numbers as the age of the universe or the age of the galaxies have nullified many of the results of early calculations on the effects of Dirac's hypothesis.

Egyed (1956); Carey (1958); Heezen (1959); Wilson (1960) amongst others have suggested that the earth has been expanding with time—a good review of these expansionist theories has been given by Dearnley (1966). In Carey's hypothesis the expansion took place mainly during the past 500 m yr, leading to an average rate of increase in the radius of the earth during this period of about 5 mm per yr; Egyed inferred a rate of increase of the earth's radius of 0·4–0·8 mm per yr. One of the mechanisms which has been suggested to account for such an expansion is Dirac's speculation that the gravitational constant G varies inversely with time. With a gradual decrease of G, the pressure would decrease inside the earth and the volume increase. Dicke (1957) showed that to account for the expansion demanded by Carey by this mechanism, it would be necessary to assume that G has been decreasing at a rate of roughly 1 part in 10^8 per yr; to meet the expansion of Egyed, the rate of decrease of G would have to have been one part in 10^9 per yr.

If G has been decreasing with time, the rate of radiation of the sun would have been higher in the past and hence asteroids and meteorite bodies would have been warmer, possibly leading to loss of argon from the material of the meteorites. From the observed K–Ar ages of meteorites, Peebles and Dicke (1962) have shown that G cannot have been decreasing by more than about 1 part in 10^{10} per yr. This rules out the possibility that a decrease in G could lead to the large expansions required by Carey and Egyed: the limit of 1 part in 10^{10} per yr in the variation in G leads to an upper limit in the rate of increase in the earth's radius of about 0·05 mm per yr. Further deductions by Dicke (1966) on various geophysical effects indicate that only a very small decrease in G is possible.

Egyed (1960) first suggested that palaeomagnetic data might be used to calculate probable ancient radii of the earth. In the expansion hypothesis of Carey, Heezen and Egyed, the continents do not increase in area, so that the distance between any two points on a stable part of one continent remains the same. Thus if the earth's radius increases, the geocentric angle between the two points decreases. Assuming the earth's ancient magnetic field to be dipolar, the earth's ancient radius R may be found from the formula (*cf* equation 5.7),

$$R = \frac{d}{\cot^{-1}\left(\frac{1}{2}\tan I_1\right) - \cot^{-1}\left(\frac{1}{2}\tan I_2\right)} \tag{1.5}$$

where I_1 and I_2 are the inclinations of contemporaneous palaeomagnetic data from two localities on the same stable continental block and d is the distance measured along the earth's surface between the two geomagnetic latitudes (Egyed (1960)). Cox and Doell (1961) have used this method on Permian data from Europe and Siberia to obtain an estimated Permian radius of 0·99 times the present radius. Ward (1963) generalized Egyed's method of calculation and applied it to Devonian and Triassic data as well as Permian from Europe and Siberia. He obtained estimated earth radii for these periods of 1·12, 0·94 and 0·99 times the present radius respectively and considered none of these estimates to be significantly different from the present radius.

Using the latest high pressure shock wave data, Birch (1968) showed that the increase in radius for an earth having a chemically distinct mantle and core would only be about 370 km for a decrease in the gravitational constant from $2G$ to its present value of G. A larger increase in radius would be possible if Ramsey's hypothesis (see §10.5) were true. Birch showed, however, that Ramsey's hypothesis is extremely unlikely to hold and concluded that if the mass of the earth has remained constant, changes in the earth's radius are unlikely to exceed 100 km. Thus all evidence to date indicates that any large expansion of the earth has not taken place and that the upper limit to any rate of change of G is about 1 part in 10^{10} per yr.

Recently there has been a fairly rapid increase in the number of observations of 'discrepant redshifts'. Hoyle and Narlikar (1971) have given a possible explanation in a theory which incorporates a gravitational constant G that is decreasing with time. In a later paper (1972) they show that their theory implies that the radius of the earth has increased at a rate of about 0·1 mm per yr: they also speculate on the possible geophysical consequences of such an expansion. Shapiro *et al.* (1971) have placed an observational upper limit of 4×10^{-10} yr^{-1} on $\dot{G}/G$. Taking Hubble's constant as 5×10^{-11} yr, the variation expected would be 10^{-10} yr^{-1} in close agreement with Dicke's (1962) estimate.

In his presidential address to the Royal Astronomical Society in 1972 on 'The history of the Earth', Hoyle devoted considerable time to the question of a varying gravitational constant. Hoyle appears to find the sum total of evidence in favour of the hypothesis that G has decreased with time. The result of this would be higher temperatures in the past, the mean sea level temperature of the earth being about 70°C 2000 m yr ago. The most serious objection to a variable G is, in Hoyle's opinion, the Precambrian glaciation of the Canadian shield which has been estimated to have occurred 2500 m yr ago. The geophysical implications of modern cosmologies in which G is variable have recently been discussed by Wesson (1973).

1.5 Meteorites

Our knowledge of the composition of the earth's interior and of the evolution and origin of the solar system is greatly increased by examination of extra-terrestrial material. Until Apollo 11 landed in Mare Tranquillitatis on July 20th, 1969 and brought back lunar samples to the earth, the only extra-terrestrial material available for study was from meteorites. A discussion of the results of the analyses of lunar samples will be given in the next chapter.

Meteorites are pieces of cosmic matter that have reached the earth's surface from outer space before being completely vaporized in passing through the earth's atmosphere. About 1500 meteorites with a mass greater than 10 kg hit the earth every year, although only about five to ten are recovered, the majority of them landing in the oceans and uninhabited regions. Meteorites may be classified according to their composition and structure (see e.g. Mason (1962)). Those that consist almost entirely of nickel-iron are called *siderites*, or simply iron meteorites; those that are chiefly composed of iron and magnesium silicates (mainly olivine and pyroxene) with comparatively little nickel-iron are called *aerolites*, or stony meteorites. There is also a third class called *siderolites* or stony-irons which contain about 50 per cent nickel-iron and 50 per cent silicates. None resembling sedimentary or metamorphic rocks have ever been found, but many meteorites contain small quantities of a peculiar iron sulphide mineral (FeS) called troilite and a few contain carbonaceous material.

Stony meteorites may be subdivided into two classes: *chondrites* and *achondrites*. In all, 85 per cent of all witnessed meteorite falls and 95 per cent of all stones are *chondrites*. Although nearly 50 different minerals have been identified in meteorites they generally consist of a few major constituents. Chondrites are composed of olivine, pyroxene, oligoclase, troilite and flecks of nickel-iron. When viewed through a microscope, a thin section shows these minerals in a disorderly array, quite unlike any terrestrial rock, with the fragments less than a fraction of a millimetre across. The distinguishing feature is the chondrule, a round inclusion, typically 1 mm in diameter (see Fig. 1.2). Wood (1963a, b) has given an excellent description of chondrules and presents some evidence for glass in many of them. They appear to be frozen liquid droplets mainly of silicate with some metallic iron, and may be primordial planetary matter.

Major advances in the number and quality of chondrite analyses (e.g. Ahrens *et al.* (1969)) have made it possible, on chemical grounds, to divide the chondrites into seven groups. All these groups (except enstatite chondrites) are remarkably homogeneous with respect to the major non-volatile constituents, though all but the carbonaceous chondrites vary widely in texture and volatile content. Van Schmus and Wood (1967) have divided each chemical group into six petrologic types on the basis of petrographic and mineralogic criteria. Most of the textural and mineralogical variations among members of each chemical group are consistent with the view that they have experienced different degrees of thermal metamorphism (Dodd (1969)). Systematic decreases in volatile constituents with advancing recrystallization also support this view: the specific pattern of volatile

13

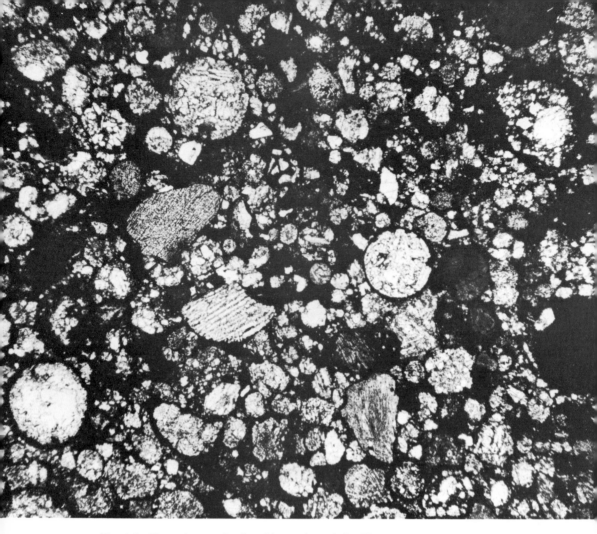

Fig. 1.2 Photomicrograph of a thin section of the Tieschatz (Czechoslovakia) chondrite by transmitted unpolarized light. Relatively transparent minerals appear light; opaque substances are darker. The rounded structures are chondrules. The field of view is approximately 1 cm. (After Wood (1968))

depletion, in particular the correlated depletion of dissimilar elements, seems, however, to be inconsistent with metamorphism. The data for volatile trace elements have still not been satisfactorily explained.

Rb–Sr ages for several members of each chondrite group suggest that all chondrites were formed during a very brief period (less than 200 m yr) about 4600 m yr ago (Kaushal and Wetherill (1970)). Xe–I studies (Podosek (1970)) suggest a slightly greater age and a still shorter period of formation. The origin of chondrules has not yet been settled; possibilities include condensation of nebular gas, fusion of nebular dust by lightning, hypervelocity impacts on extended bodies and volcanism. Larimer and Anders (1967, 1970) have shown that the condensation model can account for most of the chemical trends within and among chondrite groups, and such a model is the most widely accepted.

The *achondrites* do not contain chondrules; they are very similar to terrestrial rocks such as gabbro, the most common minerals being pyroxene and plagioclase. The achondrites are considered to show the effects of metamorphism and are thought to have crystallized from a magma in the same way that terrestrial rocks were produced—they could be derived from chondritic material that has been melted and subsequently recrystallized.

About two-thirds of all meteorites found are irons, although over 90 per cent of all observed to fall are stones. The reason for this apparent paradox is that irons are easily recognized as meteorites, whereas stony meteorites, unless observed to fall, would easily be overlooked as such and iron meteorites are much more resistant to weathering than stones. A better indication of the relative elemental abundances of meteoritic matter is thus obtained from those observed to fall. The proportions of silicate, troilite and nickel-iron in average meteoritic matter are about 13:1:3.

It will be shown later that the earth has a core consisting mainly of nickel-iron surrounded by a silicate mantle and it is thus natural to suppose that iron meteorites come from the core of a fair-sized planet and that stony meteorites come from the surrounding mantle. It has been suggested that meteorites may be asteroids that have been deflected from the asteroidal zone between Mars and Jupiter, and that the asteroids themselves may be fragments of a former planet. It is not possible in this book to consider the problem of the formation of the asteroids—Anders (1965) has given a good review of this subject. New developments have been discussed by Alfvén and Arrhenius (1970); Safronov and Zvjagina (1969) and Sonett *et al.* (1970).

Iron meteorites show crystal patterns and were undoubtedly formed from a solid, not liquid, core. The nickel-iron composition of iron meteorites can be explained by phase chemistry in terms of differentiation within a single parent body, although other data favour several (four or more) parent bodies (Anders (1964)). The size of these parent bodies was probably between 50 and 250 km in diameter. Cosmic ray exposure ages have been determined for a large number of meteorites. Typical exposure ages of stones range from 1–100 m yr and of irons from 10–1000 m yr and indicate that such ages represent the time since the break-up of much larger bodies shielded from cosmic rays and are distinct from radiogenic or formation ages. Any strong groupings of exposure ages of a particular meteorite class would imply that all such meteorites came from a single collision of a particular object, and that most meteorite types recovered on earth originated in a limited number of such parent objects.

The measurement of isotopic abundances of the noble gases in meteorites (and other extraterrestrial samples) has become a very active field of research during the past decade, and studies of the five stable noble gases have yielded much information on the early history of the solar system. Some primitive gases have become 'trapped' and retained in meteorites since their existence as solid bodies. These gases are distinct from noble gases produced in meteorites by radioactive decay and various induced nuclear reactions. Two major classes of trapped gases are found in meteorites: solar or unfractionated gases and planetary or fractionated gases. The solar gases, which are clearly found in only a small number of meteorites, represent a sampling of the solar wind (see §2.3) that has become embedded in the surfaces of individual mineral grains. Presumably this solar wind irradiation occurred early in the history of the solar system when portions of these meteorites were still finely dispersed in space. Lunar surface material also contains solar wind gases of a similar composition. The other class of trapped noble gases, the planetary component, is found in large concentrations in carbonaceous meteorites and in much smaller amounts in many ordinary chondrites. These gases represent the minute remaining fraction of noble gases that existed in the planetary nebula when the meteoritic bodies formed and, as a consequence, show elemental and isotopic fractionation.

The three principal radioactive methods used for dating rocks can also be applied to meteorites (see e.g. Anders (1962); Reynolds (1967)). Towards the end of the 1960s it was well established that an age for the earth of 4550 ± 150 m yr could be calculated by assuming that terrestrial oceanic lead evolved from the primordial lead found in troilite from some iron meteorites (Patterson (1956)). Rb^{87}–Sr^{87} analyses of chondrites gave ages of about 4600 m yr (see e.g. Gast (1962)). The maximum K–Ar ages of stone meteorites (mostly chondrite analyses) were 4500–4600 m yr, but there was a wide spread with some ages of less than 1000 m yr, presumably because of diffusive Ar^{40} loss (see e.g. Kirsten et al. (1963)). Silicate inclusions of a number of iron meteorites gave K–Ar and Rb–Sr ages in the range 4400–4700 m yr (Bogard et al. (1968); Burnett and Wasserburg (1967)). The presence of radiogenic Xe^{129} in chondrites due to in situ decay of I^{129} showed that the age of the chondrites was essentially that of the solar system. The major advance during the last few years is the increase in accuracy of the Rb^{87}–Sr^{87} method. This has come about from high sensitivity (and particularly high precision) mass spectrometry and the chemical extractions of Rb and Sr with subnanogram contamination levels.

Green II et al. (1971) believe that the Allende meteorite which fell in the Mexican state of Chihuahua in 1969 contains crystalline material older than the solar system. The meteorite consists of a fine-grained, carbon-rich matrix in which are embedded many chondrules: both the matrix and the chondrules are composed principally of olivine (Mg_2SiO_4). A detailed study of the fine structure of the matrix and chondrules by Green et al., using a high-voltage transmission electron microscope, revealed that the olivine crystals in several of the chondrules had a peculiar fine-scale, black-spot structure. This fine structure was interpreted as evidence of extensive radiation damage. By comparing the substructure with that of other meteorites, lunar basalts and terrestrial rocks, the authors concluded that the observed irradiation of the chondrules preceded the cold accretion of matter that took place during the early stages of formation of the solar system. The radiation, they believe, originated in the proto-sun: the presence of radiation damage in the chondrules and its absence in the matrix indicates that the irradiation occurred after the chondrule solidified but before the parent meteorite body was formed. The authors thus conclude that 'the Allende meteorite consists of virgin planetary material'. Banerjee and Hargraves (1972) carried out a study of the natural remanent magnetization (NRM) of three carbonaceous chondrites and concluded that the intensity of the ambient magnetic field during the formation of these meteorites was between 0·2 and 1·1 Oe. The latter figure (1·1 Oe) was for the Allende meteorite; Butler (1972) obtained the same value for the magnetic field intensity at the time of formation of the NRM in Allende. Such a value of the intensity of the primordial magnetic field is about 10^4 times greater than that of the present interplanetary field and favours those cosmological theories which require very large magnetic fields to have been present during the formation of the solar system. Alternatively if Allende was magnetized by the body from which it derived, that parent body must have possessed a strong magnetic moment.

One particular class of meteorites, *tektites*, deserves special mention, since their origin has been hotly debated for many years. Tektites are small, glassy objects a few centimetres in size found in a few special areas of the world. Such tektite 'strewnfields' have different physical and chemical characteristics and different ages. The Australasian, Ivory Coast, Czechoslovakian and North American strewnfields have approximate ages of 0·7, 1·0, 15·0 and 35·0 m yr, respectively. Some tektites have very distinctive shapes such as buttons or dumb-bells (see Fig. 1.3), and wind tunnel experiments have shown that these could have been caused by rapid transit through the earth's atmosphere. The intense heating would have melted the surface and the material would have flowed back in the jet stream. In 1967 microtektites (< 1 mm in size) were found in deep sea cores from the Indian and Pacific Oceans, establishing a geographic connection between the Australian, south-east Asian and Philippine fields—a single event, 700 000 years ago, spread glass over a range of at least 10 000 km. A detailed account of the properties of tektites can be found in the book *Tektites* (edited by O'Keefe (1963)), in his review paper (1966) and in the reports of the Third International Tektite Symposium held at Corning, New York, in 1969, published in *J. geophys. Res.*, **74**, 6722–6852, and *Geochim. cosmochim. Acta*, **33**, 1011–1147.

The essential argument is whether or not tektites are extra-terrestrial material. Material might have been fused into a glass during the impact of a large meteorite with the earth or the moon. Some scientists believe that tektites were originally lunar material thrown off the moon when a large meteorite struck its surface. The fused material solidified in space, some of it reaching the earth (Fleischer *et al.* (1965) have set a limit of 300 years in space, basing their conclusions on the absence of cosmic-ray-induced fission tracks in tektites). It seems just as reasonable, however, to suppose that a meteorite collided with the earth, not the moon, tektites being fused terrestrial material which was shot into the air before falling back to the earth's surface. Several tektite fields are now known to be close to large craters on the earth and there is convincing evidence that the Ries Kessel crater at Nordlinger, Germany, is related to the Czechoslovakian tektite field and the Bosumtwi crater in Ghana to the Ivory Coast group of tektites.

From the study of the physical and chemical differences among the Austra-

Fig. 1.3 Tektites, showing the variety of forms. (The American Museum of Natural History)

lasian tektites, Chapman (1971) divided them into groups whose geographic distribution gives the pattern of their flight trajectories. Chapman maintains that the trajectories are incompatible with a terrestrial origin but agree in detail with material ejected in a specific direction from one lunar crater. He carried out a computer analysis of possible lunar sources; various azimuth and elevation angles from a number of large, young craters were investigated. The only large crater that could satisfy the data was Tycho, the azimuth agreeing exactly with one of the strongest rays from Tycho (the 'Rosse' ray). Since tektites have not spent much time in space, Chapman's suggestion can be definitely settled when an accurate age has been obtained for Tycho. An age based on crater counts (which may be grossly in error) indicates that Tycho is at least a hundred times older than the Australasian tektites. Again, although small glassy particles resembling microtektites have been found in some of the lunar samples, Durrani (1971) has collected all available data and concludes that the chemical composition of tektites, microtektites and general lunar material obtained by Apollo 11 and 12 are not similar: the resemblance of the exceptional lunar sample 12013 to two untypical Java tektites is not by itself conclusive evidence for a lunar origin. Taylor and Epstein (1970) have also shown that although the sample 12013 has an SiO_2 content that is in the same range as that of some tektites, the oxygen isotope values are completely outside the range of values for tektites. All the evidence suggests that tektites do not come from the moon, but their origin still remains an open question.

Glass and Heezen (1967) and Cassidy et al. (1969) have pointed out that the age of the Australasian microtektites coincides with that of the reversal of the earth's magnetic field from the normal (Brunhes) epoch to the reversed (Matuyama) epoch about 0·7 m yr ago (see §6.3). Durrani and Khan (1971) have also suggested an association between the Ivory Coast microtektites and the Jaramillo reversal event which occurred about 0·9 m yr ago. However, there appears to be no record of tektites or microtektites associated with the times of other field reversals (there were more than seventy during the last 20 m yr) and it is difficult to visualize how the impact of even quite a large mass (it has been estimated as about 100 million tons for the Australasian tektites) could cause a reversal of the earth's magnetic field, whose origin lies within the core (see §5.5). Such a correlation is almost certainly due to chance and it is unlikely that there can be any physical connection between the two events.

It has also been suggested that sudden changes in fauna are associated with geomagnetic reversals (see §6.3). If tektites are produced by cometary impacts as Urey (1957, 1962) has suggested, then gases such as frozen ammonia and methane contained in cometary nuclei might have appreciable ecological effects when rapidly introduced into the atmosphere and oceans. However, the evidence is not convincing for any physical association between reversals of the earth's magnetic field and either tektites or faunal changes.

Craters are the dominant surface feature on the moon and on Mars and it is natural to ask whether the earth has also been subject to meteorite bombardment. On the moon, which lacks water and an atmosphere, and on Mars, with probably no water and a tenuous atmosphere, meteorite craters have remained unchanged, except for the impact of later meteorites. On the earth, on the other hand, craters would not persist for long: erosion would wear away their rims, tectonic processes alter their crater shapes and sedimentation fill them in. The most famous terrestrial crater is the Barringer crater in Arizona, which is about 1300 m in diameter and 175 m deep. Its age has been estimated as about 25 000 yr; climatic conditions in the arid desert have helped preserve its shape. Detailed study of the earth's surface, particularly from air photographs, has identified many faint crater features as fossil meteorite craters (see Fig. 1.4 as an example). Dietz (1963) has coined the word *astroblemes* for these ancient craters.

It is not always easy to determine whether a crater has been caused by volcanic activity or is the result of meteoritic impact. Meteorite fragments in surficial deposits are unlikely to be preserved for more than 100 000 yr at the most. Deep drilling is unlikely to prove helpful in this respect since intense shock metamor-

Fig. 1.4   Brent Crater in Algonquin Park, Ontario. Diameter of visible circle 2·9 km. Estimated original diameter of crater 4 km. Depth of sedimentary fill 260 m. Depth of underlying breccia lens 610 m in centre. Age: middle Ordovician, about 450 ± 30 m yr. Estimated energy of formation about 10^{25} ergs. (Department of Energy, Mines and Resources, Earth Physics Branch, Ottawa, Canada, Photograph No. 5015)

Fig. 1.5(a) Shatter cone measurements in Mereenie Sandstone. The cone segments near the instrument describe two-thirds of the complete cone; striations on segments in the upper right corner complete the cone. (Photographed by D. J. Milton)

phism and mixing with pulverized and melted rock would make recognition almost impossible. Drilling and geophysical surveys can, however, provide some information, e.g. evidence of shock metamorphism which would demand far higher pressures than could be released by volcanic explosion, or the lack of volcanic roots—crater deformation should die out at depth. Two further clues have been provided in recent years: *shatter cones* and the discovery of *coesite* and *stishovite*, high pressure polymorphs of quartz.

Shatter cones are conical fragments of rock characterized by striations radiating from an apex (see Fig. 1.5). A recent summary of worldwide observations has been given by Dietz (1968). Their heights vary from less than 1 cm to about 12 m. The apex is never perfectly formed and the apical angles are usually around 90°. Shatter cones may be found singly or in clusters, their axes being parallel. They

Fig. 1.5(b) Shatter-coned 'pipe rock' in Mereenie Sandstone. Coning was initiated along Scolithus tubes, casts of burrows normal to the bedding made by organisms dwelling in the sand. (Photographed by D. J. Milton)

are a rare phenomenon caused by intense stress waves from a shock source. They are not found in the immediate vicinity of the shock centre but at depth beneath the explosion source (a nuclear test or small meteorite crater) or around the periphery of the crater (as in large structures such as the Vredefort Ring in South Africa). The orientation of shatter cones records the direction of advance of the shock front and the assemblage of cone axes should point to a focus that corresponds to the centre of shock propagation. A dynamic mechanism for the formation of shatter cones has been given by Gash (1971). He proposed that they are formed beneath the shock source, but beyond its immediate domain, by brittle or semi-brittle tensile fracturing (no shear is involved) along the direction of the greatest principal stress within a region of low resultant stress.

Coesite and stishovite are high pressure forms of silica. Coesite was first formed by Coes Jr. (1953) and stishovite by Stishov and Popova (1961). Very high pressures are required to transform quartz to coesite (above 20 kbar) and coesite to stishovite (above 80 kbar). The density of coesite is about 2·92 g/cm³ and that of stishovite about 4·28 g/cm³. This large increase in density is a result of the change in coordination of silicon from 4 to 6. Akimoto (1972) has recently reviewed the experimental results on the quartz–coesite–stishovite transformations (see Fig. 1.6) and has also given the most recent estimates of their elastic properties. Davies (1972) has determined the equations of state and phase equilibria of

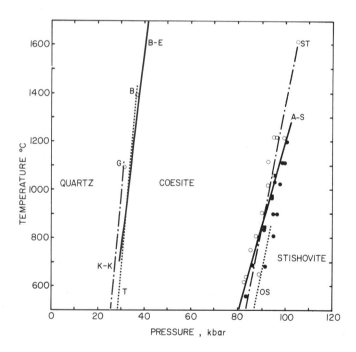

Fig. 1.6   Phase diagram for quartz–coesite–stishovite transition of SiO_2 (after Akimoto (1972). This reference contains details of the different experimental results)

stishovite and a coesite–like phase from shock wave and other data. Coesite and stishovite have been identified in samples of highly sheared and fused Coconino sandstone at the Barringer crater, Arizona (Chao *et al.* (1962)), and coesite in the Ries Kessel Caldera, an ancient basin some 25 km in diameter in southern Germany which had long been supposed to be of volcanic origin. It would appear that natural coesite and stishovite can only result from high pressure impacts (see e.g. Gigl and Dachille (1968)). Gosses Bluff in Central Australia, 160 km west of Alice Springs, has now been definitely confirmed as an eroded crater formed by impact, probably by a comet (Milton *et al.* (1972)). Although shatter cones were found at Gosses Bluff, neither coesite nor stishovite have been identified. Perhaps their absence may be due to re-equilibration during slow cooling near the base of the crater fill. It has recently been suggested that Darwin Glass may be the result of an impact event (Ford (1972)).

The Lonar crater in the Deccan Traps, India, is a unique feature whose origin has been a matter of controversy for well over a century. The results of recent detailed studies, which include the discovery of shock-metamorphosed material, provide definite evidence that it is an impact crater (Fredriksson *et al.* (1973)). As the only known terrestrial impact crater in basalt, it provides a unique opportunity for comparison with lunar craters—in particular microbreccias and glass spherules from Lonar crater have analogues among the Apollo samples.

An extremely interesting discovery is the Campo del Cielo, Argentina, meteorite and crater field (Cassidy *et al.* (1965)). At least nine impact craters were found, all of which lie close to a straight line extending for 17·5 km; the meteorite strewn field is at least 75 km long. The time of impact has been estimated as 5800 ± 200 years ago (this is the carbon-14 age of sizeable charcoal fragments found in and outside the rim of one of the craters). The authors suggest a high-altitude break-up of the parent meteorite in a very flat trajectory as the explanation of the great extent and extreme narrowness of both the crater and meteorite strewn fields. In a near-circular orbit a revolution would take about 88 min, during which time the earth would have rotated about 22° eastward. About 1000 km to the north-west of Campo del Cielo there is another strewn field in northern Chile of the same type of meteorites (hexahedrites), and Cassidy *et al.* suggest that the two strewn fields may be related, the north Chilean hexahedrites being fragments of the same body which made one more revolution before reaching the ground.

The Ries Kessel and the Steinheim Basin are two well-established meteorite craters in southern Germany. About 20 km north-east of the Ries and in line with these two craters is another circular structure, the Stopfenheim Kuppel. Storzer *et al.* (1971) suggest that this is also due to meteoritic impact and that these three impact structures, which lie in a straight line, are related. They suggest that all three were formed simultaneously by three different fragments of the same meteorite which broke up above the earth.

References

Ahrens, L. H., H. von Michaelis, A. J. Erlank and J. P. Willis (1969) Fractionation of some abundant lithophile element ratios in chondrites, in: *Meteorite Research* (ed. P. M. Millman) (D. Reidel Publ. Co., Holland).

Akimoto, S.-I. (1972) The system $MgO-FeO-SiO_2$ at high pressures and temperatures— phase equilibria and elastic properties, *Tectonophysics*, **13**, 161.

Alfvén, H. (1942–45) On the cosmogony of the solar system, *Stockholms Obs. Ann.*, **14**, No.'s 2, 5, 9.

Alfvén, H. (1954) *On the Origin of the Solar System* (Clarendon Press, Oxford).

Alfvén, H. and G. Arrhenius (1970) Structure and evolutionary history of the solar system 1, *Astrophys. Space Sci.*, **8**, 338.

Anders, E. (1962) Meteorite ages, *Rev. mod. Phys.*, **34**, 287.

Anders, E. (1964) Origin, age and composition of meteorites, *Space Sci. Rev.*, **3**, 583.

Anders, E. (1965) Fragmentation history of asteroids, *Icarus*, **4**, 399.

Anderson, D. L. and T. C. Hanks (1972) Formation of the earth's core, *Nature*, **237**, 387.

Banerjee, S. K. and R. B. Hargraves (1972) Natural remanent magnetizations of carbonaceous chondrites and the magnetic field in the early solar system, *Earth Planet. Sci. Letters*, **17**, 110.

Birch, F. (1968) On the possibility of large changes in the earth's volume, *Phys. Earth Planet. Int.*, **1**, 141.

Blake, J. B., Typhoon Lee and D. N. Schramm (1973) Chronometer for s-process nucleosynthesis, *Nature Phys. Sci.*, **242**, 98.

Blake, J. B. and D. N. Schramm (1973) ^{247}Cm as a short-lived r-process chronometer, *Nature Phys. Sci.*, **243**, 138.

Bogard, D., D. Burnett, P. Eberhardt and G. J. Wasserburg (1968) $^{40}Ar-^{40}K$ ages of silicate inclusions in iron meteorites, *Earth Planet. Sci. Letters*, **3**, 275.

Brans, C. and R. H. Dicke (1961) Mach's principle and a relativistic theory of gravitation, *Phys. Rev.*, **124**, 925.

Burnett, D. S. and G. J. Wasserburg (1967) $^{87}Rb-^{87}Sr$ ages of silicate inclusions in iron meteorites, *Earth Planet. Sci. Letters*, **2**, 397.

Butler, R. F. (1972) Natural remanent magnetization and thermomagnetic properties of the Allende meteorite, *Earth Planet. Sci. Letters*, **17**, 120.

Carey, S. W. (1958) A tectonic approach to continental drift, in: *Continental Drift, a symposium, Univ. Tasmania, Hobart*.

Cassidy, W. A., L. M. Villar, T. E. Bunch, T. P. Kohman and D. J. Milton (1965) Meteorites and craters of Campo del Cielo, Argentina, *Science*, **149**, 1055.

Cassidy, W. A., B. Glass and B. C. Heezen (1969) Physical and chemical properties of Australasian microtektites, *J. geophys. Res.*, **74**, 1008.

Chao, E. C. T., J. J. Fahey, J. Littler and D. J. Milton (1962) Stishovite, SiO_2, a very high pressure new mineral from meteor crater, Arizona, *J. geophys. Res.*, **67**, 419.

Chapman, D. R. (1971) Australasian tektite geographic pattern, crater and ray of origin, and theory of tektite events, *J. geophys. Res.*, **76**, 6309.

Clark, S. P., K. K. Turekian and L. Grossman (1972) Model for the early history of the earth, in: *The Nature of the Solid Earth* (ed. E. E. Robertson) (McGraw-Hill).

Coes, L. Jr. (1953) A new dense crystalline silica, *Science*, **118**, 131.

Cox, A. and R. R. Doell (1961) Palaeomagnetic evidence relevant to a change in the earth's radius, *Nature*, **189**, 45.

Davies, G. F. (1972) Equations of state and phase equilibria of stishovite and a coesitelike phase from shock wave and other data, *J. geophys. Res.*, **77**, 4920.

Dearnley, R. (1966) Orogenic fold-belts and a hypothesis of earth evolution, in: *Physics and Chemistry of the Earth*, **7**, 1 (Pergamon Press).

Dicke, R. H. (1957) Principle of equivalence and the weak interactions, *Rev. mod. Phys.*, **29**, 355.

Dicke, R. H. (1962) Implications for cosmology of stellar and galactic evolution rates, *Rev. mod. Phys.*, **34**, 110.

Dicke, R. H. (1966) The secular acceleration of the earth's rotation and cosmology, in: *The Earth–Moon System* (ed. B. G. Marsden and A. G. W. Cameron) (Plenum Press).

Dietz, R. S. (1963) Astroblemes; ancient meteorite impact structures on the earth, in: *The Moon, Meteorites and Comets, The Solar System*, Vol. IV (ed. B. M. Middlehurst and G. P. Kuiper) (Univ. Chicago Press).

Dietz, R. S. (1968) Shock metamorphism of natural materials, *Proc. First Conf. NASA, Goddard Space Flight Center, April 14–16, 1968*, Mono, Baltimore.

Dirac, P. A. M. (1938) A new basis for cosmology, *Proc. roy. Soc.* **A165**, 199.

Dodd, R. T. (1969) Metamorphism of the ordinary chondrites: a review, *Geochim. cosmochim. Acta*, **33**, 161.

Durrani, S. A. (1971) Origin and ages of tektites, *Phys. Earth Planet. Int.*, **4**, 251.

Durrani, S. A. and H. A. Khan (1971) Ivory Coast microtektites: fission track age and geomagnetic reversals, *Nature*, **232**, 320.

Egyed, L. (1956) A new theory on the internal constitution of the earth and its geological-geophysical consequences, *Acta Geol. Acad. Sci Hung.*, **4**, 43.

Egyed, L. (1960) Some remarks on continental drift, *Geofis, Pura Applic.*, **45**, 115.

Fleischer, R. L., C. W. Naeser, P. B. Price, R. M. Walker and M. Maurette (1965) Cosmic ray exposure ages of tektites by the fission-track techniques, *J. geophys. Res.*, **70**, 1491.

Fleischer, R. L. and C. W. Naeser (1972) Search for plutonium-244 tracks in Mountain Pass bastnaesite, *Nature*, **240**, 465.

Ford, R. J. (1972) A possible impact crater associated with Darwin Glass, *Earth Planet. Sci. Letters*, **16**, 228.

Fowler, W. A. and W. E. Stephens (1968) Origin of the elements, Resource Letter OE-1, *Amer. J. Phys.* **36**, 289.

Fredriksson, K., A. Dube, D. J. Milton and M. S. Balasundaram (1973) Lonar Lake, India: an impact crater in basalt, *Science*, **180**, 862.

Gash, P. J. Syme (1971) Dynamic mechanism for the formation of shatter cones, *Nature Phys. Sci.*, **230**, 32.

Gast, P. W. (1962) The isotopic composition of strontium and the age of stone meteorites, 1, *Geochim. cosmochim. Acta*, **26**, 927.

Gigl, P. D. and F. Dachille (1968) Effect of pressure and temperature on the reversal transitions of stishovite, *Meteoritics*, **4**, 123.

Glass, B. and B. C. Heezen (1967) Tektites and geomagnetic reversals, *Nature*, **214**, 372.

Green, H. W. II, S. V. Radcliffe and A. H. Heuer (1971) Allende meteorite: a high voltage electron petrographic study, *Science*, **172**, 936.

Heezen, B. C. (1959) Palaeomagnetism, continental displacements and the origin of submarine topography, *Inter. Ocean. Cong.*

Herczeg, T. (1968) *Planetary cosmogonics*, in *Vistas in Astronomy* (ed. A. Beer) **10**, 175 (Pergamon Press).

Hoffman, D. C., F. O. Lawrence, V. L. Merwherter and F. M. Rourke (1971) Detection of plutonium-244 in nature, *Nature*, **234**, 132.

Hoyle, F. (1955) *Frontiers of Astronomy* (Heinemann).

Hoyle, F. (1960) On the origin of the solar nebula, *Q. J. roy. Astr. Soc.*, **1**, 28, 1960.

Hoyle, F. (1972) The history of the earth, *Q. J. roy. Astr. Soc.*, **13**, 328.

Hoyle, F. and J. V. Narlikar (1971) On the nature of mass, *Nature*, **233**, 41.

Hoyle, F. and J. V. Narlikar (1972) Cosmological models in a conformally invariant gravitational theory—II A new model, *Mon. Not. roy. Astr. Soc.*, **155**, 323.

Kaushal, S. K. and G. W. Wetherill (1970) Rb^{87}–Sr^{87} age of carbonaceous chondrites, *J. geophys. Res.*, **75**, 463.

Kirsten, T., D. Kranknowsky and J. Zähringer (1963) Edelgas und Kalium Bestimmungen an einer grösseren Zahl von Steinmeteoriten, *Geochim. cosmochim. Acta*, **27**, 13.

Kohman, T. P. (1954) Geochronological significance of extinct natural radioactivity, *Science*, **119**, 851.

Larimer, J. W. (1967) Chemical fractionations in meteorites 1. Condensations of the elements, *Geochim. cosmochim. Acta*, **31**, 1215.

Larimer, J. W. and E. Anders (1967) Chemical fractionations in meteorites, 2, Abundance patterns and their interpretation, *Geochim. cosmochim. Acta*, **31**, 1239.

Larimer, J. W. and E. Anders (1970) Chemical fractionations in meteorites, 3, Major element fractionations in chondrites, *Geochim. cosmochim. Acta*, **34**, 367.

Mason, B. (1962) *Meteorites* (John Wiley and Sons Inc.).

McCrae, W. H. (1960) The origin of the solar system, *Proc. roy. Soc.* **A 256**, 245.

Milton, D. J., B. C. Barlow, R. Brett, A. R. Brown, A. Y. Glikson, E. A. Manwaring, F. J. Moss, E. C. E. Sedmik, J. Van Son, G. A. Young (1972) Gosses Bluff impact structure, Australia, *Science*, **175**, 1199.

Nölcke F. (1930) *Der Entwicklungsgang unseres Planetensystems*, (Bonn).

Notsu, K., H. Mabuchi, O. Yoshioka, J. Matsuda and M. Ozima (1973) Evidence of the extinct nuclide ^{146}Sm in 'Juvinas' achondrite, *Earth Planet. Sci. Letters*, **19**, 29.

O'Keefe, J. A. (1963) The origin of tektites, in: *Tektites* (ed. J. A. O'Keefe), Univ. Chicago Press.

O'Keefe, J. A. (1966) The origin of tektites, *Space Sci. Rev.*, **6**, 174.

Papanastassiou, D. A. and G. J. Wasserburg (1969) Initial strontium isotopic abundances and the resolution of small time differences in the formation of planetary objects, *Earth Planet. Sci. Letters*, **5**, 361.

Patterson, C. (1956) Age of meteorites and the earth, *Geochim. cosmochim. Acta*, **10**, 230.

Peebles, J. and R. H. Dicke (1962) The temperature of meteorites, Dirac's cosmology and Mach's principle, *J. geophys. Res.*, **67**, 4063.

Podosek, F. A. (1970) Dating of meteorites by the high-temperature release of iodine-correlated Xe^{129}, *Geochim. cosmochim. Acta*, **34**, 341.

Reynolds, J. H. (1963) Xenology, *J. geophys. Res.*, **68**, 2939.

Reynolds, J. H. (1967) Isotopic abundance anomalies in the solar system, *Ann. Rev. Nucl. Sci.*, **17**, 253.

Rowe, M. W. and P. K. Kuroda (1965) Fissiogenic xenon from the Passamonte meteorite, *J. geophys. Res.*, **70**, 709.

Safronov, V. S. and E. V. Zvjagina (1969) Relative sizes of the largest bodies during the accumulation of planets, *Icarus*, **10**, 109.

Shapiro, I. I., W. B. Smith, M. E. Ash, R. P. Ingalls and G. H. Pettengill (1971) Gravitational constant: experimental bound on its time variation, *Phys. Rev. Letters*, **26**, 27.

Sonett, C. P., D. S. Colburn, K. Schwartz and K. Keil (1970) The melting of asteroidal-sized bodies by unipolar dynamo induction from a primordial T Tauri sun, *Astrophys. space Sci.*, **7**, 446.

Spitzer, L. (1939) The dissipation of planetary filaments, *Astrophys. J.*, **90**, 675.

Stishov, S. M. and S. V. Popova (1961) New dense polymorphic modification of silica, *Geokhimiya*, **10**, 837.

Storzer, D., W. Gentner and F. Steinbrunn (1971) Stopfenheim Kuppel, Ries Kessel and Steinheim basin: a triplet cratering event, *Earth Planet. Sci. Letters*, **13**, 76.

Taylor, H. P. and S. Epstein (1970) Oxygen and silicon isotope ratios of lunar rock 12013, *Earth Planet. Sci. Letters*, **9**, 208.

ter Haar, D. (1967) On the origin of the solar system, in: *Annual Review of Astronomy and Astrophysics*, **5**, 267.

Urey, H. C. (1957) Origin of tektites, *Nature*, **179**, 556.

Urey, H. C. (1962) Origin of tektites, *Science*, **137**, 746.

Van Schmus, W. R. and J. A. Wood (1967) A chemical-petrologic classification for the chondritic meteorites, *Geochim. cosmochim. Acta*, **31**, 747.

Ward, M. A. (1963) On detecting changes in the earth's radius, *Geophys. J.*, **8**, 217.

Wesson, P. S. (1973) The implications for geophysics of modern cosmologies in which G is variable, *Q. J. roy. Astr. Soc.*, **14**, 9.

Williams, I. P. and A. W. Cremin (1968) A survey of theories relating to the origin of the solar system, *Q. J. roy. Astr. Soc.*, **9**, 40.

Wilson, J. T. (1960) Some consequences of expansion of the earth, *Nature*, **185**, 880.

Wood, J. A. (1963a) Physics and chemistry of meteorites, in: *The Moon, Meteorites and Comets, The Solar System*, Vol. IV (ed. B. M. Middlehurst and G. P. Kuiper) (Univ. Chicago Press).

Wood, J. A. (1963b) On the origin of chondrules and chondrites, *Icarus*, **2**, 152.

Wood, J. A. (1968) *Meteorites and the Origin of Planets* (McGraw-Hill).

York, D. and R. M. Farquhar (1971) *The Earth's Age and Geochronology* (Pergamon Press).

2

Mercury, the Moon, Mars and Venus

2.1 Introduction

One might think that there have been no changes over the last few years in our knowledge of the general characteristics of the solar system, such as the mean density, mean radius and period of rotation of the planets. Apart from increased precision in some of the measured quantities, there have also been some radical changes, such as in the rotation rate of Venus and Mercury. These have come about through the development of radar astronomy, and will be discussed below. The Apollo landings on the moon and the Mariner flights to Mars and Venus have also added enormously to our knowledge of these bodies, and the results of these space flights will also be discussed in this chapter. For the outer planets, our estimate of the radius of Neptune has been increased 10 per cent (Bixby and van Flandern (1969))—this has reduced the average density of the planet from 2·23 to 1·65 g/cm^3; a tenth satellite of Saturn was discovered by Dollfus in 1966; Ash *et al.* (1971) have concluded that Pluto's mass (and hence its average density) cannot be determined reliably from existing data—present views on planet and satellite formation from the solar nebula indicate that it is unlikely that a body formed as far away as Pluto could have a density much higher than 3 g/cm^3. Much higher values have been suggested in the past. A survey of representative dynamical data for the major planets and satellites has been given by Message (1972), and by Kovalevsky (1972).

2.2 Planetary radar astronomy: the rotation rate of Mercury

Planetary radar astronomy has essentially developed only during the last decade but has already made many major contributions to our knowledge of the solar system. It has given us new information on planetary motions (both orbital and spin), planetary distances, planetary surface characteristics and the properties of the intervening medium. Radar measurements have improved by almost five orders of magnitude the accuracy relating the astronomical unit (AU) to a terrestrial unit of length, but it is in the determination of planetary rotations (particularly of Mercury and Venus) that the most spectacular advances have been made. A good review of the dynamics of planetary rotations has been given by Goldreich and Peale (1968).

Because of its small size, proximity to the sun and low reflectivity, Mercury is very difficult to observe telescopically and to photograph. However, since the 1880s most astronomers have believed that Mercury was rotating slowly with a period equal to its orbital period of 88 days. It was not until 1965 (Pettengill and Dyce (1965)) that delay Doppler maps of the surface, obtained using radar, showed the rotation period to be 59 ± 3 days. Following these radar observations, several groups have independently re-examined the visual determinations of the rotation period and concluded that the optical data are consistent with a rotation period of about 59 days—in fact the most accurate rotation rate for Mercury (58·66 days) has been obtained from visual data (Smith and Rees (1968); Chapman (1968)).

Peale and Gold (1965) attempted to explain this new result in terms of solar tidal torques only. For stability the average torque exerted on Mercury must vanish. Because of the large eccentricity of Mercury's orbit, the tidal torque could change sign twice during an orbital revolution and could thus have a zero average over this interval. Colombo (1965) first noticed that the observed sidereal spin period T_s was nearly two-thirds of the 88 day orbital period T_o and suggested that the axial rotation might be 'locked' to the orbital motion in a three-halves resonance state by the additional solar torque exerted on an axial asymmetry in Mercury's inertia ellipsoid. Colombo and Shapiro (1966) investigated a two-dimensional model of sun–Mercury interactions and showed that resonances occur when a planet makes an integral number of half-rotations during one orbital revolution, i.e. when

$$T_s = 2T_o/k \qquad (k \text{ an integer}) \qquad (2.1)$$

Thus the solar torque exerted on the permanent asymmetry could cause Mercury to be trapped into a $k=3$ resonance spin state (see Fig. 2.1; and Goldreich and Peale (1966)). For reasonable assumptions about the magnitude of the tidal torque, the asymmetry in the inertia ellipsoid required for stability in the $k=3$ state need only be several orders of magnitude less than the observed asymmetry of the moon.

The stability conditions for the rotational motion of any celestial object that interacts with the primary about which it orbits have been investigated in detail by Bellomo et al. (1967). These authors also considered the long-term evolution of Mercury's spin rate. If initially Mercury was spinning much faster in either a direct or retrograde sense, then in order to have reached its present rotation rate it would have had to avoid being captured into either the $k=4$ resonance state or the synchronous $k=2$ state. Penetration of the higher resonance is not unlikely because of the high orbital eccentricity required for capture, for which the resonance state would not be stable. Conditions for penetration of the $k=2$

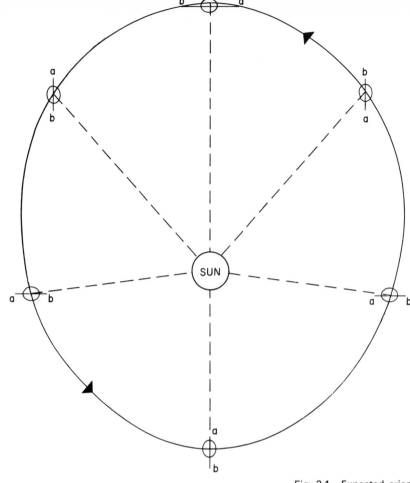

Fig. 2.1 Expected orientation of Mercury's axis of minimum moment of inertia as a function of orbital position for the $k = 3$ resonance spin state (after Colombo (1965))

resonance are more difficult to establish, but it appears that Mercury may have evolved to its present spin state from an initially retrograde axial rotation. The rotation of Venus poses an even more interesting problem, which is discussed in §2.6.

2.3 The solar wind

Satellites have now shown that the interplanetary medium in the vicinity of the earth is not just empty space but instead is filled with a highly tenuous plasma (ionized gas) which is being continuously blown radially out from the sun at speeds of the order of 300–500 km/s. This gas has been called the *solar wind*. Biermann (1953) presented evidence for the existence of a continuous solar wind from observations of comet tails. Impacts between solar wind particles and gaseous particles in the tail of a comet cause the latter to always point away from the sun. Biermann (1971) has later given a comprehensive review of the interaction of the solar wind with a comet. Ejection of particles from the sun to form a 'wind' is probably the result of an accleration process involving changing magnetic or electric fields, i.e. the sun acts as a giant particle accelerator. More details about the solar wind can be found in the book by Brandt (1970), and in Sonett *et al.* (1972).

The solar wind is very 'gusty', showing fluctuations in energy and density in times of the order of hours. The solar plasma consists primarily of ionized hydrogen (protons and electrons) and is electrically neutral; the density is of the order of 10 ions/cm^3. Embedded in the solar wind is an interplanetary magnetic field whose strength is of the order of 5 γ* during quiet solar periods but increases to many times this value during periods of high solar activity. Its energy density is much smaller (approximately 1 per cent of that of the solar wind), so that it is carried along by the solar wind.

Plasmas and magnetic fields tend to confine each other. If a streaming plasma encounters a magnetic body such as a magnetized sphere, the plasma will confine the magnetic field to a limited region around the body. The body in turn will tend to exclude the plasma, thus creating a hole or cavity. The size of the cavity is determined by the energy density of the streaming plasma and the degree of magnetization of the body. In addition, if the velocity of the plasma is great enough to be highly supersonic, a detached shock wave may be produced in the region ahead of the cavity boundary. This is analogous to the formation of the detached shock front of an aerodynamic object travelling at hypersonic speed (above a Mach number of about 5). The analogy, however, is by no means perfect. In aerodynamics, the shock wave results from collisions of particles and is about one mean free path thick. In the solar wind, a Coulomb collision mean free path (approximately 10^{14} cm) is so large that collisions play no part in the observed shock wave. This 'collisionless' shock is produced by the action of the magnetic field; the characteristic dimension is the cyclotron radius, not the mean free path. A 1 keV proton in the interplanetary magnetic field of 5 γ at IAU has a cyclotron radius of about 1000 km.

As the solar wind flows through the solar system, it interacts with comets, the moon, the planets, interplanetary dust and cosmic rays. If a planet possesses a magnetic field, the solar wind is arrested on the sunward side of the planet at a radial distance R such that the dynamic pressure of the wind is equal to the opposing pressure of the (compressed) magnetic field B of the planet, i.e. when

$$knmv^2 = B^2/8\pi \qquad (2.2)$$

where n is the number density of protons of mass m and velocity v, and k is a numerical factor between 1 and 2 (Beard (1960)). If M is the magnetic dipole moment of the planet, $B \propto M/R^3$. A necessary condition for the existence of a

---

* 1 $\gamma = 10^{-5}$ Γ. See also §5.1.

trapped radiation belt around a planet is that the stagnation point given by equation (2.2) is far enough above the surface for the atmospheric density to be sufficiently small (see Van Allen (1966) for a more quantitative statement of this condition). The interaction of the solar wind with the moon is very different from that with the earth (see Fig. 5.9) since the moon has little or no magnetic field (see §2.4). Thus the solar wind particles reach the lunar surface and are absorbed there. This absorption creates a cavity in the solar wind but there is no lunar bow shock. The electrical conductivity of the lunar interior appears to be low enough to allow the interplanetary field to pass through the moon without significant change.

Measurements of the intrinsic magnetic fields of Mars and Venus by fly-by spacecraft indicate that their magnetic moments are very small (see §§2.5 and 2.6). Hence the solar wind interaction is with the solid planet or its atmosphere. If Mars and Venus possessed ionospheres similar to that of the earth, they would form an obstacle to the solar wind and cause a bow shock to be formed. A schematic picture (see Fig. 2.2) of such an interaction has been given by Dessler (1968). This should be compared with Fig. 5.9. There are at present no observa-

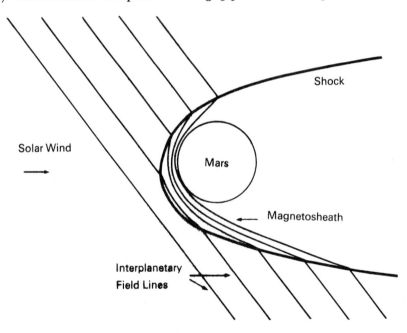

Fig. 2.2 Schematic diagram of the interaction of the solar wind with Mars (after Dessler (1968))

tions on Mars to test this theoretical model; however, measurements made during the Mariner 5 fly-by of Venus indicate a bow shock close to that planet (Bridge et al. (1967)). Further theoretical models of the interaction of the solar wind with planetary atmospheres have been given by Elco (1969); Johnson and Midgley (1969); Spreiter *et al.* (1970); and Beard (1973).

2.4 The moon

Our knowledge of the moon has increased enormously from data obtained by US and USSR space craft. The US Ranger, Surveyor, Orbiter and Apollo missions successively carried out reconnaissance, unmanned and manned missions, and the published data run into thousands of pages. It is impossible to give a detailed summary of all the findings to date, or even to give a comprehensive bibliography. Instead a few selected topics will be discussed: the origin of the moon, its magnetic field, its seismic response and its temperature. The moon's gravitational field, with its large positive gravity anomalies (mascons) is discussed in §4.5. Figs. 2.3 and 2.4 show some of the varied surface features observed by Apollo 15.

In our solar system the earth–moon system occupies a unique position. It is the only planet–satellite system in which the satellite has a larger angular momentum than that of the planet. Again, while there are other bodies in the

Fig. 2.3   Landing site of Apollo 15 between the jutting bend of Hadley Rille and the Apennines (NASA 15-0993)

Fig. 2.4 Meandering rilles near the Prinz crater. They may have been formed by flowing lava or jets of gas from the interior. The paper-clip-shaped trough was formed either by collapse of a lava tube or subsidence between parallel faults (NASA 15-11978)

solar system (e.g. Io, one of the satellites of Jupiter) which have approximately lunar mass or larger, the earth's moon has the largest fractional mass, the ratio being about $\frac{1}{81}$: 1. Among the other terrestrial planets, only Mars has satellites and their masses are of the order of 10^{-8} to 10^{-9} of that of the planet.

2.4.1   *The origin of the moon*   There are three main classes of theories of the origin of the moon: that the moon was captured by the earth; that the moon and earth were originally one body, the moon being later ejected from the earth by some fission process; and that the moon and earth were formed together as à binary system. All of these theories are open to serious objections although their proponents have found evidence in the analyses of the lunar rocks brought back by the Apollo flights to support their own particular views.

The origin of the earth–moon system is a unique event and it is tempting to search for other unique features of the earth and try to relate them, e.g. the existence of a magnetic field, the only one among the terrestrial planets, the existence of an oxygen atmosphere and the existence of life. If the moon and earth were once very close together, it may be conjectured that the intense stresses and heating caused the rapid evolution of an atmosphere and oceans through intensive and rapid defluidization of the surface rocks of the earth. These events then perhaps set the stage for the evolution of life on the earth. If such events took place early in the earth's history, subsequent evolution of the earth's surface could have removed all traces from the geologic record. The moon, on the other hand, may still retain evidence of such a catastrophic event.

Let us now consider in some detail the three main classes of theories of the origin of the moon. The plane of the moon's orbit is at present inclined at about 5° to the ecliptic and the earth's equator is inclined by about 23·5° to the ecliptic. By integrating the moon's orbit backwards in time, Gerstenkorn (1955) showed that the inclination of the moon's orbit and the obliquity change rapidly as the moon approaches the earth.* If the moon were to approach the earth in a retrograde orbit (i.e. moving in the opposite sense to its present motion and to the rotation of the earth) it would be drawn in by the tides and not pushed out as at present. As it came closer to the earth the torque on the plane of its orbit would tip the orbit across the poles and down the other side. The effect of this would be to reverse the direction of the moon's orbit around the earth, and once in a prograde orbit the tides on the earth would begin to push the moon outwards as at present. One of the chief difficulties with such an evolution and similar theories lies in the time scale and the catastrophic events that would have accompanied the moon's closest approach to earth (about 1700 m yr ago). The geologic record goes back at least 3000 m yr and would be expected to contain some evidence of the moon's capture. It is more than possible that the oceans would have evaporated and the earth's surface melted during the moon's approach. Associated with all capture theories there is also the implausibility of the process which requires the velocity of an initially unbound moon to be reduced by some kind of dissipation process.

Singer (1968, 1970a) is one of the supporters of an origin of the moon by capture, but capture as it passed near the earth in a direct (prograde) orbit, shortly after the formation of the earth and moon about 4500 m yr ago, and not from an initially retrograde orbit as proposed by Gerstenkorn (1955), MacDonald (1964) and others. Singer's calculations are based on a frequency-dependent tidal perturbation. He believes that the effects of lunar capture on the earth would have been cataclysmic, leading to intensive heating of its interior, to volcanism and to the immediate formation of an atmosphere and hydrosphere— the capture of the moon may then have given rise to the unique properties of the earth (in the solar system) and to the early evolution of life. For a number of reasons, Singer favours a time scale of about 4500 m yr ago for the capture of the moon rather than about 2000 m yr.

The capture of the moon could have initiated the formation of a core in the

---

* The past history of the lunar orbit has been re-calculated by Goldreich (1966); the dynamical aspects of lunar origin have recently been reviewed by Kaula (1971).

earth (see §10.5), which took place within the first hundred million years of the earth's existence. Again the great age (over 4000 m yr) of igneous rocks from the lunar surface indicates that capture must be at least that old. The earlier age for the capture of the moon is based on the assumption that the elastic parameters and dissipation constants have not changed throughout geologic time. Some support for this assumption has been provided by Wells (1963), who suggested that the yearly growth bands on fossil corals are made up of daily growth ridges. By counting the number of daily ridges per yearly band he estimated the number of days per year back to mid-Devonian times, some 380 m yr ago (see §4.4). The length of the day was then about 22 h, consistent with a constant phase lag of 3° between the tidal bulge on the earth and the earth-moon line of centres. Remarkable as the coral data are, they extend back less than 10 per cent of geologic time. Most of the tidal energy is dissipated in the shallow seas and thus depends quite critically on the details of ocean–land boundaries. Evidence of continental drift and sea-floor spreading (see §6.5) indicates a very different configuration of the oceans and continents as recently as 500 m yr ago, and ocean–tide dissipation could thus have been substantially less in earlier times.

Herz (1969) pointed out that massifs of anorthosite globally distributed around the world have a common age of around 1300 m yr. Although these massifs are at present randomly distributed throughout the world, when plotted on a reconstruction of the continents in their pre-drift positions, they become aligned into two broad belts, one of which crosses the supercontinent of Laurasia in the northern hemisphere and the other Gondwanaland in the southern hemisphere. Anorthosite is a material which must have originated at high temperatures and pressures and, because of the coincidence of the massifs in both space and time, Herz suggested that they may have been thrust up from the mantle during a cataclysmic event which overtook the earth some 1300 m yr ago. In a recent study of lead isotope data from young mantle-derived volcanics, Ulrych (1969) proposed that the mantle is a two-stage system. From a consideration of the global variations in lead isotope ratios, he obtained an age of 1300 m yr for the formation of the second stage. This coincides with the age of the anorthosite event, and Ulrych suggested that both these events may have come about as a result of the circumstances which led to the birth of the moon. However, it seems more than likely that such catastrophic events as would have occurred following the birth of the moon would have obliterated the geologic record of earlier times —and the geologic record certainly extends back earlier than 1300 m yr.

Let us consider now the possible origin of the moon by fission from the earth. If a rotationally unstable body breaks up into two separate components Lyttleton (1953) has shown, from considerations of energy and angular momentum, that the mass ratio of the two parts must be about 8 to 1. There is no possibility of a mass breaking into two equal or nearly equal pieces, nor is there any possibility of a very large mass ratio, such as about 100 to 1 as is the case for the present earth–moon system. Again, the length of the day would have been about 5 h when the moon had just split off from the earth, whereas an initial rotation period close to 2·6 h is necessary to induce rotational instability. Lyttleton also

36

showed that the type of instability that brings about disruption of a rotating mass of approximately uniform density is such that the two resulting components would separate to infinity, i.e. the smaller cannot be put into close orbit around the larger. It thus appears that it is dynamically impossible for the moon by itself to have been ejected from the earth through rotational instability. Lyttleton (1960) later suggested that perhaps the earth could throw off a mass equal to about one-tenth of its own mass and that at the same time a much smaller mass, amounting to about one-hundredth of the whole, could be left intermediate between the two separating main pieces. He showed that if such an intermediate small droplet were formed it could be left in orbit around the earth, the smaller mass (about a tenth of the whole) carrying away in hyperbolic orbit the excess angular momentum that rendered the original mass unstable. Lyttleton suggested that the smaller mass (about a ninth that of the earth) might be Mars. Since the velocity of escape from the earth is about 11 km/s, a body projected away from the earth with sufficient speed for complete escape would soon reach a great distance with a residual relative speed of this order of magnitude. However, the velocity of escape from the sun at the general distance of the earth is more than 42 km/s and thus the smaller mass (Mars), ejected in the proposed break-up, would not be able to escape from the sun.

The initial orbit around the earth of a large satellite formed in this way would be very different from the present lunar orbit since its least distance would only be about twice the earth's radius. The orbital period would exceed 7 h while the earth would have a rotation period considerably less than this. Thus tidal forces would begin at once to drive the moon further out and the effect would be far stronger at this close separation than it is today. As in the case of capture theories, the earth–moon system should still show some effects of the moon's birth had the event taken place during geologic time.

Lyttleton (1960) also showed that it is dynamically possible for Mercury and Venus to have at one time formed one body. As the planets exist now there is no simple relationship between their densities and their masses; nor is there any trend in their densities with distances from the sun. McCrae (1969) has shown, however, that with the most recent determinations of the masses and radii of the planets, the systems earth–moon–Mars and Mercury–Venus could have resulted from the break-up of two unstable planetary bodies of identical chemical composition. In order to start with a common supply of raw material and to finish up with several planets possessed of a variety of masses and densities, showing no simple correlation among themselves, it seems reasonable to suppose that one or more bodies were formed as an intermediate stage, that segregation of material occurred within them and that these bodies then broke up in a variety of ways to yield the existing planets. The asteroids may be the debris left behind from one such break-up, while similar considerations applied to Saturn, Uranus and Neptune suggest that they may each be the remaining, larger portion of a single primitive planet broken up by rotational instability.

O'Keefe (1970) believes that the analyses of the lunar material brought back by Apollo 11 support the idea that the moon was formed by the break-up of the

37

earth, and that after separation strong tidal interaction between the two bodies heated both of them (especially the moon), causing the loss of large amounts of mass and angular momentum from the moon. If over half of the moon's mass was boiled away at this time, then we have simultaneously an explanation for the deficiency of volatile elements and the enhancement of refractory elements in the moon (as indicated by the Apollo 11 data), as well as for the loss of angular momentum which is required for the formation of the moon by fission. (The loss of about one-half of the initial angular momentum of the earth–moon system is a difficulty that all fission hypotheses face.) In point of fact, theories in which fission of the earth was followed by a period of very strong heating in the moon had been suggested (e.g. Wise (1969); O'Keefe (1969)) before the Apollo results became available.

Preliminary analyses of lunar samples returned by Apollo 11, Apollo 12 and Luna 16 indicate that the major surface rock-forming units of the maria are basalts (LSPET (1969, 1970); Vinogradov (1971)). This conclusion is based both on the mineralogy and on typical igneous textures. The basalts are neither

Fig. 2.5 Log scale plot of chemical-element abundances in lunar basalt lavas (Apollo 11 mean) versus Type 1 carbonaceous chondrites. Elements above the line are relatively enriched, and below the line relatively depleted in the lunar rocks (adapted from Mason and Melson (1970))

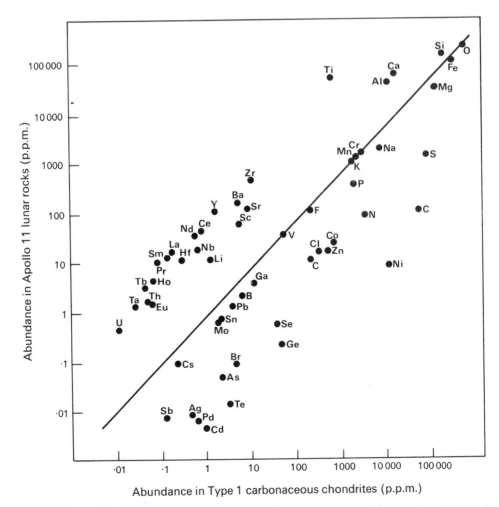

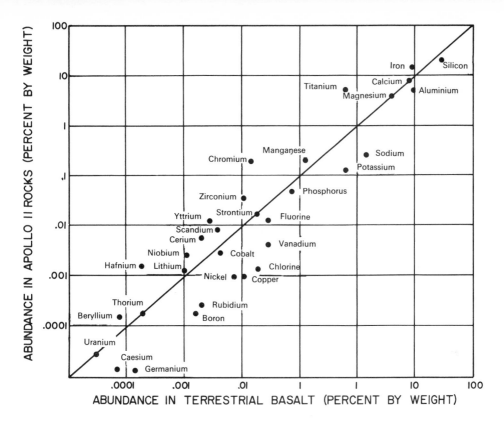

Fig. 2.6   Log scale plot of chemical-element abundances in lunar basalt lavas (Apollo 11 mean) versus terrestrial basalts (after Mason (1971))

compositionally nor texturally uniform within a given mare site. The chemical element abundances in lunar basalts are compared to those in Type I carbonaceous chondrites in Fig. 2.5, and to terrestrial basalts in Fig. 2.6. It can be seen that relative to both cosmic and terrestrial abundances, lunar basalts are depleted in volatiles and enriched in certain refractory elements, including titanium and especially the rare earth elements. The problem is when and where did the lunar material undergo this relative loss and gain of elements. Although the high vacuum conditions at the lunar surface would favour the loss of volatiles from cooling lava pools, evidence indicates that the lunar interior source material was already depleted before the lavas formed. Perhaps there always has been a depletion of volatiles and enrichment of refractories in the outer regions of much of the moon. It has been pointed out that lunar basalt abundances appear to be more closely related to terrestrial than to cosmic abundances (see Figs. 2.5 and 2.6), thereby indicating a close relationship between the origin of the earth and the moon. However, as Ringwood (1970a) has stressed, there are significant differences between terrestrial and lunar basalts and between their inferred source regions which preclude the possibility that the moon was once fissioned from the earth's mantle, in which case lunar basalts would be essentially identical

39

to terrestrial basalts. Finally, the fission hypothesis cannot easily explain the present 5° inclination of the lunar orbit to the plane of the ecliptic. Integration of the moon's orbit backwards in time shows that it would have been inclined at least 10° to the earth's equator when the two bodies were close together. Fission would result in an initially equatorial orbit and this orbit would have evolved into one that lay in the plane of the ecliptic at the present earth–moon separation (Goldreich (1966)).

The history of the tidal evolution of the earth–moon system indicates that at one time the moon was only 2·7 earth radii from the earth. This is almost identical with the Roche limit (see Appendix A) and, as Ringwood (1970a) has pointed out, it does not seem likely that the agreement between these two distances is just a coincidence (as is implied by the capture hypothesis). It is more plausible that the material now on the moon originally came from within the Roche limit where it existed in the form of a 'sediment ring' of planetesimals—in fact Öpik in 1961 proposed that the moon was formed by the aggregation of just such a sediment ring. This hypothesis clearly suggests a genetic relationship between the material now in the moon and the earth. Öpik did not attempt to explain the origin of the sediment ring nor its composition. A possible explanation has been given by Ringwood (1970a).

According to Ringwood's 'precipitation' hypothesis, in the later stages of accretion of the earth a massive primitive atmosphere developed, the temperature being sufficiently high to evaporate selectively a substantial proportion of the silicates falling upon the earth. The primitive atmosphere was subsequently dissipated as the sun passed through a T-Tauri phase and, on cooling, the silicates were precipitated to form a ring of planetesimals orbiting the earth. Within the above framework there is a great deal of flexibility concerning the detailed physical and chemical processes. This precipitation hypothesis of Ringwood (which is supported by Cameron (1970)) is closely related to the fission hypothesis since it maintains that the material now in the moon was ultimately derived from the earth, not from the mantle but from the earth's massive, primitive atmosphere. Further discussion of this hypothesis has been given by Ringwood (1970b, 1971, 1972) and Singer (1971a).

Finally let us consider the hypothesis that the earth and moon formed as a binary system—a hypothesis which faces no real dynamical problems. The main difficulty is that if the earth and moon formed from the same dust cloud, they would be expected to have a similar composition, yet the moon's average density is only 3·34 g/cm³ compared with 5·5 g/cm³ for the earth. Anderson (1972) has recently attempted to answer this question. The inclination of the orbit of the moon must have been greater in the past, decreasing with time owing to gas drag and tidal interactions. The composition of the solar nebula depends not only on distance from the sun, but also on distance from its median plane. The special features of the moon's composition are thus presumed to arise because of the strong dependence of condensation temperature on pressure in the nebula and the rapid decrease in pressure away from the median plane of the disc. Anderson suggests that the composition of the moon can be understood if it

40

formed relatively late, after most of the condensable material near the median plane had been swept up by the earth. The earth was probably more than 50 per cent assembled before the moon started and will have swept up most of the iron. The bulk of the moon is thus composed of material that condensed at much higher temperatures and lower pressures than the other terrestrial planets and chondritic meteorites. In a later paper (1973), Anderson develops these ideas in more detail—he is also able to account for the moon's depletion in volatiles and enrichment in refractories.

In spite of the difficulties associated with the gross chemical differences between the earth and the moon, there are a number of advantages for theories in which the moon formed from material in orbit around the earth. Ganapathy et al. (1970), from an analysis of trace elements in Apollo 11 lunar samples, and Tatsumoto (1970), from an isotopic study of the U–Th–Pb systematics of Apollo 11 samples, both favour a simultaneous origin for the earth and moon as a planet–satellite system.

None of the three traditional theories of lunar origin are completely satisfactory; perhaps some more complicated sequence of events involving several bodies is necessary. A moon formed by the accretion of planetesimals orbiting the earth and other space debris seems more likely than a process demanding fission from the earth or capture. Although future analyses of lunar samples will greatly increase our knowledge of the moon, they will probably also bring fresh problems, and the question of the origin of the moon is likely to remain unanswered for a long time yet.

2.4.2 *The magnetic field of the moon* Data from Explorer 35 give an upper limit to the overall dipole moment of the moon of 10^{20} emu which would create a surface field of $< 2\,\gamma$ (Sonett et al. (1967)). However, there appear to be significant local magnetic fields. The lunar surface magnetometer at the Apollo 12 site recorded a field of $36\,\gamma$ (Dyal et al. (1970)), while at the Apollo 14 site, fields of $103\,\gamma$ and $43\,\gamma$ were measured at stations over 1 km apart by the lunar portable magnetometer (Dyal et al. (1971)). Although differences in the magnetic susceptibility of rocks can produce magnetic field variations, there is at present no global lunar field of sufficient strength to cause such variations.

Strangway et al. (1971) have shown that the remanent magnetization of lunar samples brought back by Apollo 11 and 12 consists, in most cases, of two distinct components: an unstable component which is easily removed by a.c. demagnetization in fields of < 100 Oe and which is considered to be an isothermal remanence* acquired during or after return to earth, and a second component which is unaltered by demagnetization in fields up to 400 Oe. This very stable component is as strong as that found in almost any igneous terrestrial rocks; it is probably a thermal remanent magnetization due to cooling from above 800°C in the presence of a field of a few 1000 γ. If we accept this interpretation it implies that the moon had a magnetic field at the time of formation of the Apollo 11 and

---

* See §6.2 for an account of the magnetization of rocks.

41

12 samples, i.e. during the period from 3000–3800 m yr ago. If we also accept that breccias acquired their magnetization in the same way, the period during which a magnetic field existed can be extended to cover the time of breccia formation which could be much younger. The origin of the ancient lunar field is unknown although a number of possibilities can be suggested:

(i)   The moon had at one time a liquid core and generated its own field by a self-exciting dynamo action (see §5.5).
(ii) When the moon was close to the earth the surface rocks became magnetized by the earth's magnetic field.
(iii) There was an early magnetic field present in the solar system.

The first possibility seems the most probable. This would imply that the moon has an iron core which was above the melting point over 3000 m yr ago. If the internal temperature has fallen to below the melting point of iron since then, it is not difficult to understand why the moon has lost its internal magnetic field: the free-decay time is only a few thousand years. A difficulty with the second possibility, i.e. that the lunar magnetic field was acquired when the moon was much closer to the earth, is that the range of ages of the magnetic lunar material returned to earth is at least 500 m yr, i.e. the moon would have had to have been very close to the earth for an improbably long time, and also at a distance close to the Roche limit.

In addition to permanent magnetic fields, magnetometers at the Apollo 12 and 14 sites have recorded transient fields due to electric currents generated in the interior of the moon. These transient fields are induced in the moon by changes in the magnetic field associated with the solar wind. Fig. 2.7 shows the effects of a solar storm on the lunar magnetic field at the Apollo 12 site and on the magnetic field at a site on earth. The top curve shows the rise in solar wind

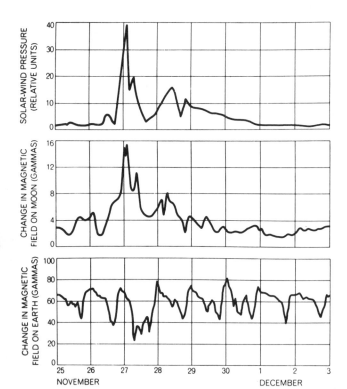

Fig. 2.7 Effects of a solar storm on the lunar magnetic field at the Apollo 12 site and on the terrestrial magnetic field at a site on earth. The top curve shows the rise in solar wind pressure accompanying the arrival of solar particles in the vicinity of the earth–moon system. The middle curve shows that the horizontal component of the steady field recorded by Apollo 12 rises in direct proportion to solar wind pressure. The bottom curve shows that the intensity of the terrestrial field exhibits a corresponding decrease during the solar storm (after Dyal and Parkin (1971))

42

pressure accompanying the arrival of solar particles in the vicinity of the earth–moon system; the middle curve shows that the horizontal component of the steady field recorded by Apollo 12 rises in direct proportion to the solar wind pressure; while the bottom curve shows that the intensity of the terrestrial field exhibits a corresponding decrease during the solar storm. This decrease in the earth's magnetic field is believed to be caused by the formation of a ring current involving charged particles trapped in the earth's magnetosphere (see §5.7). No known mechanism exists for the formation of an analogous ring current around the moon.

Analyses of these transient magnetic fields enable estimates to be made of the electrical conductivity and temperature of the lunar interior. Dyal and Parkin (1971) estimated that the temperature at a depth half-way to the centre is only about 1000°C. Such low lunar temperatures are consistent with the low seismic activity recorded by seismometers that have been placed on the moon. Seismic data indicate that the interior of the moon is not hot and molten but relatively cool, and that its surface has been modified more by meteoritic impact than by volcanic activity (see also §§2.4.3 and 2.4.4).

Sonett et al. (1971) found a layer of high electrical conductivity in the moon at a depth of about 250 km. Rama Murthy and Hall (1970) have suggested that one of the first major chemical differentiations in a planetary body would be the segregation of an Fe–FeS liquid, because the eutectic temperature of this system is much lower than the temperature required to melt silicates. All indications are that the surface of the moon has been melted and silicates differentiated about 4600 m yr ago (see, e.g., Papanastassiou et al. (1970)). Any Fe–S present would melt before the silicates and, because of its greater density, would sink into the interior. Since the interior of the moon would have been cold at the time of this melting, the sinking Fe–S liquid would be trapped at some depth. Rama Murthy et al. (1971) suggest that the layer of high electrical conductivity found by Sonett et al. is the trapped Fe–S liquid layer (Fig. 2.8). They further suggest that variations in the thickness of the Fe–S layer due to differential gravitational effects is the reason for the moon's three unequal moments of inertia.

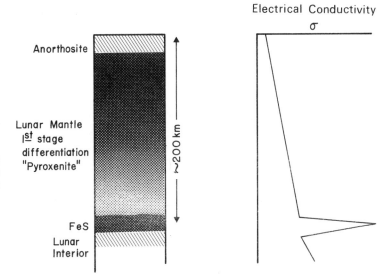

Fig. 2.8 Position of the Fe–S layer in the moon about 4500 m yr ago (after Rama Murthy *et al.* (1971))

43

2.4.3 *The seismicity of the moon* Records from the lunar seismic stations set up at the Apollo 11 and 12 sites have been analysed by Latham and his colleagues in a series of papers (see, e.g., Latham *et al.* (1970)). They found that continuous seismic signals, analagous to microseisms on earth, are not detectable on the lunar surface. Thus seismometers can operate there at a much higher sensitivity than can normally be used on earth. Seismic signals from 208 events of natural origin were identified on the records of the first nine months of operation of the Apollo 12 station (on the south-eastern edge of Oceanus Procellarum), i.e. about one event occurred every one and a half days. Signals from two artificial impacts, the impact of the Apollo 12 landing module ascent stage, and the third stage of Apollo 13 Saturn booster, produced records which were utterly unlike any observed on earth (see Fig. 2.9). The lunar maria are thus clearly very different from typical regions of the earth's crust.

Lunar seismic signals have very long durations, their intensity increasing and decreasing gradually with little correlation between any two components of

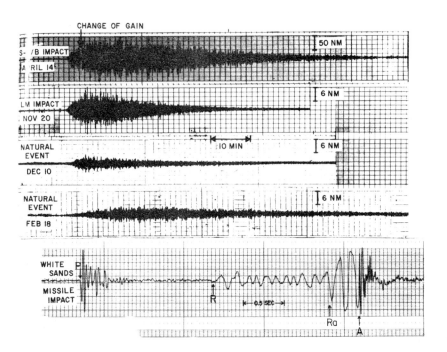

Fig. 2.9 ·Signals from the impacts of the Apollo 12 landing module (LM) ascent stage and the third stage (SIVB) of the Apollo 13 Saturn booster and from two of the largest natural events recorded on the lunar surface to date. All signals were recorded on the long period vertical component seismometer. A record of the seismic signal from a missile impact recorded at the White Sands Missile Range is shown for comparison. For the White Sands record, P indicates P wave; R, Rayleigh wave; A, atmospheric acoustic arrival. Note that the time scale of the White Sands impact signal is greatly expanded relative to that of the lunar signals (after Latham *et al.* (1970))

44

ground motion. Arrivals of compressional and shear waves can be identified in the early parts of wave-trains but are less distinct than on records of normal earthquakes. The predominant signal frequency remains relatively constant throughout a given wave train but differs for different events.

Ewing *et al.* (1970) suggested that the lunar seismic reverberations are the result of intensive scattering (and possibly dispersion) of surface waves in the upper 2–3 km of the moon. Absorption of seismic energy in this material is extremely low compared with that in typical earth crustal materials. Seismic wave velocities are extremely low near the surface, increasing rapidly with depth to approximately 5–6 km/s (for compressional waves) at a depth of about 20 km, in close agreement with values predicted from measurements of elastic properties of lunar samples as a function of pressure (Kanamori *et al.* (1970); Schreiber *et al.* (1970)). It does not follow automatically, however, that the basaltic rock material found at the surface actually extends to a depth of 20 km or more. Again although almost all the Apollo rock samples are basaltic in character, high pressure studies of this material have confirmed the pre-Apollo observation of Wetherill (1968) that the moon as a whole cannot have this composition. It would invert at relatively modest depths to an eclogite assemblage of such high density as to be inconsistent with the overall density of the moon.

Six artificial seismic events caused by the impact of spent rocket stages and lunar modules have now been analysed by Toksöz *et al.* Their results indicate that the moon has a layered crust some 85 km thick. The upper 1 ~ 2 km appears to consist of broken rock and rock fragments. Below this, to a depth of about 25 km, the velocity of compressional waves agrees with that found in laboratory experiments on mare basaltic rocks. A second layer, extending down to about 85 km, appears to be of different composition with higher seismic velocities similar to those observed in plagioclase-rich materials. The seismic evidence for the existence of a layered crust with apparent compositional differences is an additional indication that the moon has undergone differentiation.

Natural seismic signals on the moon are due to moonquakes or meteoritic impacts. With few exceptions, moonquakes occur at monthly intervals at or near the times of closest approach of the earth and moon (perigee) during the monthly orbital revolution of the moon about the earth. This strongly suggests that moon-quakes are induced by tidal strains. All moonquakes are small, the largest having magnitudes between 1 and 2 on the Richter scale (see §3.1). The total energy released by moonquakes is thus exceedingly small compared to that released by earthquakes: the outer shell of the moon must be relatively cold and tectonically stable compared with the outer regions of the earth.

The records from the four Apollo seismograph stations now operating on the moon show remarkably consistent signatures. First motions are always in the same sense and direction. Only a few of the sources have been definitely pin-pointed—but they all lie at depths between 600 and 1000 km, the average depth being about 800 km. The interior of the moon at these depths is thus rigid enough to support considerable stress. There is also some evidence that these foci tend to lie below the rims of those maria that are associated with mascons. At a recent

45

meeting on geophysical and geochemical exploration of the moon and planets held at Houston, Texas, on January 10th–12th, 1973, it was suggested that this seismicity might be the result of mascons dropping back into the moon, the whole process being triggered by the tidal effects of the earth (Runcorn, private communication).

An explanation of the relative aseismicity of the moon has been put forward by Thomsen (1972). He suggests that the moon is aseismic because its pressure–temperature (p, T) curve lies in the (p, T) field where rock fails by stable-sliding rather than by stick-slipping, i.e. the moon at depth is too warm (i.e. soft) rather than too cold. It need not, of course, be hot, since the transition zone to stable-sliding is probably well below the solidus. Some support for the suggestion that the moon may have a soft region below about 800 km comes from the observations following the fall of a large meteorite on the far side of the moon near the Sea of Moscow on day 134, 1972: no S waves were observed below about 800 km. (Ewing, private communication).

2.4.4 *The temperature of the moon* At the third lunar science conference held in Houston in January 1972, some of the results from Apollo 14 and 15 and Luna 16 were presented. The general consensus of opinion then was that the moon is now a relatively cold and inactive body, but that it has had a complex thermal history. The maria basalts acquired their gross composition 4500–4700 m yr ago during a primitive melting and crystallization process, their composition being slightly modified during a second (partial) melting about 3100–4000 m yr ago (Urey and Marti (1972)). There appears to be as yet no resolution of the conflict between geochemical data which demand an initially cold moon in which partial melting of its outer layers occurred and the magnetic evidence which can best be explained by a hot moon with a molten iron core for most of the first 1500 m yr of its history.

Hays (1972) has tried to resolve this question by carrying out a series of calculations on the moon's thermal history. He found that, for an initially cold moon with a uniform distribution of U and Th and no K, the maximum concentration of U is 23×10^{-9} if the low temperatures ($\sim 800°C$ at a depth of 900 km), inferred by Sonett *et al.* (1971) from lunar electrical conductivity profiles, are correct. This value is only slightly raised by increasing the thermal conductivity, while a higher initial temperature or allowance for K leads to a still lower U content. This value is an order of magnitude less than that found for the least radioactive lunar samples. One explanation is that radioactive heat sources in the moon are preferentially concentrated near the surface, but this would probably require differentiation through partial melting (conflicting with a cold moon theory and the temperature distribution of Sonett *et al.*), or a primary stratification during accretion (which would require U and Th to behave differently from K). The calculations of Hays thus appear to have only increased the dilemma—he concludes that either the temperature distribution of Sonett *et al.* is incorrect or else the moon has a most 'unusual' geochemistry.

That the temperature distribution of Sonett *et al.* may be incorrect is indicated by recent work of Duba and Ringwood (1973). Sonett *et al.* obtained their temperature distribution by inverting the electrical conductivity distribution (obtained from an analysis of the interaction of the moon with the solar wind), using experimental data on the relationship between electrical conductivity and temperature in olivine. It is these last data that Duba and Ringwood question. The electrical conductivity of terrestrial olivines varies over a very wide range (for similar temperatures), being possibly controlled by redox conditions. Duba and Ringwood thus measured the electrical conductivity of several olivines and pyroxenes (probably the most abundant minerals in the lunar interior) over a wide range of temperatures and redox conditions. Using their results obtained under redox conditions similar to those that exist in the moon, together with Sonett *et al.*'s electrical conductivity distribution, Duba and Ringwood estimate the temperature at a depth of 600 km in the moon to be between 950°C and 1560°C. Temperatures in the depth range 500–900 km are much higher than those obtained by Sonett *et al.* and are at or near the solidus, i.e. the lunar interior is hot, not cold. Duba and Ringwood also argue that the moon accreted under relatively high temperature conditions. Large molten bodies of iron–nickel–sulphide formed during the early near surface differentiation would thus be able to sink through the lunar mantle to form a core which may generate a magnetic field.

Anderson and Hanks (1972) have summarized the conflicting evidence for a hot or a cold moon. Arguments in favour of a cold moon are the temperatures deduced from the lunar electrical conductivity profile, the non-hydrostatic shape of the moon, the existence of mascons, the aseismicity of the moon and the absence of present-day volcanism. Some of these points have already been discussed. Arguments in favour of a hot origin and high initial temperatures are the evidence for rapid differentiation of the moon some 4600 m yr ago, the extensive igneous period resulting in mare formation 3700–2800 m yr ago, the depletion of the moon in volatiles and its enrichment in calcium, aluminium and the trace refractory elements. In a critical appraisal of all these data, Anderson and Hanks conclude that the most reasonable model of the present lunar interior has a cool, rigid lithosphere several hundred kilometres thick overlying a hot interior.

2.5 Mars

Our knowledge of Mars increased enormously with the launching of the series of Mariner spacecraft. Mariner 4 was launched from Cape Kennedy on November 28th, 1964, and passed within about 13 200 km of Mars on July 14–15th, 1965. During the close encounter with the planet, measurements were made of the magnetic field and various particle fluxes, all of which indicated that Mars had at most a very weak magnetic field.

We do not have a quantitative theory of the origin of the earth's radiation belts, and it is not possible to predict the nature of such belts for a planet at arbitrary distance from the sun; however, the planet must have a sufficiently strong magnetic field and be exposed to the solar wind. Since the distance of Mars from the sun is intermediate between that of the earth and Jupiter, both of which have intense radiation belts, it is reasonable to assume that Mars also would have radiation belts provided that it is a sufficiently magnetized body. The criterion for sufficiency is given approximately by equation (2.2). A system of sensitive particle detectors on Mariner 4 indicated the presence of electrons of energy > 40 keV out to a radial distance of 165 000 km ($\simeq 25 R_\mathrm{e}$) on the morning side of the earth, yet failed to detect any such electrons during the close encounter with Mars (Van Allen et $al.$ (1965)). This implies (from equation (2.2)) that the magnetic dipole moment of Mars is less than 0·001 that of the earth, i.e. the upper limit on the equatorial magnetic field at the surface of Mars is about 200 γ.

Similar results have been reported by O'Gallagher and Simpson (1965). The Mariner 4 carried a solid-state charged-particle telescope capable of detecting electrons with energies greater than 40 keV and protons with energies greater than 1 MeV. The trajectory of Mariner 4 would have carried it through a bow shock, transition region and magnetospheric boundary had these existed. No evidence of charged particle radiation was found in any of these regions. Again a planet with even a very small magnetic field might be expected to produce a wake in the anti-solar direction. Mariner 4 passed sufficiently close to Mars to have detected such a wake, had it existed; no escape of electrons, as would be expected in such a wake, was observed. O'Gallagher and Simpson place the same upper limit on any Martian magnetic field, namely 0·1 per cent of that of the earth.

Mariner 4 also carried a magnetometer during the close encounter with Mars; no magnetic effects were observed that could be definitely associated with a Martian magnetic field. E. J. Smith et $al.$ (1965a) put an upper limit for a Martian magnetic moment of 3×10^{-4} that of the earth. Since the rotational period of Mars is approximately the same as that of the earth, its much weaker magnetic moment would suggest that it has at most a very small fluid, electrically conducting core.

Dolginov et $al.$ (1972), in an analysis of the magnetograms transmitted from the USSR Mars 2 and 3 orbiting probes, claim that Mars also possesses an internal magnetic field. The magnetograms indicate a field intensity some seven to ten times greater than that of the interplanetary field at the distance of the orbit of Mars. The authors conclude that Mars possesses a dipole moment of $2 \cdot 4 \times 10^{22}$ emu and an intensity of ~ 60 γ at the magnetic equator. These values are still less than the upper limit deduced from the analysis of the Mariner 4 data.

In addition to the above experiments, Mariner 4 also successfully transmitted to earth 22 photographs of the surface taken at distances of 17 000 to 12 000 km. These revealed the unexpected result that the Martian surface is heavily cratered, comparable to the bright upland areas of the moon. About 1 per cent of the Martian surface was photographed and these showed more than 70 craters, ranging in diameter from 4 to 200 km. No mountain chains, valleys or ocean basins were identified nor were there any traces of 'canals', although the line of flight crossed several positions of these hypothetical features. Leighton et al. (1965) argued, by analogy with the moon, that much of the heavily cratered surface of Mars must be very old, perhaps 2000–5000 m yr, and that the state of preservation indicated that no atmosphere significantly denser than the present very tenuous one has ever existed. They also inferred the absence of any appreciable amounts of free water on Mars since the time the surface was formed, and that, unlike the earth, Mars has long been inactive. These conclusions have been challenged by Anders and Arnold (1965), Witting et al. (1965) and Baldwin (1965). Leighton et al. assumed that cratering rates on Mars and the moon are equal. On the other hand, Anders and Arnold estimated that the rate of crater formation on Mars is about 25 times higher than that on the moon, taking into consideration both asteroidal and cometary impacts. In this case the crater density observed by Mariner 4 indicates an age only one-sixth that of the lunar maria, i.e. the Martian craters are of the order of 300–800 m yr old. Such a low age implies a significant erosion rate; Anders and Arnold suggested dust storms as a possible agent. Witting et al. concluded that the average age of the Martian craters may be less than 300 m yr, and Baldwin suggested an age between about 340–680 m yr. In a later paper Hartmann (1966) disagreed with all the above analyses, and gave a different interpretation of the Mariner 4 photographs. He showed that the difference in impact velocities on Mars and the moon, which was neglected by the above authors, has an appreciable effect. Hartmann showed that the age of a surface layer capable of retaining the large Martian craters (diameter greater than 50 km) is about 4000 m yr (within a factor of 2). He stressed that this is not the age of the surface of Mars, but the duration of stable conditions in a crustal layer that has retained all craters of diameter greater than 50 km. If any form of erosion is present, smaller craters would have shorter lifetimes. 'Crater retention age' is thus a function of crater diameter. A 1 km crater is estimated to have a lifetime of the order of 100 m yr. Hartmann suggested that perhaps the Martian crustal regime is analogous to the stable shield areas of the earth. In the Canadian shield, where there has been no orogenic activity for about 2000 m yr, large impact craters seen on air photographs are of great age (see §1.6), although small craters are younger; in unstable areas of the earth, on the other hand, which are subject to orogeny and erosion, large craters are not found because the length of time the surface has been exposed has been too short to record such infrequent major impacts.

Far more details of the Martian surface were revealed when Mariners 6 and 7 successfully flew past Mars on July 31st and August 5th, 1969, respectively. The television cameras on Mariner 6 recorded 75 pictures (50 far-encounter and 25 near-encounter) and those on Mariner 7 126 pictures (93 far-encounter and 33 near-encounter). While confirming the earlier evidence of a moon-like cratered surface for much of Mars, the photographs from Mariners 6 and 7 indicated at least three distinctive terrains (Leighton et al. (1969)): cratered, chaotic and featureless. Those parts of the Martian surface where craters are the dominant

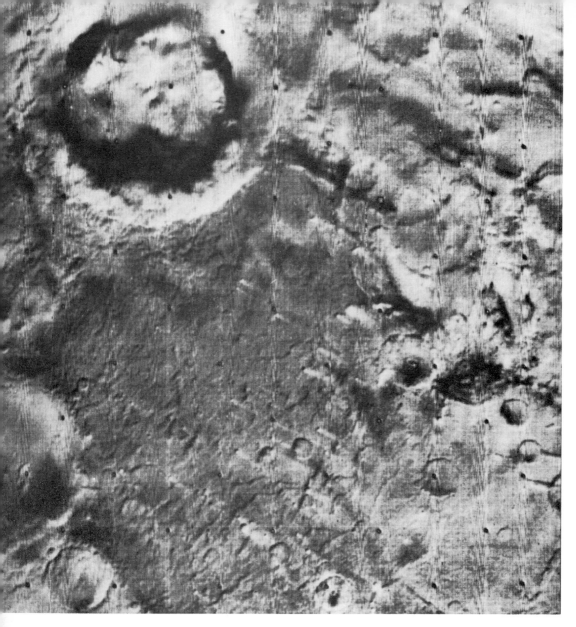

Fig. 2.10 Cratered terrain on Mars with numerous small channels. One interpretation is that this channel system is due to running water; another hypothesis is that they are of volcanic origin. This photograph was taken by the Mariner 9 space craft on the 210th orbit of Mars. (NASA 72-H-816; Mariner 9 Orbit 210, 4244-56, 211-4351)

topographic form have been called cratered terrains (see Figs. 2.10 and 2.11). There appears to be two distinctive types of craters: large and flat-bottomed, and small and bowl-shaped. The latter resemble lunar primary impact craters. Diameters range from a few to a few hundred km. Although many features of large lunar impact craters can be seen on the Martian surface, other features such as rays and secondary crater swarms are conspicuously absent. The 'missing' features are those most easily destroyed or hidden by erosion or blanketing. No earth-like tectonic forms have been recognized. Some of the photographs show a relatively smooth cratered surface giving way abruptly to areas of 'chaotically jumbled ridges'. Chaotic terrain appears to be essentially devoid of craters. Finally there are large, featureless areas on Mars (such as the floor of the circular

Fig. 2.11 Cratered terrain modified by many small channels. Several proposals have been made to account for the channels; water erosion, wind erosion, lava channels. This photograph was taken by the Mariner 9 space craft on the 423rd orbit of Mars. (NASA 72-H-818; Mariner 9 Orbit 423, 4288-28, 211-4349)

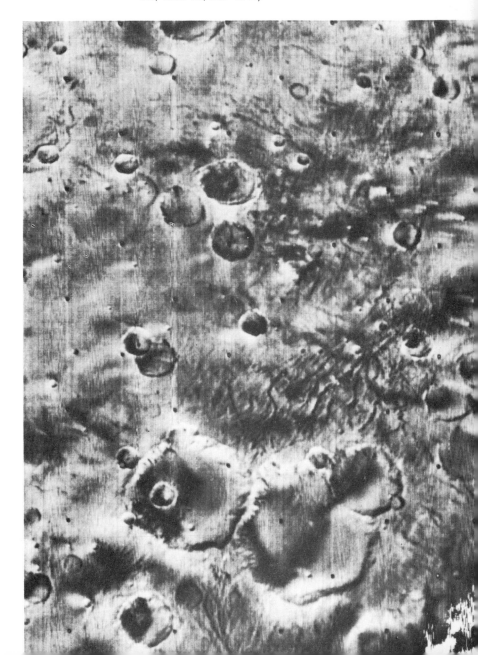

'desert', Hellas; no area of comparable size and smoothness has been found on the moon). Perhaps local dust storms occur very frequently in Hellas, obscuring the surface and any craters that may be present.

A far more detailed interpretation of the photographs returned by Mariners 6 and 7 has been given in a series of papers published together in *J. geophys. Res.*, 1971, **76**, 293–472. In a later paper, Oberbeck and Aoyagi (1972) have made a study of Martian craters and found that many occur in clusters or as isolated crater doublets. They showed that many of these Martian craters could not have resulted from random single-body impacts and suggested that they might have been caused by meteorite break-up resulting from stresses induced in the meteorite by the gravitational field of Mars. The Clearwater Lakes craters and the Ries Kessel and Steinheim basin may be examples of terrestrial doublets that were produced by the same mechanism. If this hypothesis is true, the distribution of large craters on different planetary bodies cannot be compared directly for interpretations of the geologic history of one planetary surface relative to another. Many inferences which have been drawn from a comparison of the relative crater density and size on Mars and on the moon may thus not be valid.

No account will be given of the findings on the atmosphere of Mars. However, large-scale variations in Martian topography can be found from the relative abundance of CO_2 (the main constituent of the Martian atmosphere). Such spectroscopic mapping (Wells (1969)) is in good agreement with previous range-gated radar data along 21°N which indicated variations in elevation of the order of 12 km. The S band occultation experiment carried out by Mariner 9 (Kliore *et al.* (1972)) also revealed overall elevation differences of about 13 km.

The latest information on Mars has come from Mariner 9, which was launched from Cape Kennedy on May 30th, 1971, and put into orbit around Mars on November 14th. The height of the orbit above the surface ranged from 1650 to 17 100 km. Unlike the other Mariner flights which flew past Mars, Mariner 9 was the first man-made satellite to orbit the planet. (Two Soviet spacecraft have subsequently been put into orbit around Mars and a short lived instrument capsule landed on the surface.) When Mariner 9 went into orbit, a severe dust storm which was first observed from earth on September 22nd was still raging, and photography was not possible for many weeks. Dust seems to be far more prevalent than had been realized and dust, rather than biological phenomena, may be the cause of the seasonal and secular changes in markings on the planet as viewed from the Earth.

Since October 27th, 1972, when its instrumentation was turned off, more than 7000 photographs have been obtained from Mariner 9, providing more than a hundred times the amount of information accumulated by all previous missions to Mars. The most spectacular new features revealed were four large volcanic mountains, larger than any such features on earth (see Fig. 2.12), and a vast system of canyons, tributary gullies and sinuous channels (see Fig. 2.13).

Evidence now of past volcanic activity, faulting and geochemical differentiation indicates that Mars has many features of recent origin (on a geological time scale) and is far from being a dead planet. Masursky *et al.* (1972) found some of the

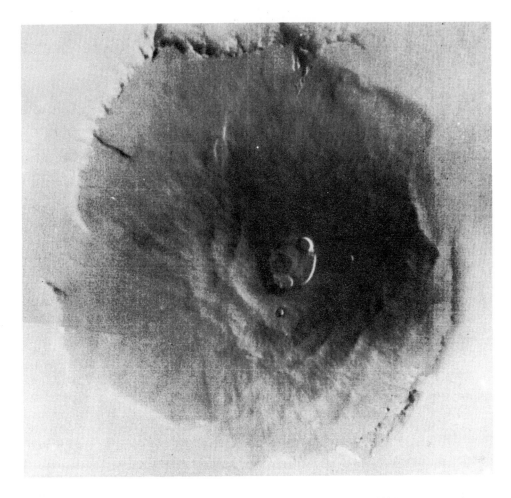

Fig. 2.12 Nix Olympica, a gigantic volcanic mountain on Mars. It is 500 km across at the base; the main crater at the summit is 65 km in diameter. The mountain is more than twice as broad as the most massive volcanic pile on earth. (The mountain that forms the Hawaiian Island is 225 km. across and rises 9 km. from the floor of the Pacific to the summit crater, Mauna Loa.) This photograph was taken by the Mariner 9 spacecraft on the 146th orbit of Mars (NASA 72-H-716, JPL P-13074)

largest craters to be volcanic in origin, resembling collapsed volcanic craters or calderas on earth. Surrounding the high caldera uplift area is an array of grabens, horsts and a rift valley system extending for 80° of Martian longitude and comparable in extent to the East African rift system. Most of these features are relatively uneroded, and indicate significant tectonic activity on Mars in geologically recent times. Equally surprising are a number of dendritic rilles which do not have craters at the sources of their tributaries. They are unlike collapsed lava tubes and the Hadley Rille in the lunar Apennines, and are difficult to

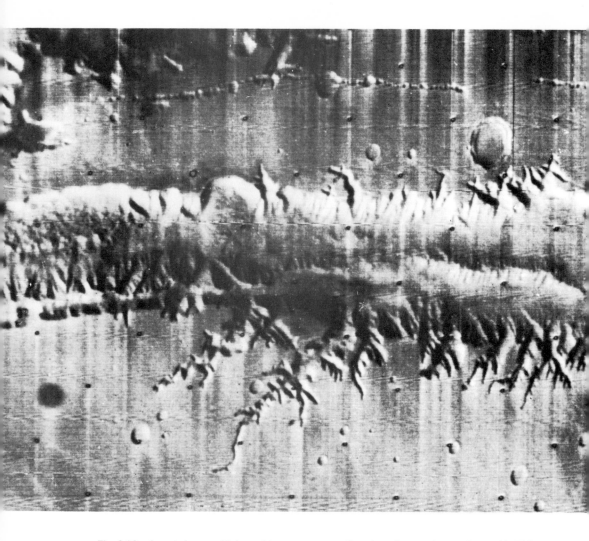

Fig. 2.13   A vast chasm with branching canyons eroding the adjacent plateau, located in Tithonius Lacus, 300 miles south of the equator. This photograph was taken by the Mariner 9 spacecraft on the 119th orbit of Mars from a height of 1977 km and covers an area 376 by 480 km. (NASA 72-H-43, JPL P-12732)

explain without invoking liquid water on Mars. This is difficult to accept since the low pressures on Mars are on the wrong side of the triple point for liquid water to exist. Additional evidence that Mars has undergone substantial geochemical differentiation has come from the infra-red spectroscopy data from Mariner 9 (Hanel *et al.* (1972))—many of the thermal emission spectra are significantly affected by dust with a SiO_2 content approximately corresponding to that of intermediate igneous rocks.

54

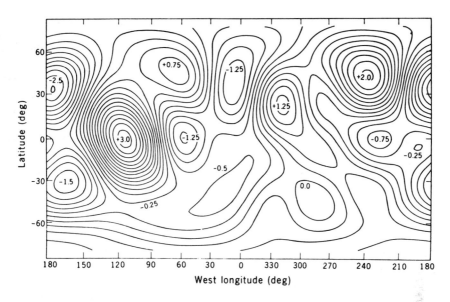

Fig. 2.14 Contours of equivalent surface heights deduced from a sixth-degree solution for the Martian gravity field. These contours represent the deviations from sphericity of a uniformly dense body with external potential given by the first sixth-degree solution, with the effect of J_2 omitted. Contours are labelled in km; the contour interval is 250 m (after Lorell *et al.* (1972))

Preliminary results on the gravitational field of Mars have been obtained from Mariner 9 radio-tracking data (Lorell *et al.* (1972)). Contours of constant equivalent surface heights deduced from a sixth degree solution of the Martian gravity field are shown in Fig. 2.14. The Martian gravity field is rougher than that of either the earth or the moon: deviations from a spheroid vary from about $-2\cdot5$ to $+3$ km, implying substantial stress. An analysis of high frequency gravity variations may possibly reveal the presence of 'mascons' (see §4.5).

Mariner 9 also obtained the first ever close-up pictures of the two Martian satellites Phobos and Deimos. Phobos is about $25\ (\pm 5) \times 21\ (\pm 1)$ km and Deimos about $13\cdot5\ (\pm 2) \times 12\cdot0\ (\pm 0\cdot5)$ km. Both have irregular shapes (as expected for bodies too small for their gravitational fields to demand spherical shapes). A large indentation can be seen on the right-hand side of Phobos (Fig. 2.15). The two satellites are highly cratered, some of the craters showing raised rims, consistent with an impact origin. The crater near the centre in Fig. 2.15 is about $5\cdot3$ km in diameter; the impact that produced this crater must have been close to the largest impact Phobos could have sustained without disruption. The whole of the July 10th, 1973 issue of *J. geophys. Res.* (pp. 4007–4440) is devoted to a discussion of the scientific results of the Mariner 9 mission.

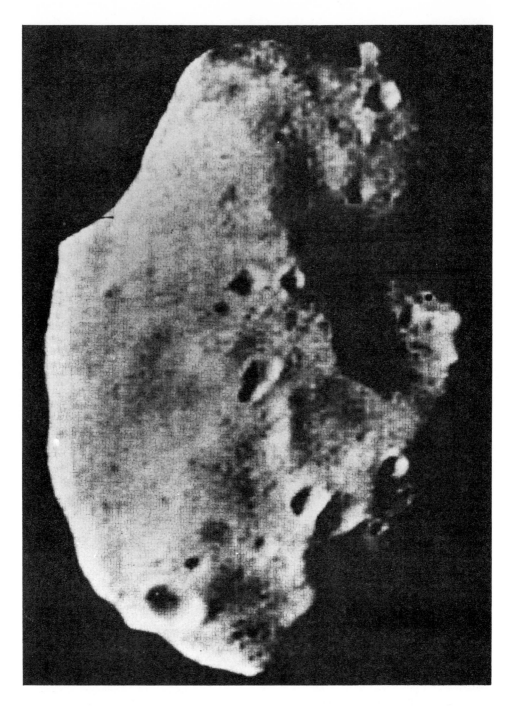

Fig. 2.15 The first close up view ever obtained of the Martian satellite Phobos. This picture was taken at a distance of 5540 km by Mariner 9. The large number of craters suggests that Phobos is very old and possesses considerable strength. (JPL P-12694, Mariner 9 Orbit 34)

2.6 Venus

Because of its dense cloud cover, no reliable estimates of the rotation rate of Venus were possible prior to radar measurements in 1962 (Carpenter (1964); Goldstein (1964)). These led to the surprising result that the rotation rate is slow (about 250 days) and retrograde. More recent determinations indicate the period to be within a day or two of 243·15 days, for which period Venus would present the same face to the earth at every inferior conjunction—to an observer on earth Venus would appear to rotate exactly four times between close approaches. The spin of Venus may thus be controlled by the earth. The most recent and accurate determination by Carpenter (1970) of the rotation period of Venus (242·98 days) from radar tracking of features and Doppler broadening differs significantly from the earth's synchronous value, and it is now very uncertain whether Venus is in synodic spin resonance. A sidereal period of 250·5 days (retrograde) would imply that Venus presents the same face to Jupiter at every close approach of these two planets. However, the average gravitational torques exerted by Jupiter on Venus are about an order of magnitude less than those exerted by the earth so that Jupiter is far less likely to be the controlling influence. (Venus would present the same face to Mercury at alternate close approaches for a spin period of 238 days. Mercury, however, exerts an even smaller average torque than Jupiter.)

Two questions concerning the anomalous rotation of Venus need to be answered: how did its axial rotation reach its present state and is a spin state that is locked to the relative orbital motion of the earth and Venus stable? Bellomo *et al.* (1967) have attempted to answer these difficult questions and concluded that the existence of a stable earth–Venus resonance rotation would be 'a most surprising phenomenon'. For the rotation to be stable, the fractional difference in its principal equatorial moments of inertia would have to be quite large, greater than about 10^{-3} (the corresponding difference for the moon is 2×10^{-4}). The anomalous rotation of Venus remains largely unexplained, and is probably bound up with the origin of the solar system. Goldreich and Peale (1970), from their theory of tidal evolution, find that Venus must have a liquid core and solid mantle to maintain its retrograde rotation.

Singer (1970b) believes that, like the other planets, Venus initially had a prograde spin with a period between 10 and 20 h. This immediately raises the question of how this angular momentum was lost; solar tidal friction is completely inadequate. Singer suggests that it was accomplished by the capture of a moon-like object by Venus from an initially retrograde orbit. This would despin Venus, transforming the planet's rotational kinetic energy into internal heat. This in turn would lead to widespread volcanism and the liberation of large amounts of volatiles. Singer further suggests that the 'moon' would finally crash into the surface of Venus. French (1971) is of the opinion, however, that if such were the case, the impact would destroy Venus. Singer (1971b) still believes that his hypothesis is possible, the essence of his theory being that the moon was first *captured* into an orbit, retrograde, with respect to Venus. About one-half of the moon's initial angular momentum would be transmitted to Venus through (solid body) tidal interactions before any impact. Also the orbit of the captured moon would gradually approach Venus, and once inside the Roche limit (see Appendix A) the moon would break into a number of pieces, the remainder of its angular momentum being transmitted to Venus through a *series* of impacts of smaller bodies.

The Mariner 5 space craft was launched from Cape Kennedy on June 14th, 1967 and passed about 10 150 km from the centre of Venus on October 19th. A previous attempt by Mariner 2 on December 14th, 1962 to observe the interaction between Venus and the solar wind was unsuccessful (E. J. Smith *et al.* (1965b)), partly because the space craft passed on the sunward side of the planet and came no closer than 40 000 km. Mariner 5 observed abrupt changes in the amplitude of magnetic fluctuations, in field strength and in plasma properties in the vicinity of Venus. Data from both the magnetometer and plasma probe show clear evidence for the existence of a bow shock around Venus, similar to, but much smaller than, that near the earth (Bridge *et al.* (1967)). The solar wind appears to flow around the planet without striking it, in contrast to the case of the moon. It appears to be deflected by a dense ionosphere on the sunlit side of Venus. Its high electrical conductivity prevents the passage of the incoming magnetic field and the 'pile up' of the field alters the plasma flow. Strong support for this interpretation comes from the observation of the Mariner Stanford Group (1967) that the upper boundary of the daylight ionosphere is very sharp and is pushed down to within about 500 km of the planet by the momentum of the solar wind. In the case of the moon, the plasma ions are absorbed by the lunar surface and no shock develops (Lyon *et al.* (1967)). The moon appears to be a sufficiently good insulator to allow the interplanetary field to be convected through it essentially unaltered (Colburn *et al.* (1967); Ness *et al.* (1967)). Although the shock around Venus resembles that around the earth (except in scale) conditions inside are quite different: in the case of Venus it is the ionosphere that deflects the solar wind; in the case of the earth, it is its magnetic field.

The above observations also place an upper bound on the magnetic dipole moment of Venus. It cannot exceed 10^{-3} times that of the earth (within a factor of two). Similar conclusions were drawn by Van Allen *et al.* (1967) from the absence of energetic electrons (> 45 keV) and protons (> 320 keV) during the fly-by of Mariner 5. They concluded that the dipole moment of Venus is almost certainly less than 0·01 and probably less than 0·001 that of the earth. Since Venus has about the same size and average density as the earth, its very small magnetic dipole moment is probably a consequence of its very slow rotation.

References

Anders, E. and J. R. Arnold (1965) Age of craters on Mars, *Science*, **149**, 1494.
Anderson, D. L. (1972) The origin of the moon, *Nature*, **239**, 263.

Anderson, D. L. (1973) The composition and origin of the moon, *Earth Planet. Sci. Letters*, **18**, 301.

Anderson, D. L. and T. C. Hanks (1972) Is the moon hot or cold?, *Science*, **178**, 1245.

Ash, M. E., I. I. Shapiro and W. B. Smith (1971) The system of planetary masses, *Science*, **174**, 551.

Baldwin, R. B. (1965) Mars: an estimate of the age of its surface, *Science*, **149**, 1498.

Beard, D. B. (1960) The interaction of the terrestrial magnetic field with the solar corpuscular radiation, *J. geophys. Res.*, **65**, 3559.

Beard, D. B. (1973) The interactions of the solar wind with planetary magnetic fields: basic principles and observations, *Planet. Space Sci.*, **21**, 1475.

Bellomo, E., G. Colombo and I. I. Shapiro (1967) Theory of the axial rotations of Mercury and Venus, in: *Mantles of the Earth and Terrestrial Planets* (ed. S. K. Runcorn) (Interscience Publ.).

Biermann, L. (1953) Physical processes in comet tails and their relation to solar activity, *Soc. Roy. Sci. de Liege*, **13**, 291.

Biermann, L. (1971) Comets and their interaction with the solar wind, *Q. J. roy. Astr. Soc.*, **12**, 417.

Bixby, J. E. and T. C. Van Flandern (1969) The diameter of Neptune, *Astron. J.*, **74**, 1220.

Brandt, J. C. (1970) *Introduction to the Solar Wind* (W. H. Freeman and Co., San Francisco).

Bridge, H. S., A. J. Lazarus, C. W. Snyder, E. J. Smith, L. Davis Jr., P. J. Coleman Jr. and D. E. Jones (1967) Mariner V: plasma and magnetic fields observed near Venus, *Science*, **158**, 1669.

Cameron, A. G. W. (1970) Formation of the earth–moon system, *Trans. Amer geophys. Union EθS*, **51**, 628.

Carpenter, R. L. (1964) Study of Venus by CW radar, *Astron. J.*, **69**, 2.

Carpenter, R. L. (1970) A radar determination of the rotation of Venus, *Astron. J.* **75**, 61.

Chapman, C. R. (1968) Optical evidence on the rotation of Mercury, *Earth Planet. Sci. Letters*, **3**, 351.

Colburn, D. S., R. G. Currie, J. D. Michalov and C. P. Sonett (1967) Diamagnetic solar-wind cavity discovered behind the moon, *Science*, **158**, 1040.

Colombo, G. (1965) Rotational period of the planet Mercury, *Nature*, **208**, 575.

Colombo, G. and I. Shapiro (1966) The rotation of the planet Mercury, *Astrophys. J.*, **145**, 296.

Dessler, A. J. (1968) Ionizing plasma flux in the Martian upper atmosphere, in: *Atmospheres of Venus and Mars* (ed. J. C. Brandt and M. B. McElroy) (Gordon and Breach, New York).

Dolginov S. S., E. G. Eroshenko and L. N. Zhuggov (1972) *Dokl. Akad. Nauk SSSR*, **207**, 1296.

Duba, A. and A. E. Ringwood (1973) Temperatures in the lunar interior and some implications, *Earth Planet. Sci. Letters*, **18**, 158.

Dyal, P., C. W. Parkin and C. P. Sonett (1970) Apollo 12 magnetometer measurement of a steady magnetic field on the surface of the moon, *Science*, **169**, 762.

Dyal, P. and C. W. Parkin (1971) The Apollo 12 magnetometer experiment: internal lunar properties from transient and steady magnetic field measurements, *Proc. Sec. Lunar Sci. Conf., Houston, Texas, Jan. 1971* (M.I.T. Press).

Dyal, P., C. W. Parkin, C. P. Sonett, R. L. Dubois and G. Simmons (1971) Lunar portable magnetometer experiment, Apollo 14, *Preliminary Sci. Rep. NASA SP-272*, 227.

Elco, R. A. (1969) Interaction of the solar wind with planetary atmospheres, *J. geophys. Res.*, **74**, 5073.

Ewing, M., G. Latham, F. Press, G. Sutton, J. Dorman, Y. Nakamura, R. Meissner, F. Duennebier and R. Kovach (1970) Seismology of the moon and implications on internal structure, origin and evolution, in: *Highlights of Astronomy* (ed. De Jager), p. 155 (D. Reidel Publ. Co., Holland).

French, B. M. (1971) How did Venus lose its angular momentum? *Science*, **173**, 169.

Ganapathy, R. R., R. R. Keays, J. C. Land and E. Anders (1970) Trace elements in Apollo 11 lunar rocks: implications for meteorite influx and origin of moon, *Proc. Apollo 11 Lunar Sci. Conf. Houston, Texas, Jan. 1970.* Vol. 2, p. 1117 (Pergamon Press).

Gerstenkorn, H. (1955) Über Gezeitenreibung beim Zweikörper Problem, *Z. Astrophys.*, **36**, 245.

Goldreich, P. (1966) History of the lunar orbit, *Rev. Geophys.*, **4**, 411.

Goldreich, P. and S. J. Peale (1966) Resonant spin states in the solar system, *Nature*, **209**, 1078.

Goldreich, P. and S. J. Peale (1968) The dynamics of planetary rotations, *Ann. Rev. Astron. Astrophys.*, **6**, 287.

Goldreich, P. and S. J. Peale (1970) The obliquity of Venus, *Astron. J.*, **75**, 273.

Goldstein, R. M. (1964) Venus characteristics by earth-based radar, *Astron. J.*, **69**, 12.

Hanel, R. A., B. J. Conrath, W. A. Hovis, V. G. Kunde, P. D. Lowman, J. C. Pearl, C. Prabhakara, B. Schlachman and G. V. Levin (1972) Infrared spectroscopy experiment on the Mariner 9 mission: preliminary results, *Science*, **175**, 305.

Hartmann, W. K. (1966) Martian cratering, *Icarus*, **5**, 565.

Hays, J. F. (1972) Radioactive heat sources in the lunar interior, *Phys. Earth Planet. Int.*, **5**, 77.

Herz, N. (1969) Anorthosite belts, continental drift, and the anorthosite event, *Science*, **164**, 944.

Johnson, F. S. and J. E. Midgley (1969) Induced magnetosphere of Venus, *Space Res.*, **9**, 760.

Kanamori, H., A. Nur, D. Chung, D. Wones and G. Simmons (1970) Elastic wave velocities of lunar samples at high pressures and their geophysical implications, *Science*, **167**, 732.

Kaula, W. M. (1971) Dynamical aspects of lunar origin, *Rev. Geophys. Space Phys.*, **9**, 217.

Kliore, A. J., D. L. Cain, G. Fjeldbo, B. L. Seidel and S. I. Rasool (1972) Mariner S-band Martian occultation experiment: initial results on the atmosphere and topography of Mars, *Science*, **175**, 313.

Kovalesky, J. (1972) A system of planetary masses and related quantities, *Phys. Earth Planet. Int.*, **6**, 29.

Latham, G., M. Ewing, J. Dorman, F. Press, N. Toksöz, G. Sutton, R. Meissner, F. Duennebier, Y. Nakamura, R. Kovach and M. Yates (1970) Seismic data from man-made impacts on the moon, *Science*, **170**, 620.

Leighton, R. B., B. C. Murray, R. P. Sharp, J. D. Allen and R. K. Sloan (1965) Mariner IV photography of Mars: initial results, *Science*, **149**, 627.

Leighton, R. B., N. H. Horowitz, B. C. Murray, R. P. Sharp, A. H. Herriman, A. T. Young, B. A. Smith, M. E. Davies and C. B. Leovy (1969) Mariner 6 and 7 television pictures: preliminary analysis, *Science*, **166**, 49.

Lorell, J., G. H. Born, E. J. Christensen, J. F. Jorden, P. A. Laing, W. L. Martin, W. L. Sjogren, I. I. Shapiro, R. D. Reasenberg and G. L. Slater (1972) Mariner 9 celestial mechanics experiment: gravity field and pole direction of Mars, *Science*, **175**, 317.

LSPET (1969) Preliminary examination of lunar samples from Apollo 11, *Science*, **165**, 1211.

LSPET (1970) Preliminary examination of lunar samples from Apollo 12, *Science*, **167**, 1325.

Lyon, E. F., H. S. Bridge and J. H. Binsack (1967) Explorer 35 plasma measurements in the vicinity of the moon, *J. geophys. Res.*, **72**, 6113.

Lyttleton, R. A. (1953) *The Stability of Rotating Liquid Masses* (Camb. Univ Press).

Lyttleton, R. A. (1960) Dynamical calculations relating to the origin of the solar system, *Mon. Not. Roy. Astr. Soc.*, **121**, 551.

MacDonald, G. J. F. (1964) Tidal friction, *Rev. Geophys.*, **2**, 467.

Mariner Stanford Group (1967) Venus: ionosphere and atmosphere as measured by dual-frequency radio occultation of Mariner V, *Science*, **158**, 1678.

Mason, B. (1971) The lunar rocks, *Sci. Amer.* (Oct.).

Mason, B. and W. G. Melson (1970) *The Lunar Rocks* (Wiley-Interscience).

Masursky, H., R. M. Batson, J. F. McCauley, L. A. Soderblom, R. L. Wildey, M. H. Carr, D. J. Milton, D. E. Wilhelms, B. A. Smith, T. B. Kirby, J. C. Robinson, C. B. Leovy, G. A. Briggs, T. C. Duxbury, C. H. Acton Jr., B. C. Murray, J. A. Cutts, R. P. Sharp, S. Smith, R. B. Leighton, C. Sagan, J. Veverka, M. Noland, J. Lederberg, E. Levinthal, J. B. Pollack, J. T. Moore Jr., W. K. Hartmann, E. N. Shipley, G. de Vaucouleurs and M. E. Davies (1972) Mariner 9 television reconnaissance of Mars and its satellites: preliminary results, *Science*, **175**, 294.

McCrae, W. H. (1969) Densities of the terrestrial planets, *Nature*, **224**, 28.

Message, P. J. (1972) A survey of dynamical data for the major planets and satellites, *Phys. Earth Planet. Int.*, **6**, 17.

Murthy, V. Rama and H. T. Hall (1970) The chemical composition of the earth's core: possibility of sulphur in the core, *Phys. Earth Planet. Int.*, **2**, 276.

Murthy, V. Rama, N. M. Evenson and H. T. Hall (1971) Model of early lunar differentiation, *Nature*, **234**, 267.

Ness, N. F., K. W. Behannon, C. S. Scearce and S. C. Cantarano (1967) Early results from the magnetic field experiment on Lunar Explorer 35, *J. geophys. Res.*, **72**, 5769.

Neugebauer, M. and C. W. Snyder (1965) Solar wind measurements near Venus, *J. geophys. Res.*, **70**, 1587.

Oberbeck, V. R. and M. Aoyagi (1972) Martian doublet craters, *J. geophys. Res.*, **77**, 2419.

O'Gallagher, J. J. and J. A. Simpson (1695) Search for trapped electrons and a magnetic moment at Mars by Mariner IV, *Science*, **149**, 1233.

O'Keefe, J. A. (1969) Origin of the moon, *J. geophys. Res.*, **74**, 2758.

O'Keefe, J. A. (1970) Apollo 11: implications for the early history of the solar system, *Trans. Amer. Geophys. Union. EθS*, **51**, 633.

Öpik, E. J. (1961) Tidal deformations and the origin of the moon, *Astron. J.*, **66**, 60.

Papanastassiou, D. A., G. J. Wasserburg and D. S. Burnett (1970) Rb–Sr ages of lunar rocks from the Sea of Tranquillity, *Earth Planet. Sci. Letters*, **8**, 1.

Peale, S. J. and T. Gold (1965) Rotation of the planet Mercury, *Nature*, **206**, 1241.

Pettengill, G. H. and R. B. Dyce (1965) A radar determination of the rotation of the planet Mercury, *Nature*, **206**, 1240.

Pettengill, G. H., C. C. Counselman, L. P. Rainville and I. I. Shapiro (1969) Radar measurements of Martian topography, *Astron. J.* **74**, 461.

Ringwood, A. E. (1970a) Origin of the moon: the precipitation hypothesis, *Earth Planet. Sci. Letters*, **8**, 131.

Ringwood, A. E. (1970b) Petrogenesis of Apollo 11 basalts and implications for lunar origin, *J. geophys. Res.*, **75**, 6453.

Ringwood, A. E. (1971) Reply to S. F. Singer—discussion of paper by A. E. Ringwood, 'Petrogenesis of Apollo 11 basalts and implications of lunar origin', *J. geophys. Res.*, **76**, 8075.

Ringwood, A. E. (1972) Some comparative aspects of lunar origin, *Phys. Earth Planet. Int.*, **6**, 366.

Schreiber, E., O. Anderson, N. Soga, N. Warren and C. Scholz (1970) Sound velocity and compressibility for lunar rocks 17 and 46, and for glass spheres from the lunar soil, *Science*, **167**, 732.

Singer, S. F. (1968) The origin of the moon and geophysical consequences, *Geophys. J.*, **15**, 205.

Singer, S. F. (1970a) Origin of the moon by capture and its consequences, *Trans. Amer. Geophys. Union. EθS*, **51**, 637.

Singer, S. F. (1970b) How did Venus lose its angular momentum? *Science*, **170**, 1196.

Singer, S. F. (1971a) Discussion of paper by A. E. Ringwood 'Petrogenesis of Apollo 11 basalts and implications for lunar origin', *J. geophys. Res.*, **76**, 8071.

Singer, S. F. (1971b) How did Venus lose its angular momentum? *Science*, **173**, 170.

Smith, B. A. and E. J. Rees (1968) Mercury's rotation period—photographic confirmation, *Science*, **162**, 1275.

Smith, E. J., L. Davis Jr., P. J. Coleman Jr. and D. E. Jones (1965a) Magnetic field measurements near Mars, *Science*, **149**, 1241.

Smith, E. J., L. Davis Jr., P. J. Coleman Jr. and C. P. Sonett (1965b) Magnetic measurements near Venus, *J. geophys. Res.*, **70**, 1571.

Sonett, C. P., D. S. Colburn and R. G. Currie (1967) The intrinsic magnetic field of the moon, *J. geophys. Res.*, **72**, 5503.

Sonett, C. P., D. S. Colburn, P. Dyal, C. W. Parkin, B. F. Smith, G. Schubert and K. Schwartz (1971) Lunar electrical conductivity profile, *Nature*, **230**, 359.

Sonett, C. P., P. J. Coleman, Jr. and J. M. Wilcox (1972) (Eds.) Solar Wind, *Proc. NASA Conf.*, March 1971, U.S. Gov. Printing Off., Washington.

Spreiter, J. R., A. L. Summers and A. W. Rizzi (1970) Solar wind flow past non-magnetic planets—Venus and Mars, *Planet. Space Sci.*, **18**, 1281.

Strangway, D. W., G. W. Pearce, W. A. Gose and R. W. Timme (1971) Remanent magnetization of lunar samples, *Earth Planet. Sci. Letters*, **13**, 43.

Tatsumoto, M. (1970) Age of the moon: an isotopic study of U–Th–Pb systematics of Apollo 11 lunar samples, *Proc. Apollo 11 Lunar Sci. Conf. Houston, Texas, Jan. 1970*, Vol. 2, p. 1595 (Pergamon Press).

Thomsen, L. (1972) Implications of lunar aseismicity, *Nature*, **240**, 94.

Ulrych, T. J. (1969) Lead isotopes, lunar capture and mantle evolution, *Nature*, **224**, 766.

Urey, H. C. and K. Marti (1972) Lunar basalts, *Science*, **175**, 118.

Van Allen, J. A. (1966) Some general aspects of geomagnetically trapped radiation, in: *Radiation trapped in the earth's magnetic field* (ed. B. M. McCormac) (D. Reidel Publishing Co., Holland 65).

Van Allen, J. A., L. A. Frank, S. M. Krimigis and H. K. Hills (1965) Absence of Martian radiation belts and implications thereof, *Science*, **149**, 1228.

Van Allen, J. A., S. M. Krimigis, L. A. Frank and T. P. Armstrong (1967) Venus: an upper limit on intrinsic magnetic dipole moment based on absence of a radiation belt, *Science*, **158**, 1673.

Vinogradov, A. P. (1971) Preliminary data on lunar ground brought to earth by automatic probe 'Lunar-16', Apollo 12, *Lunar Sci. Conf., Houston, Texas, Jan. 1971*.

Wells, J. W. (1963) Coral growth and geochronometry, *Nature*, **197**, 948.

Wells, R. A. (1969) Martian topography: large-scale variations, *Science*, **166**, 862.

Wetherill, G. W. (1968) Lunar interior—constraint on basalt composition, *Science*, **160**, 1256.

Wise, D. U. (1969) Origin of the moon from the earth: some new mechanisms and comparisons, *J. geophys. Res.*, **74**, 6034.

Witting, J., F. Narin and C. A. Stone (1965) Mars: age of its craters, *Science*, **149**, 1496.

3

Seismology and the Physics of the Earth's Interior

3.1 The seismicity of the earth

The earth is continually undergoing deformation due to stresses that are set up within it. If the stresses are not too great, elastic or plastic deformation will occur. However if the stresses continue to build up over a long time, fracture may take place. This involves a sudden release of energy, part of which takes the form of elastic waves which travel through the earth. These waves emanate from a confined region below the surface of the earth, called the *focus* of an earthquake. The point on the surface of the earth vertically above the focus is called the *epicentre*. If the disturbance is large enough, it may be picked up by any of the seismological stations scattered across the world. Most of our information about the interior of the earth has been derived from observations of the propagation of elastic waves generated by earthquakes.

Gutenberg and Richter (1954) have classified earthquakes according to their depth of focus as shallow (0–70 km), intermediate (70–300 km) and deep (below 300 km). The deepest earthquakes occur at a depth of about 700 km. Figures 3.1 and 3.2 show the global distribution of earthquakes in the depth range 0–100 km and 100–700 km, respectively. The very distinctive pattern of seismic activity plays a key role in the development of what has been called the new global tectonics or plate theory, and will be discussed in Chapter 8. It will be shown in §3.2 that three types of waves can be propagated: P and S waves, both of which are body waves, and surface waves. P waves are condensation-rarefaction waves, and S waves are shear waves.

Two different scales have been established to describe the 'bigness' of an earthquake: a magnitude scale and an intensity scale. The intensity scale is a measure of the observed effects of earthquake damage at a particular place on the earth's surface. It is not based on actual measurements of any kind. To classify an earthquake in a manner more representative of its tectonic importance

63

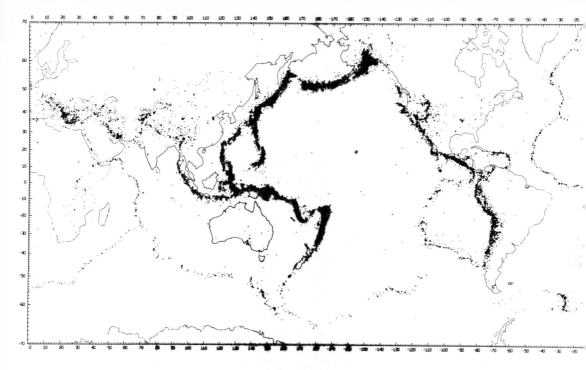

Fig. 3.1 Distribution of earthquakes in the depth range 0–100 km for 1961–1967 (after Barazangi and Dorman (1969)). Note the continuity of the seismic belts that outline large stable blocks, the narrowness of the belts and the low level of activity in zones of divergence and the greater width and increased activity in zones of convergence

Fig. 3.2 Distribution of earthquakes in the depth range 100–700 km for 1961–1967 (after Barazangi and Dorman (1969))

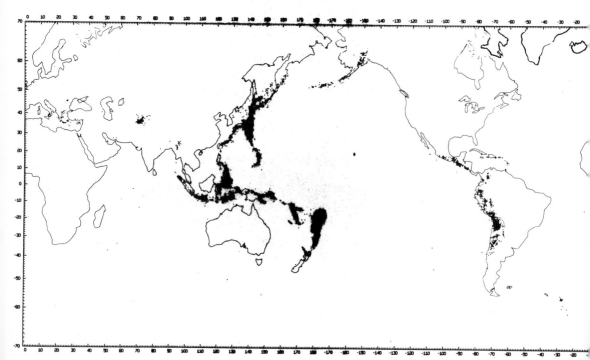

and energy released, Richter introduced in 1935 the concept of earthquake magnitude. The historical development of these two scales has been summarized by Richter (1958) and the whole question of magnitudes reviewed by Bath (1966a). Richter suggested the empirical formula

$$\log_{10} E = A + BM \tag{3.1}$$

where M is the magnitude, E the total radiated energy and A and B constants. The most recent values of the constants are $A = 12 \cdot 24$ and $B = 1 \cdot 44$ when the energy E is measured in ergs (Bath (1966b)). The magnitude scale is thus logarithmic and is similar to the term 'order of magnitude' as used by the physicist or astronomer. The actual range of magnitudes is enormous, the largest events representing motions a hundred million times as large as those of small events (nine magnitudes). The scale is open ended at both ends; it does not have a maximum value of ten, as is often reported in the press, and negative magnitudes are routinely measured by seismologists studying micro-earthquakes.

One of the most widely accepted explanations of shallow earthquakes is Reid's (1911) elastic-rebound theory, which attributes earthquakes to the progressive accumulation of strain energy in tectonic regions and the sudden release of this energy by faulting when the fracture strength is exceeded (see Fig. 3.3). Strain

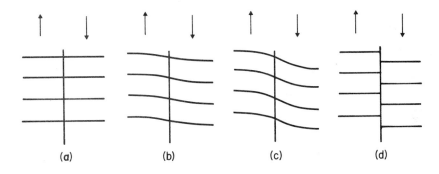

Fig. 3.3 Sketch illustrating Reid's elastic rebound earthquake source mechanism. Due to regional shearing movement, elastic strain is slowly built up from the unstrained state (a) through (b) to state (c), at which it is suddenly released across the fault, producing displacements as in (d)

release at the focus may be investigated by observing the pattern of ground motion at a distance from an earthquake. Details of the method may be found in Stauder (1962) and Benioff (1964).

There has been considerable controversy concerning faulting as the mechanism for earthquakes. A confined seismic source that is in equilibrium before and after an earthquake must, in the low frequency limit, be a superposition of zero-order monopoles and dipoles with no net moment. It was originally believed that the elastic rebound mechanism could be represented by a single couple model. However, as Honda pointed out in 1962, such a model has a net un-balanced torque. Honda also showed that while the radiation pattern of P waves from both single and double couple models is identical, the two models can be distinguished by the S wave radiation pattern—a single couple model gives rise to two shear wave lobes while a double couple model shows four lobes (see Fig. 3.4). The high quality data now available for source mechanism studies have

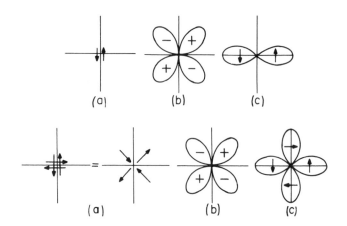

Fig. 3.4 Single couple (Top) and double couple (Bottom) fault mechanisms (a), with their radiation patterns of first motion for P waves (b) and S waves (c)

led to almost universal acceptance of the double couple model as the appropriate low frequency source representation of earthquakes.

The possibility that intermediate and deep-focus earthquakes may be caused by phase changes has been suggested by a number of authors (Bridgman (1945); Dennis and Walker (1965); Evison (1967); Ringwood (1967)). There have been two main difficulties with such suggestions. It is hard to understand how a phase transition can occur sufficiently rapidly in the mantle to cause an earthquake, and first motion studies indicate a double couple mechanism rather than the implosion or explosion mechanisms believed to be required on a phase change hypothesis. Ringwood (1972) has re-examined the whole question and shown that in a dynamic (rather than a static) mantle both the above objections may be overcome, and that it is quite likely that deep-focus earthquakes are caused by failure under stresses generated directly or indirectly by phase changes. Again, Knopoff and Randall (1970) and Randall and Knopoff (1970) in a recent study on earthquake source mechanisms often found an additional component in the seismic radiation field which was not of the double couple type but which is consistent with a phase transformation origin.

The concept of fracture followed by frictional sliding across a fault face is plausible only for shallow shocks. The coefficient of friction between dry rock faces is of the order of unity, and at depths below a few kilometres the overburden

66

pressure exceeds the shear strength of the rock, and dry frictional sliding is impossible. If a pore fluid pressure is introduced, the effective normal stress across the fault is reduced; this has the effect of lubricating it, allowing sliding to occur at greater depths.

In 1966 a series of small earthquakes was felt in Denver, Colorado, which was attributed, by Evans (1966), to the injection of fluid into a waste disposal well. This interpretation has been challenged by Simon (1970), who pointed out that the area was seismically active before the fluid injection and that earthquakes continued to occur after stoppage of injection. Other authors (e.g. Healy et al. (1968)) remain convinced that these earthquakes were at least triggered by the initiation of fluid injection. Healy and Raleigh are now carrying out experiments in the Rangely oil field in western Colorado. The fluid pressure in the field was raised by as much as 60 per cent above the normal hydrostatic pressure by fluid injection and small earthquakes occurred along a fault system at a rate of 15–20 per week. Seismic activity ceased when water was pumped out of the wells. Pumping has since been reversed in order to see whether earthquakes begin again as the pore fluid pressure builds up.

Earthquakes may also be triggered by underground nuclear explosions (Ryall and Boucher (1970)) and by the filling of large reservoirs. At present there is no reliable method of predicting when and where earthquakes will occur. Preliminary results from the experiments in the Rangely oil field have indicated that under certain circumstances earthquakes may be controlled, but it is not known whether the same technique could be applied to major fault zones such as the San Andreas fault. Johnson et al. (1973) have shown that creep occurring in the central part of the San Andreas fault is related to changes in the pore pressure of water in a well 150 m deep near Hollister. Both the maximum pore pressure and the offset in water level in the well are linearly related to the total creep in motions which occur within hours of anomalous changes in water level: it is not possible, however, to say whether the transient pore pressure is produced by creep or the creep by changes in fluid levels.

Release of tectonic strain by underground nuclear explosions has been intensively studied, but not enough is known of the processes involved to use such a method for earthquake control. An attempt by Aki et al. (1970) to measure 'tectonic' stress with small explosions on the San Jacinto fault, California, was unsuccessful. McKenzie and Brune (1972) have shown that moderate or large earthquakes could melt the rock within about a centimetre of their fault planes. This could be important since the molten film of rock may act as a lubricant and, in the case of large earthquakes, allow the release of almost all the elastic strain in the region of the shock. McKenzie and Brune suggest that bore holes should be drilled through the fault plane following a large earthquake to check their calculations.

A meeting was held in January 1972 at the Massachusetts Institute of Technology to 'review scientific capability to detect, locate and identify earthquakes and underground nuclear explosions'. The papers presented at this meeting have been published in a triple issue of the *Geophysical Journal* (Vol. 31, Nos. 1–3 (1972)).

A summary of present-day capabilities in distinguishing between nuclear explosions and earthquakes has been given recently by Davies (1973).

Semonov (1969) found that before small to moderate sized earthquakes there were significant changes in the ratio V_P/V_S of the velocities of compressional and shear waves. His observations were made on micro earthquakes in the Garm region of the USSR. He found that the ratio V_P/V_S remained fairly constant for considerable periods of time before it began to decrease to a broad minimum followed by a sharp increase before a larger earthquake. Confirmation of these results has since been reported by Aggarwal et al. (1973) in a different region of the earth—the Blue Mountain Lake area, New York. They found the decrease in the V_P/V_S ratio to be as much as 13 per cent on occasions and the time duration of the order of a few days. The time duration (but not the size) of the decrease appears to be longer the larger the magnitude of the earthquake. These results were confirmed by Whitcomb et al. (1973) for the San Fernando earthquake of February 9th, 1971. The lead time of the decrease was about $3\frac{1}{2}$ years for this much larger event (magnitude 6·4). The data of this earthquake clearly indicate, however, that the variation of V_P/V_S is primarily due to large changes in V_P and not V_S as had been proposed by Savarensky (1968) and Aggarwal et al. (1973). Whitcomb et al. explain their findings as a result of rock dilatancy and its effects on rock strength and fluid saturation. Dilatancy is an increase of volume due to a change in shape. In the field, the change in shape is thought to be due to shear stresses associated with regional tectonic strain. If the pores and cracks in the rocks are initially saturated with fluid, the volume increase will lead to a decrease in pore fluid pressure. If the dilatancy is large enough, the pore and crack volume will exceed the fluid volume. In this undersaturated state, the rock voids will partially contain vapour, resulting in a reduction of the overall bulk modulus. This will cause a large drop in V_P with but little change in V_S (cf. equations (3.9) and (3.12)). The effect of the reduction in pore fluid pressure is to increase the fracture strength of the rocks and thus delay the onset of an earthquake. Fluid flow from outside the dilatant volume, however, gradually restores the permeable rock to its saturated state, producing a gradual increase in V_P/V_S and a gradual decrease in fracture strength until an earthquake occurs. The above explanation also accounts for the observed relation between earthquake magnitude and precursor time interval (the time taken for V_P/V_S to return to its normal value) through the time needed for fluid flow to restore undersaturated rock in the dilatant volume to the saturated state—a longer precursor time interval implies a larger dilatant volume which in turn leads to a larger earthquake. The size of the decrease in V_P/V_S does not appear to be related to the magnitude of the subsequent earthquake. This is also consistent with the above explanation that the observed seismic velocities depend on whether or not the rock is saturated with water rather than on the volume of the dilatant region and hence on the magnitude of the earthquake.

3.2 Theory of elastic waves

Consider a homogeneous, isotropic, perfectly elastic medium. Using the infinitesimal strain theory of elasticity, the equation of motion satisfied by the displacement vector $\mathbf{D}$ is

$$(\lambda+\mu) \text{ grad div } \mathbf{D}+\mu\nabla^2\mathbf{D}+\rho(\mathbf{F}-\mathbf{f}) = 0 \qquad (3.2)$$

where λ, μ are the Lamé stress constants, ρ the density, $\mathbf{F}$ the external force per unit mass, and $\mathbf{f}$ the acceleration. Since the displacement and velocity are always small in elasticity (so that second powers of the displacements and their derivatives may be neglected), $\mathbf{f}=d^2\mathbf{D}/dt^2$ may be replaced by $\partial^2\mathbf{D}/\partial t^2$. The elastic (restoring) forces are also very much greater than gravitational forces (except for very long waves, i.e. those with little strain), so that we may take $\mathbf{F}=0$. With these assumptions, equation (3.2) becomes

$$(\lambda+\mu) \text{ grad div } \mathbf{D}+\mu\nabla^2\mathbf{D}-\rho\frac{\partial^2\mathbf{D}}{\partial^2 t} = 0 \qquad (3.3)$$

Taking the divergence of equation (3.3)

$$(\lambda+2\mu)\nabla^2 \text{ div } \mathbf{D}-\rho\frac{\partial^2}{\partial t^2} \text{ div } \mathbf{D} = 0 \qquad (3.4)$$

which shows that the cubical dilatation $\delta = \text{div } \mathbf{D}$ satisfies the wave equation

$$\left(\nabla^2-\frac{\rho}{\lambda+2\mu}\cdot\frac{\partial^2}{\partial t^2}\right)\delta = 0 \qquad (3.5)$$

The disturbance is transmitted with speed

$$V_P = \sqrt{\left(\frac{\lambda+2\mu}{\rho}\right)} \qquad (3.6)$$

Again taking the curl of equation (3.3),

$$\mu\nabla^2 \text{ curl } \mathbf{D}-\rho\frac{\partial^2}{\partial t^2} \text{ curl } \mathbf{D} = 0 \qquad (3.7)$$

which shows that the rotation $\boldsymbol{\omega}=\tfrac{1}{2} \text{ curl } \mathbf{D}$ satisfies the wave equation

$$\left(\nabla^2-\frac{\rho}{\mu}\cdot\frac{\partial^2}{\partial t^2}\right)\boldsymbol{\omega} = 0 \qquad (3.8)$$

the disturbance being transmitted with speed

$$V_S = \sqrt{\left(\frac{\mu}{\rho}\right)} \qquad (3.9)$$

In seismology these two waves are called P and S waves respectively. A P wave is a condensation–rarefaction wave involving change of volume. Motion of the medium is longitudinal so that there is no polarization of a P wave. An S wave is a shear wave in which there is distortion without change of volume. S waves, being transverse, are polarized and it is thus necessary to disinguish between vertical and horizontal polarizations, termed SV and SH waves, respectively. Both wave velocities depend only on the elastic parameters and the density of the medium.

In particular, if the rigidity μ is zero, $V_S=0$, i.e. shear waves cannot be transmitted through a material of zero rigidity. If k_S is the adiabatic incompressibility (or bulk modulus) defined by the equation

$$\frac{1}{k_S} = \frac{1}{\rho}\left(\frac{\partial \rho}{\partial p}\right)_S \tag{3.10}$$

where p is the pressure and S entropy, it can be shown that

$$k_S = \lambda + \tfrac{2}{3}\mu \tag{3.11}$$

Thus the velocity of P waves may be written alternatively as

$$V_P = \sqrt{\left(\frac{k_S + \tfrac{4}{3}\mu}{\rho}\right)} \tag{3.12}$$

It follows from equations (3.9) and (3.12) that

$$\phi \equiv \frac{k_S}{\rho} = V_P{}^2 - \frac{4}{3}V_S{}^2 \tag{3.13}$$

P and S waves are body waves and travel throughout the earth. There is another class of elastic waves (surface waves) which are guided by density and velocity layering in the outer regions of the earth. They are important in studying the structure of these regions and are discussed in § 3.6. The validity of the assumptions of isotropy and homogeneity are considered in later chapters dealing with the detailed structure and composition of the earth.

3.3 The effect of boundaries

Consider for the moment an earthquake as a generator of elastic waves and a seismic station as a receiver, and suppose also that the earthquake occurs close to the earth's surface so that the source is effectively at the epicentre. When an elastic wave meets a sharp boundary between two media of different properties, part of it will be reflected and part refracted, and laws of reflection and refraction analogous to those of geometrical optics apply. The case of elastic waves is more complicated, however, since waves of both P and S type may be reflected and refracted. The theory is based on Fermat's principle according to which an elastic wave takes the quickest path between any two points. This does not imply that there is only one path—there may be a number of alternative paths—but each path must involve a minimum transit time relative to small deviations in the path. When an SH wave meets a horizontal boundary, there is just simple refraction and reflection, but an SV wave, which has a component of motion normal to the boundary, generates P waves in addition to refracted and reflected SV waves, as shown in Fig. 3.5.

Snell's law is equally applicable to seismic waves as in optics. In Fig. 3.5, the

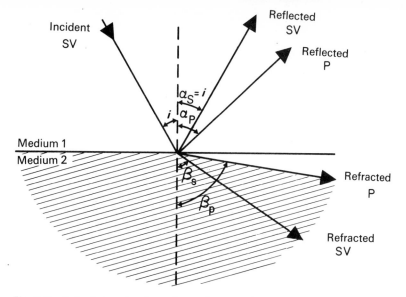

Fig. 3.5 Reflected and refracted rays from an SV ray incident on a plane
boundary between two solids

angles of reflection and refraction are thus related to the angle of incidence by
the wave velocities V_P and V_S of P and S waves in the media 1 and 2:

$$\frac{\sin i}{V_{S_1}} = \frac{\sin \alpha_S}{V_{S_1}} = \frac{\sin \alpha_P}{V_{P_1}} = \frac{\sin \beta_S}{V_{S_2}} = \frac{\sin \beta_P}{V_{P_2}} \qquad (3.14)$$

Similar equations can be written down for reflections and refractions of an
incident P wave and for an SH wave (in which case only reflected and refracted
SH occur).

Fig. 3.6 illustrates some of the many possible reflections and refractions of
elastic waves at discontinuities within the earth. The figure also illustrates the

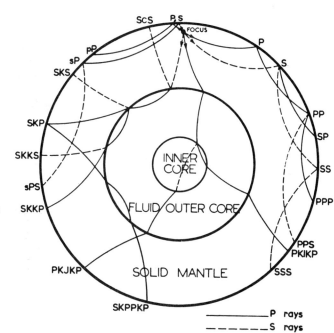

Fig. 3.6 Representative seismic rays
through the earth (after Bullen (1954))

terminology used to designate the different phases. A P wave reflected from the earth's surface can give rise to both a P wave and an S wave (called PP and PS waves, respectively). Likewise an S wave reflected from the surface can give both P and S waves (called SP and SS waves respectively). The letters c and i denote reflections from the outer and inner core boundaries, respectively, while the letters K and I indicate P waves through the outer and inner core. Thus ScS is an S wave that has travelled down to the core boundary and been reflected as an S wave. Similarly SKP is an S wave that has been refracted into the outer core (necessarily as a P wave) and refracted back into the mantle as a P wave. P^1 is often written for PKP. S waves have never been observed in the outer core, which is thus considered to be fluid. On the other hand, it has often been conjectured that the inner core is solid (see §7.4) and the symbol J was proposed for paths of S waves (if they exist) in the inner core. It was not until 1972 (Julian *et al.*) that the phase PKJKP was identified on seismograms and the solidity of the inner core definitely established. Multiple (n) internal reflections from the core mantle boundary into the outer core are indicated by nK. Thus P_4KP is a P wave that has been refracted into the outer core, suffered 4 internal reflections in the core before being refracted back into the mantle as a P wave. The letter h is used to denote reflection from the surface of the F shell (see §3.5), and the letter d reflection from the underside of a layer at a depth d below the earth's surface. If d can be calculated, its value (in km) is used instead of d. Thus $P_{250}P$ is a wave that has been reflected as a P wave from a layer 250 km beneath the earth.

3.4 Travel-time and velocity-depth curves

The times at which signals from the same earthquake arrive at different seismic observatories can be recorded so that it is possible to determine the travel-time of the disturbance as a function of distance. Fig. 3.7 illustrates travel-time curves for P, S and surface waves where distance Δ is measured in degrees along the surface of the earth. The graph for the first arrival of surface waves is a straight line while the graphs for P and S waves show curves with a downward concavity, indicating that the velocity over most regions of the earth is an increasing function with depth. If the velocities of elastic waves were uniform within the earth, seismic rays would be straight lines along chords.

Consider an earth model consisting of thin shells in each of which the velocity is uniform, each shell having a velocity higher than that in the shell above it (Fig. 3.8). Then from Snell's law, a ray is successively refracted away from the normal such that

$$\frac{\sin i_1}{v_1} = \frac{\sin i'_1}{v_2}$$

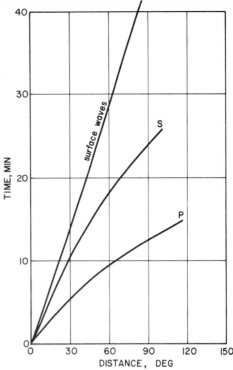

Fig. 3.7 Travel-time curves

or

$$\frac{r_1 \sin i_1}{v_1} = \frac{r_1 \sin i'_1}{v_2}$$

But from Fig. 3.8, $r_1 \sin i'_1 = r_2 \sin i_2 = d$ so that

$$\frac{r_1 \sin i_1}{v_1} = \frac{r_2 \sin i_2}{v_2} \tag{3.15}$$

and this equation can be extended to refractions at any number of boundaries or

Fig. 3.8 Successive refraction of a wave in a layered sphere

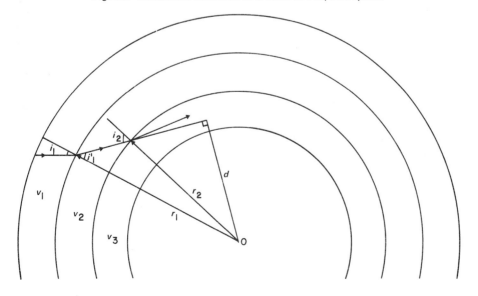

to gradual refraction in a layer of progressively increasing velocity. Hence for the particular ray considered

$$\frac{r \sin i}{v} = \text{constant} = p \qquad (3.16)$$

where i is the angle between the ray and the radius r at any point. p is a constant for all points along a given ray and is called the ray parameter. A knowledge of p gives the value of r/v at the deepest point of penetration of the ray, where $\sin i = \sin \pi/2 = 1$.

There is another interpretation of the parameter p. Let the ray emerge at Q at time T, let Q' be a neighbouring point $(r_0, \Delta + d\Delta)$ in the same plane, and let $T + dT$ be the time required to reach Q'. Draw QR perpendicular to the ray that reaches Q' (Fig. 3.9). Then to the first order of small quantities

$$RQ' = v\,dT \quad \text{and} \quad QQ' = r_0\,d\Delta$$

so that with the values of i and v at the surface in Q,

$$\sin i = \frac{v\,dT}{r_0\,d\Delta} \qquad (3.17)$$

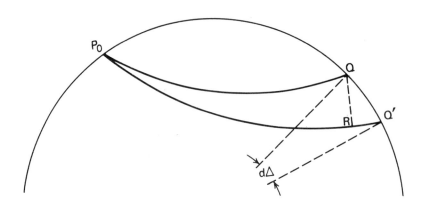

Fig. 3.9

Thus from equation (3.16), the ray parameter p may be identified as $dT/d\Delta$. Since T, Δ are known from observations, $dT/d\Delta$ can be determined, and thus p is a known function of Δ. We now have to solve the inverse problem to find v as a function of r from a knowledge of p as a function of Δ. Consider any point P on a given ray (Fig. 3.10). If s is distance measured along the ray,

$$p = \frac{r \sin i}{v} = \frac{r^2}{v}\frac{d\theta}{ds}$$

But $(ds)^2 = (dr)^2 + r^2(d\theta)^2$, so that eliminating ds,

$$\left(\frac{d\theta}{dr}\right)^2 = \frac{p^2 v^2}{r^2(r^2 - p^2 v^2)} \qquad (3.18)$$

74

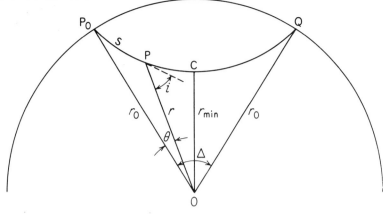

Fig. 3.10

Writing

$$\eta = r/v$$

equation (3.18) becomes

$$d\theta = \pm pr^{-1}(\eta^2 - p^2)^{-1/2}\, dr \qquad (3.19)$$

If θ begins by increasing, r will begin by decreasing and the negative sign must be taken. At the deepest point (i.e. at the midpoint) of the ray, $\sin i = 1$, and η at that point is equal to the value of p for the ray. Integrating along the ray between the outer radius and the deepest point

$$\frac{\Delta}{2} = \int_{r_{\min}}^{r_0} pr^{-1}(\eta^2 - p^2)^{-1/2}\, dr \qquad (3.20)$$

Equation (3.20) is an integral equation for η, and hence v, as a function of r, since p is a known function of Δ. The following solution, due to Rasch, was given to Jeffreys in a private communication. Rewriting equation (3.20) as an integral over η,

$$\Delta = \int_{p}^{\eta_0} 2pr^{-1}(\eta^2 - p^2)^{-1/2}\frac{dr}{d\eta}\, d\eta \qquad (3.21)$$

Let the subscript 1 denote values of the variables at the level r_1, and let Δ_1 be the value of Δ for the ray whose deepest point is at the level r_1. Multiply both sides of equation (3.21) by $(p^2 - \eta_1^2)^{-1/2}$ and integrate with respect to p over the range η_1 to η_0. Then

$$\int_{\eta_1}^{\eta_0} \Delta(p^2 - \eta_1^2)^{-1/2}\, dp = \int_{\eta_1}^{\eta_0} dp \int_{p}^{\eta_0} 2pr^{-1}[(\eta^2 - p^2)(p^2 - \eta_1^2)]^{-1/2}\frac{dr}{d\eta}\, d\eta \qquad (3.22)$$

Change the order in the double integration. The limits for p become η_1 to η and those for η run from η_1 to η_0. But if $\eta > \eta_1$,

$$\int_{\eta_1}^{\eta} \frac{p\, dp}{[(\eta^2 - p^2)(p^2 - \eta_1^2)]^{1/2}} = \frac{\pi}{2} \qquad (3.23)$$

75

so that the right-hand side of equation (3.22) becomes

$$\int_{\eta_1}^{\eta_0} \pi r^{-1} \frac{\mathrm{d}r}{\mathrm{d}\eta} \, \mathrm{d}\eta = \pi \log \left(\frac{r_0}{r_1} \right) \tag{3.24}$$

Integrating the left-hand side of equation (3.22) by parts leads to

$$\left[\Delta \cosh^{-1} \left(\frac{p}{\eta_1} \right) \right]_{\eta_1}^{\eta_0} - \int_{\eta_1}^{\eta_0} \frac{\mathrm{d}\Delta}{\mathrm{d}p} \cosh^{-1} \left(\frac{p}{\eta_1} \right) \, \mathrm{d}p = \int_0^{\Delta_1} \cosh^{-1} \left(\frac{p}{\eta_1} \right) \, \mathrm{d}\Delta \tag{3.25}$$

Thus equation (3.22) finally reduces to

$$\int_0^{\Delta_1} \cosh^{-1} \left(\frac{p}{\eta_1} \right) \, \mathrm{d}\Delta = \pi \log \left(\frac{r_0}{r_1} \right) \tag{3.26}$$

In this equation, η_1 is the value of p for the ray that emerges at Δ_1, i.e. it is the slope of the travel-time curve at Δ_1. The left-hand side can therefore be integrated numerically to any prescribed value of Δ_1 by obtaining values of p at intermediate points and evaluating $\cosh^{-1}(p/\eta_1)$. The unknown in equation (3.26), r_1, is then determined. This is the radius to the mid-point of the ray that emerges at Δ_1, and the velocity at the mid-point is given by $v_1 = r_1/\eta_1$. Inversion therefore consists of the evaluation of equation (3.26) for successively greater values of Δ_1, to obtain velocities over a range of depths.

It has been assumed in the above derivation of the velocity-depth curves that there are no discontinuities in the velocity distribution and that the velocity increases monotonically with depth. In this case, p decreases with Δ, and, in the range of integration, $p > \eta_1$. Departures from this assumption can complicate the ray structure and travel-time curves. Consider first a zone in which there is a greater than normal increase of velocity with depth. If the velocity increase is sufficiently rapid, a situation such as that shown in Fig. 3.11(a) may arise. Within a limited range of p, Δ decreases with decreasing i_0, instead of increasing. This leads to a triplication of arrivals over a certain range of Δ, corresponding to the three paths by which rays may reach a given point, and a travel-time curve such as that shown in Fig. 3.11(b).

Consider now a zone in which the velocity decreases with depth. This will cause refraction, as shown in Fig. 3.12(a), and there will be a range of depths within

Fig. 3.11 (a) Effect on seismic rays of a layer in which the velocity increases rapidly with depth in an otherwise 'normal' earth; (b) Corresponding features of the travel–time curves

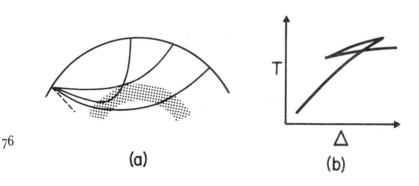

76

(a)
(b)

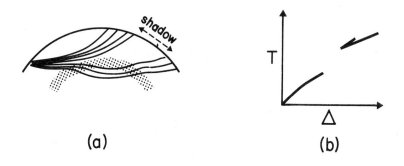

Fig. 3.12 (a) Effect on seismic rays of a layer in which the velocity decreases with depth in an otherwise 'normal' earth; (b) Corresponding features of the travel–time curves

which no rays have their deepest points and a range of Δ in which arrivals are not observed (a shadow zone). If the velocity distribution is normal below the low velocity zone, Δ decreases as η_1 further decreases and passes through a minimum before increasing again. This minimum in Δ is analogous to the case of refraction of light through a spherical lens. The resulting travel-time curve in Fig. 3.12(b) shows the duplication of arrivals over a certain range of Δ beyond the shadow zone.

3.5 Major subdivisions of the earth

The construction of travel-time curves has had a long history of successive approximation and increasing accuracy. Revision of the early travel-time curves was undertaken by Jeffreys in 1931 (see Jeffreys (1936, 1961) for complete details) using a least-squares technique. In collaboration with Bullen he produced the first J.B. Tables in 1935. Substantial refinements were incorporated in a new set of J. B. Tables first published in 1940—these gave travel-times, not only of P and S waves, but also of reflected and refracted waves. Details of the production of these tables and of the early history of the subject may be found in Bullen (1963a). Improvements in the records and the use of large artificial explosions have led to corrections in the J.B. Tables (see e.g. Herrin (1968)). However, the production of the J.B. Tables, before high-speed computers were available, was a monumental achievement. They provided the basic data from which velocity-depth curves could be constructed, which in turn have provided the information necessary for estimating the physical properties of the earth's interior (see §3.8).

Figure 3.13 shows the gross features of the velocity-depth curves in the mantle and core according to Jeffreys and Gutenberg. During 1940–1942, Bullen divided the earth into a number of regions based on these velocity-depth curves. His nomenclature has since been widely used, and, in spite of uncertainties in the positions (and realities) of the boundaries between the different regions, continues to serve as a useful basis in discussing the earth's interior. The upper

77

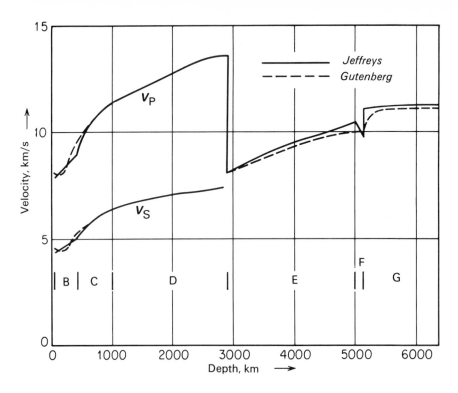

Fig. 3.13 Seismic velocities of P and S waves as a function of depth (after Birch (1952))

mantle consists of region B extending from the base of the crust (region A) to a depth of about 400 km, and region C which is a transition zone between depths of about 400 and 1000 km. The lower mantle, below a depth of about 1000 km, is called region D. Bullen later subdivided D into D' and D", the region D" being the bottom 200 km. The core is divided into an outer core E, a transition region F and an inner core G. In recent years much finer detail has been obtained, particularly in the upper mantle (regions B and C) and in the vicinity of the boundary of the inner core (region F). The structure and composition of the different regions of the earth will be discussed in Chapter 10.

The most notable feature of the velocity-depth curves is the discontinuity at a depth of about 2900 km, marking the boundary between the mantle and core. There is a discontinuous drop in the velocity of P waves across the boundary, below which S waves have never been detected. The outer core thus has negligible rigidity (equation (3.9)) and is fluid. Oldham had proposed in 1906 the existence of a core with lower P velocities than in the mantle to account for the late arrival of P waves at points antipodal to an earthquake. He also predicted the presence of a shadow zone which was later verified by Gutenberg in 1912 for P waves in the range $105° < \Delta < 143°$ (see Fig. 3.14). Gutenberg estimated the depth to the core boundary as 2900 km and predicted that both P and S waves would be reflected from the boundary. These reflected phases were not observed until much later,

78

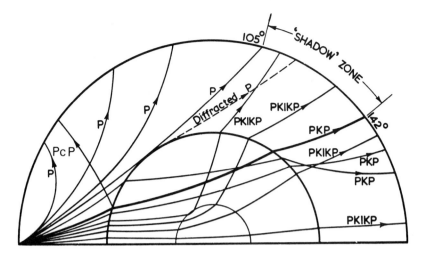

Fig. 3.14   P, PcP, PKP, PKIKP, and diffracted P rays (after Bullen (1954))

but by 1939 Jeffreys had used their travel-times to determine the depth to the core boundary as (2898 ± 4) km. This figure has only very recently been challenged (see §10.5). Some waves are recorded in the range $105° < \Delta < 143°$ so that it is not a true shadow zone. The amplitudes of such waves are much reduced and for many years their presence was attributed to diffraction round the boundary of the core. Lehmann suggested in 1935 that the waves recorded in the shadow zone had passed through an inner core (see Fig. 3.14) in which the P velocity is significantly greater than that in the outer core. Later work has corroborated her hypothesis and the existence of an inner core is well established now.

There is a second major discontinuity in the velocity-depth curves just below the earth's surface which marks the boundary between the crust and mantle. This boundary is called the Mohorovičić discontinuity after its discoverer, who identified it on a seismogram of an earthquake in Croatia in 1909. The nature of the Mohorovičić discontinuity is discussed in §10.2. The depth to the 'Moho' is about 33 km under the continents and about 5 km under the ocean floors. Although the detailed structure of the crust is extremely complex, it comprises less than 1 per cent of the mass of the whole earth and details of its structure do not affect the interpretation of the physical properties of the deeper parts of the mantle and core.

Other discontinuities within the earth are indicated by the seismic data, although none of them is so pronounced as either the core–mantle boundary or the Mohorovičić discontinuity. It is unlikely that there will be any major revisions in the velocity-depth curves, but the greatly increased number of seismograph stations, including arrays of seismometers, have yielded much more precise estimates of $p = dT/d\Delta$ leading to finer detail in the structure of the curves and evidence of lateral inhomogeneities in the earth. Two regions with high velocity gradients have been found in the upper mantle near depths of 400 and 650 km. Both of these zones are between 50 and 100 km thick and involve a velocity

79

increase of 9–10 per cent. They have been interpreted as phase transitions and are discussed in detail in §10.4. Increased velocity gradients have also been found at various depths in the lower mantle, although they are an order of magnitude less than those in the upper mantle. Finally there is an increasing body of evidence that there is a low velocity (LV) layer in the upper mantle at a depth of about 150 km. Its depth and thickness vary, not only beneath continents and oceans but also beneath different tectonic areas. The LV layer will be discussed further in §10.3.

Buchbinder (1971) has obtained a new velocity model for the lower mantle and core using travel-times and amplitudes of PKP, P2KP and higher multiple K phases from a worldwide distribution of short-period seismograms, constrained by P and PcP tables. In his model the lowest part of the mantle contains a zone of decreasing velocity and the radius of the core is about 6 km larger. In a later paper (1972a) he showed that the standard deviation of the mean of the P7KP data is 0·2 s which is the same as that of his PKP data (Buchbinder (1971)), that of Bolt's (1968) PKP data being slightly less. The repeated traverses of the outer core by the P7KP phases do not appear to contribute significantly to the scatter in the observations and Buchbinder infers that the velocity structure of the outer core is laterally very homogeneous. Such homogeneity would place constraints upon the nature of any convection cells in the outer core (see §5.6).

3.6 Surface waves

In addition to body waves there is another class of elastic waves: surface waves. These are of two kinds (Rayleigh waves and Love waves) and may be propagated around boundaries between layers of different material or in strong velocity and density gradients near the surface of the earth. A detailed treatment of the subject is beyond the scope of this book; reference may be made to the monograph by Ewing *et al.* (1957). A characteristic of surface waves which is of extreme importance in geophysics is that the energy is spread in two rather than three dimensions as is the case for body waves. This accounts for their relative prominence on seismic records.

Rayleigh waves can propagate along a free surface such as the earth's surface. Particle displacement is in the vertical plane containing the direction of propagation, the amplitude of the displacement decreasing with distance from the free surface. The motion of a particle is a retrograde ellipse with its major axis vertical. Rayleigh waves which propagate at the free surface of a homogeneous medium are non-dispersive with a velocity of about 0·92 V_s if Poisson's ratio is 0·25. On the other hand, if the elastic moduli and density vary with distance from the free surface, Rayleigh waves show dispersion. Rayleigh waves can also exist along the boundary between two elastic media whose elastic properties are not too different or if one of them is a liquid, as on the ocean floor. Such waves are known as Stonely waves.

Love waves are SH waves propagating in a wave guide, the particle motion being horizontal and transverse to the direction of propagation. The simplest structure in which Love waves can propagate is a uniform layer with one boundary at the earth's free surface. A necessary condition for their propagation is that the shear wave velocity (V_{S_1}) in the layer is less than that in the substratum (V_{S_2}). Love waves show dispersion, the phase velocity varying from V_{S_1} for very short wavelengths to V_{S_2} for very long wavelengths. They can also propagate in more complicated structures, the character of the dispersion curve reflecting the layering. The low velocity layer in the earth at a depth of about 150 km can also act as a wave guide for Love waves.

The group velocity u is related to the phase velocity c by the equation

$$u = c - \lambda \frac{\mathrm{d}c}{\mathrm{d}\lambda} \qquad (3.27)$$

where λ is the wavelength. From the dispersed train of surface waves observed at a single station, the group velocity curve may be obtained by dividing the arrival times of oscillations of different periods into distance to the source. Each oscillation is considered as a signal which has travelled from the source to the station at the appropriate group velocity. The structure of the outer layers of the earth may be investigated by comparing observed dispersion curves with theoretically computed curves for different earth models. However, the parameters of the earth control variations of phase velocity rather than group velocity. The phase velocity dispersion curve may be obtained from the group velocity dispersion curve by integration (see equation (3.27)), but this introduces an arbitrary constant. Moreover, the measurement of group velocities by the method described above assumes that all frequencies are generated simultaneously: the earthquake source mechanism is rather complex and phase shifts between the different frequencies may be introduced. More recent work on surface waves has concentrated on determining phase velocities direct; Press (1956) has shown how this may be done using an array of three or more long period seismometers and observing the velocity of a peak or trough as it crosses the array. Another approach is to compute the Fourier spectrum of the arrivals at each station in order to obtain the relative phases of different frequency components from which the phase velocity curve may be computed (see e.g. Ben-Menahem and Toksöz (1963); Anderson (1965)).

Before high speed computers were available, theoretical dispersion curves could only be constructed for relatively simple earth models, although a method for computing them for multi-layered models using matrix methods had been given by Haskell (1953). The method, readily adaptable to computers, enables theoretical dispersion curves to be calculated for realistic models of the mantle. However, as Knopoff (1961) has shown, this inversion problem, i.e. the determination of a model from observations, does not have a unique solution, i.e. different models can yield identical dispersion curves.

3.7 Free oscillations of the earth

The free oscillations of a uniform elastic sphere were first investigated more than a hundred years ago, and followed directly from the equations of elasticity. This early work has been reviewed by Stonely (1961). Calculations for the real earth taking into account its detailed structure, gravitational forces, and rotation would have been impossible without the aid of modern computers. Renewed interest in the problem developed when improved instrumentation enabled free oscillations to be actually observed, and in 1959 Alterman *et al.* carried out a series of calculations of the periods of oscillation for realistic earth models.

There are two main classes of oscillations. The first class are called torsional (or toroidal) oscillations—the dilatation is everywhere zero and there is no radial component of displacement. The second class are called spheroidal oscillations—these are coupled oscillations and involve both radial displacements as well as torsional motions. The deformation of the earth's surface is best described by spherical harmonic functions

$$ P_n^m (\cos \theta) \quad \begin{matrix} \cos m\phi \\ \sin m\phi \end{matrix} $$

(see Appendix B). Each class of vibration is characterized by the degree n of the spherical harmonic involved and by the number of nodal surfaces in the radial direction. It has become general practice to specify oscillations by the symbols $_l S_n^m$ and $_l T_n^m$ for the spheroidal and toroidal types respectively, the subscript l indicating the number of nodal surfaces. The superscript m is dropped when there is no longitude dependence.*

The fundamental spheroidal oscillation $_0 S_0$ is an alternating compression and rarefaction of the whole earth. There are an infinite series of overtones $_l S_0$ with spherical nodal surfaces within the earth. The case $n=1$ is precluded—it represents a net translation of the surface and hence of the centre of gravity and would require the application of an external force. $_0 S_2$ represents the 'football' mode. There are two nodal parallels of latitude, dividing the surface into three zones. As the sphere oscillates, it is distorted alternately into an oblate and prolate spheroid (see Fig. 3.15). Spheroidal modes $_0 S_3$, $_0 S_4$, . . ., are motions with an increasing number of subdivided zonal distributions; each of these modes also has overtones with internal nodal surfaces.

For torsional oscillations, there is no solution for $n=0$. The case $_0 T_1$ is precluded. It implies a variation in the rate of rotation of the whole earth—concentric shells within the earth execute small rigid body rotations about the polar axis. The simplest torsional mode is $_0 T_2$: in this case there is a nodal line for displacement around the equator, the two hemispheres oscillating in anti-phase (see Fig. 3.16). Higher modes arise from subdivisions of the earth into 3, 4, . . ., zones with opposite motions. There are also overtones with internal nodal surfaces.

---

* In most of the literature on the free oscillations of the earth these symbols are written $_n S_l^m$ and $_n T_l^m$ with the subscripts interchanged. We have, however, preserved the more common notation $P_n^m(\cos \theta)$ for spherical harmonics and use l to indicate the number of nodal surfaces.

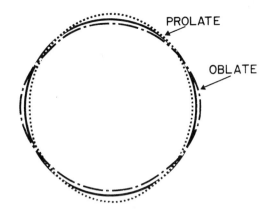

Fig. 3.15   Spheroidal 'football' mode $_0S_2$

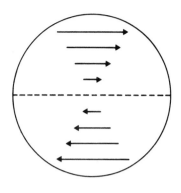

Fig. 3.16   Instantaneous motion
in toroidal mode $_0T_2$

Toroidal oscillations are, of course, only observable on horizontal component instruments. Spheroidal oscillations on the other hand, since they involve a radial displacement, may be observed on long-period vertical seismometers; they have also been detected on gravimeters set up to observe earth tides. Free oscillations of the earth were first observed in 1952, although it was not until the great Chilean earthquake of May 1960 that detailed measurements were made; the results were confirmed following the great Alaska earthquake of March 1964 (see Fig. 3.17). The records showed that the lowest frequency modes often appeared as a number of closely spaced lines in the spectrum. Pekeris *et al.* (1961) suggested that this may be due to the rotation of the earth splitting any mode having a longitude dependence into a number of frequencies. They coined the term 'terrestrial

Fig. 3.17   Power spectral density of the Alaskan and Chilean earthquakes recorded on a strain
seismometer at Isabella, California (after Smith (1966))

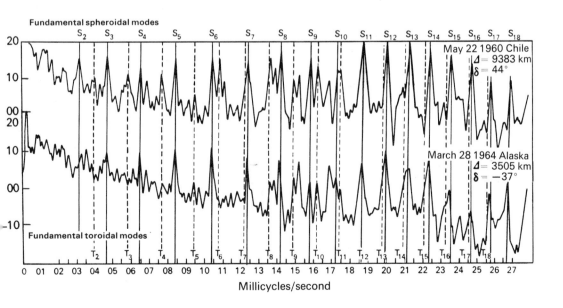

spectroscopy' for the study of free oscillation spectra, pointing out that the observed splitting of certain lines due to the earth's rotation is a mechanical analogue of the Zeeman effect. The splitting, if not resolved, produces a broadening of the spectral lines making the determination of the central frequency much more difficult. The free oscillations excited by a major earthquake last for several days, but their amplitude diminishes because the earth is not a perfectly elastic body. If spectra are computed for successive time intervals following the excitation, the damping of each mode can be determined and information on the anelasticity of the earth's interior obtained.

3.8 Variation of the physical properties within the earth

A basic problem in geophysics is the determination of the structure and constitution of the earth's interior from measurements made at the surface. This inverse problem does not have a unique solution although it is possible to put bounds on permissible solutions, and certain properties are known to a high degree of accuracy. Fundamental to the problem is the determination of the density distribution and this will be considered first.

The density ρ will depend on the pressure p, the temperature T, and an indefinite number of parameter n_i specifying the chemical composition, i.e.

$$\rho = \rho(p, T, n_i) \tag{3.28}$$

If m is the mass of material within a sphere of radius r, then, since the stress in the earth's interior is essentially equivalent to a hydrostatic pressure,

$$\frac{dp}{dr} = -g\rho \tag{3.29}$$

where

$$g = \frac{Gm}{r^2} \tag{3.30}$$

and G is the constant of gravitation. The assumption of hydrostatic pressure would be a poor approximation for the stresses in the crust, but is unlikely to be much in error in the mantle and core. If we consider for the moment a chemically homogeneous layer in which the temperature variation is adiabatic, it follows from equation (3.28) that

$$\frac{d\rho}{dr} = \frac{d\rho}{dp}\frac{dp}{dr} = \frac{-g\rho^2}{k_S} \tag{3.31}$$

where k_S is the adiabatic incompressibility defined by equation (3.10). A homogeneous region is here defined as one in which there are no significant changes of either phase or chemical composition. It follows from equations (3.30) (3.31) and (3.13) that

$$\frac{d\rho}{dr} = \frac{-Gm\rho^2}{k_s r^2} = \frac{-Gm\rho}{r^2\phi} \tag{3.32}$$

The distribution of ϕ throughout the earth is known from the velocity-depth curves.

Since $dm/dr = 4\pi\rho r^2$, a second order differential equation for $\rho = \rho(r)$ can be written down by differentiating equation (3.32), namely,

$$\frac{d^2\rho}{dr^2} - \frac{1}{\rho}\left(\frac{d\rho}{dr}\right)^2 + P(r)\frac{d\rho}{dr} + Q(r)\rho^2 = 0 \tag{3.33}$$

where

$$P(r) = \frac{2}{r} + \frac{1}{\phi}\frac{d\phi}{dr}$$

and $$\tag{3.34}$$

$$Q(r) = \frac{4\pi G}{\phi}$$

Equation (3.33), which was first obtained by Adams and Williamson in 1923, may be integrated numerically to obtain the density distribution in those regions of the earth where chemical and non-adiabatic temperature variations may be neglected.

Any density distribution must satisfy two conditions—it must yield the correct total mass of the earth and the moment of inertia about its rotational axis. Using these two conditions and a value ρ_1 of 3·32 g/cm³ for ρ at the top of layer B of the mantle (assumed to be at a depth of 33 km), Bullen applied equation (3.33) throughout the regions B, C and D. He then found that this led to a value of the moment of inertia I_c of the core greater than that of a uniform sphere of the same size and mass. This would entail the density to decrease with depth in the core and would be an unstable state in a fluid.

If the temperature gradient is not adiabatic, Birch (1952) showed that the right-hand side of equation (3.31) may be modified by the inclusion of an additional term, namely,

$$\frac{d\rho}{dr} = -\frac{g\rho}{\phi} + \alpha\rho\tau \tag{3.35}$$

where α is the volume coefficient of thermal expansion and τ the difference between the actual temperature gradient and the adiabatic gradient (see Appendix C). Allowance for the second term in equation (3.35) increased the value of I_c still more. A reasonable value for I_c could be obtained by increasing the initial value ρ_1, but only if an impossibly high value (at least 3·7 g/cm³) was chosen. Thus the assumption of chemical homogeneity must be in error; the region where this assumption is most likely to be invalid is region C where there are large changes in the slope of the velocity-depth curves. In his original earth Model A, Bullen thus used the Adams-Williamson equation (3.33) in regions B and D, while in region C he fitted a quadratic expression in r for $\rho = \rho(r)$. In the outer core (region E) equation (3.33) is likely to apply, and values of ρ down to a depth of about 5000 km can be obtained with some confidence. In the core, one boundary condition is $m = 0$ at $r = 0$, but lack of evidence on the value of the density ρ_0 at the centre of the earth leads to some indeterminacy in the density distribution in regions F and G. However, since these regions constitute only about 1 per cent of the earth's total volume, the density distribution within E can be estimated fairly precisely.

Bullen* also found that strong controls on permissible density values are exercised by various moment of inertia criteria. In particular he showed that the minimum possible value of ρ_0 is 12·3 g/cm³ and that increasing the value of ρ_0 by 5 g/cm³ affects the formally computed densities elsewhere by maximum amounts of only 0·03 g/cm³ in the mantle and 0·4 g/cm³ in the outer core. Bullen derived density distributions on two fairly extreme hypotheses:

(i) $\rho_0 = 12\cdot3$ g/cm³ and
(ii) $\rho_0 = 22\cdot3$ g/cm³ (this value being taken quite arbitrarily).

A model with density values midway between those of these two hypotheses has been called Model A. More recent evidence indicates that ρ_0 is probably much nearer its minimum value and that a model based on $\rho_0 = 12\cdot3$ (Model A-i) is more likely to be correct. There have been a number of more recent determinations of the density distribution within the earth—these will be discussed in § 3.9 after it has been shown how other physical parameters of the earth may be determined once the density distribution is known.

From equations (3.29) and (3.30), it follows that

$$\frac{dp}{dr} = \frac{-Gm\rho}{r^2} \tag{3.36}$$

Hence by numerical integration, the pressure distribution may be obtained once the density distribution has been determined. Since the density is used only to determine the pressure gradient, the pressure distribution is insensitive to small changes in the density distribution and may be determined quite accurately.

The variation of g can be calculated from equation (3.30); its value does not differ by more than 1 per cent from 990 cm/sec² until a depth of over 2400 km is reached. On the other hand, values of g deep within the earth are sensitive to changes in density, and values below 4000 km may be in error by as much as 5 per cent.

From a knowledge of the density distribution, it is easy to compute values of the elastic constants. Thus equations (3.9) and (3.13) give μ and k_s directly. From the

---

* A good account of this early work can be found in Bullen (1963a)

known relationships between the elastic constants, it is possible to compute the distribution of Young's Modulus E and Poisson's ratio σ. In particular

$$\sigma = \frac{3k_S - 2\mu}{6k_S + 2\mu} = \frac{V_P^2 - 2V_S^2}{2(V_P^2 - V_S^2)} \tag{3.37}$$

and is thus independent of the density ρ.

3.9 Density distribution within the earth; more recent determinations

Additional information on the physical properties of the earth's mantle and core has been obtained from an analysis of the free vibrations of the earth excited by the great Chilean earthquake of May 22nd, 1960. Both spheroidal and toroidal oscillations were observed; however, neither the Jeffreys nor the Gutenberg velocity distributions combined with Bullen's density distribution were consistent with the longer period free oscillation data. The higher order free oscillations also provided some of the best evidence for the existence of a low velocity zone in the upper mantle, first advocated by Gutenberg.

Landisman *et al.* (1965) considered the inverse problem, that of determining the radial distribution of ρ, V_P and V_S from the travel-time curves for P and S waves and the free periods of the earth. Their method is independent of the assumptions of homogeneity and an adiabatic temperature gradient (except in the region of the outer core). They investigated a number of earth models subject to the constraints fixed by the total mass of the earth, the revised estimate of its moment of inertia about its axis of rotation as determined from analyses of the orbits of artificial satellites (Cook, 1963), and by a density of 3·32 g/cm³ at the top of the mantle. The outer core was taken to be chemically homogeneous with an adiabatic temperature gradient and in it the Adams-Williamson equation was used. In their model M1, the lower core was considered to be chemically inhomogeneous following the model proposed by Bullen (1962, 1963b) and Bolt (1962); in their model M3, the entire core was taken to be a homogeneous, adiabatic fluid. The density in both these models is about 10 per cent lower near the base of the mantle than that predicted by the Bullen A-*i* distribution. Constant densities were also found for both models at depths between about 1600 and 2800 km. A super-adiabatic temperature of 4°–5° C/km would be required to explain such a result—this would lead to excessive temperatures at the core-mantle boundary. If the vanishing density gradient were the result of a compositional change such as the depletion of iron, a concomitant increase of almost 10 per cent would be expected in the shear velocity gradient. A super-adiabatic temperature gradient of about 2° C/km combined with some depletion of iron is a possible explanation. The central density for their model M1 is about 15·42 g/cm³ and for model M3 about 12·63 g/cm³.

Pekeris (1966) also obtained density distributions in the earth from the observed periods of the free oscillations of the earth, without using the Adams-Williamson assumption of homogeneity and adiabaticity in any region (including the outer core). The density distribution $\rho(r)$ was represented by 50 pivotal values $\rho(r_k)$ with linear variations in between, and with discontinuities at the Moho and at the core-mantle boundary. The ρ_k were varied by the method of steepest descent so as to minimize the sum of the squares of the residuals of all the observed periods. Pekeris found that a somewhat better fit to the observed spectral data was obtained for a Gutenberg-type velocity distribution (with a low velocity layer) than for the Jeffreys distribution. The density distributions for all models in the depth range 400–1200 km converge in a narrow band close to that of Bullen's model A-i. In the first 400 km, the density is nearly constant for Gutenberg-type models with a tendency towards negative density gradients in the first 200 km. All models give a density at the outer boundary of the core close to $10\cdot0$ g/cm^3; also, as is to be expected, the density distribution in the inner core has little effect on the spectrum as a whole.

Birch (1964) has estimated the density distribution in the earth based in part on his empirical observation that for silicates and oxides of about the same iron content, there is an approximate linear relationship between density and the velocity of compressional waves, namely,

$$\rho = a + bV_P \tag{3.38}$$

In the lower mantle and core Birch determined the density distribution from the Adams-Williamson equation and only used the empirical relation (3.38) in the upper mantle and transition zone; i.e. the Adams-Williamson equation was used where the change of density is most probably determined by compression alone, and the empirical rule for the upper mantle, where there are high thermal gradients, and the transition region, where there are phase changes. He considered two models. In the first (Solution I), the constant b was given the value $0\cdot328$ (g cm^{-3})/(km s^{-1}) as found for rocks and crystals of low mean atomic weight. The constant a and the mass of the core are then the only adjustable parameters and are determined by the total mass and moment of inertia of the earth. For the second model (Solution II), Birch chose the constant a so that the density at 33 km is $3\cdot32$ g/cm^3 and the adjustable parameters are then b and the mass of the core. The density in the lower mantle is about 1 per cent higher, and in the core about 1 per cent lower in the second than in the first model, and the values are very similar to Bullen's model A-i. The densities in the lower mantle are in good agreement with shock wave measurements on rocks having FeO contents in the range 10 ± 2 per cent by weight, and may be accounted for by mixtures of close-packed oxides, silica transforming to stishovite in the transition layer. A constant density in the lower mantle, combined with rising velocities as suggested by Landisman et al. (1965), is not compatible with Birch's results.

Wang (1970) has suggested that better estimates of the density in the upper mantle (to a depth of $\sim$1000 km) may be obtained by using an empirical relationship between the bulk sound velocity $c = (V_P^2 - \frac{4}{3} V_S^2)^{1/2}$ and density ρ, rather than the linear relation between V_P and ρ proposed by Birch. Such a relationship was first suggested by McQueen *et al.* (1964) and later investigated by Wang (1968, 1969). Wang (1970) found the density in the upper 200 km of the mantle to be 3·3–3·4 g/cm^3, this lower value lending support to the suggestion that the upper mantle may be composed of some variations of peridotite rather than eclogite, which would require densities in the range 3·5–3·6 g/cm^3. Wang also found the mean atomic weight of the lower mantle to be about 21·3–21·5, and suggested that perhaps the entire mantle may have a uniform iron content.

Clark and Ringwood (1964) estimated densities in the earth using petrological models of the upper mantle constructed on the assumption of an overall pyrolite (ultrabasic) composition and an eclogite composition. Their pyrolite model is almost identical with Bullen's Model A-*i* in the lower mantle. Their use of petrological arguments in the upper mantle and transition region, however, leads to densities in these regions which are systemically lower than Bullen's values. Fig. 3.18(a),(b),(c) shows the density distribution for some of the more recently proposed earth models together with Bullen's Model A-*i* for comparison. The HB$_1$ model (Haddon and Bullen, 1969) is discussed later in this section.

Wang (1972) has combined data from a number of sources to construct another earth model. Because of significant uncertainty in the shear wave velocity distribution in the upper mantle, widely different density models satisfy data on the earth's free oscillations. Thus Wang uses a petrological model for the upper mantle. In the transition region where the V_P distribution is reliable, he uses Birch's (1964) empirical relation between V_P and density. In the lower mantle, both the V_P and V_S velocity distributions are considered reliable and Wang uses his (1970) empirical relation between the bulk sound velocity and density: in the core, the Adams-Williamson relation is used. Wang's final earth model is obtained from his initial model by perturbing it to satisfy observational constraints such as total mass, moment of inertia, surface wave velocities and periods of free oscillations. His final model shows a pronounced low-velocity zone for both P and S waves between about 120 and 220 km, and satisfies the observational constraints within the widely accepted uncertainties.

Yet another earth model consistent with the free oscillation and surface wave data has been constructed by Mizutani and Abe (1972) using a trial and error method with the help of an equation of state of some rock-forming minerals. Their model has a pronounced low velocity channel between 70 and 210 km, the lid being characterized by high density ($\simeq 3\cdot5$ g/cm^3) and high shear velocity ($\simeq 4\cdot72$ km/s). The shear wave velocity has a slight negative gradient between 200 and 340 km which is probably due to the increase of the iron and pyroxene content with depth. Their model also indicates a slight discrepancy between the results of Love and Rayleigh wave data. This can be removed by introducing small ($\sim 0\cdot7$ per cent) anisotropy in the low velocity zone—such anisotropy may be interpreted in terms of partially molten material in that zone.

It should be emphasized that the overall density distribution is not drastically changed by taking into account the combined effects of the revised estimate of the

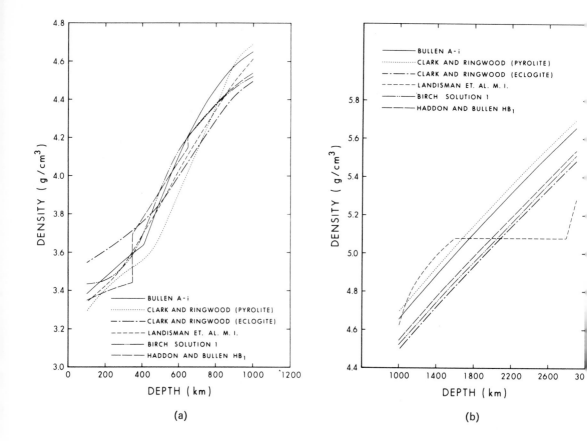

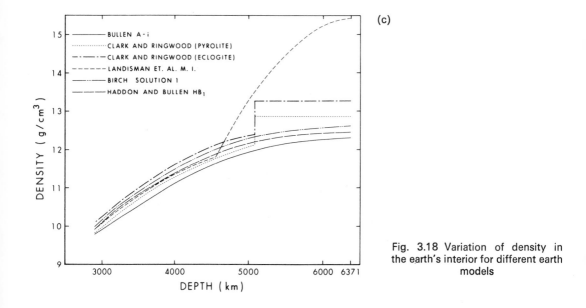

Fig. 3.18 Variation of density in the earth's interior for different earth models

moment of inertia of the earth and the observational data on its free vibrations. Inside the mantle the largest difference in ρ between Models A-i and HB_1 is only 0·15 g/cm³, and inside the core the values of ρ in Model HB_1 exceed those in Model A-i at all depths by amounts between 0·2 and 0·3 g/cm³.

Buchbinder (1968) has carried out a detailed study of PcP phases from eight explosions and three earthquakes. His results show that the initial direction of motion of PcP reverses at an epicentral distance of about 32°, at which distance the amplitudes of PcP pass through a minimum. Buchbinder maintains that these results reflect properties at the core-mantle boundary and cannot be satisfied by any of the 'conventional' earth models. He finds that the velocity at the top of the core is some 7–8 per cent lower than values usually quoted and that there is no density discontinuity across the core-mantle boundary. Buchbinder interprets his results to indicate that the bottom of the mantle is inhomogeneous, the inhomogeneity being caused by an increase in iron (or other heavy metal) content with depth, which increases the mean atomic weight and density with little change in P velocity. The drop in P velocity across the core-mantle interface may then be explained by a discontinuous increase in mean atomic weight with little change in density. It is not easy to assess Buchbinder's results—the overall evidence still appears to favour the more conventional models with a density ratio of about 1·7 at the core-mantle boundary.

Press (1968a, b) has used a Monte Carlo inversion method to obtain a number of earth models using as data 97 eigenperiods, travel times of P and S waves and the mass and moment of inertia of the earth. The Monte Carlo method uses random selection to generate large numbers of models in a computer, subjecting each model to a test against geophysical data. Only those models whose properties fit the data within prescribed limits are retained. This procedure has the advantage of finding models without bias from preconceived or oversimplified ideas of earth structure. Monte Carlo methods also offer the advantage of exploring the range of possible solutions and indicate the degree of uniqueness obtainable with currently available geophysical data.

Press was later able (1970a) to speed up considerably his Monte Carlo procedures. Using new and more accurate data he was able to find a large number of successful models; Fig. 3.19 and Fig. 3.20 show 27 successful density distributions in the mantle and core, respectively. Of the millions of earth models generated, every successful model showed a low velocity zone for shear waves in the sub-oceanic mantle centred at depths between 150 and 250 km. Moreover, although density values in the mantle just below the Moho cover the entire permissible range, in the vicinity of 100 km all values fall in the narrow band 3·5–3·6 g/cm³. This value is so high that it narrows the range of possible compositions there to an eclogite facies. Large density and velocity gradients were found in the transition zone in the mantle; in particular high density gradients are localized near 350–450 and 550–800 km. These are interpreted as phase transitions and are discussed later (see § 10.4). Upper mantle models which are more complex, with large fluctuations in the shear velocity and density, fit better than 'standard' models. Such complexity might be expected if the mantle is

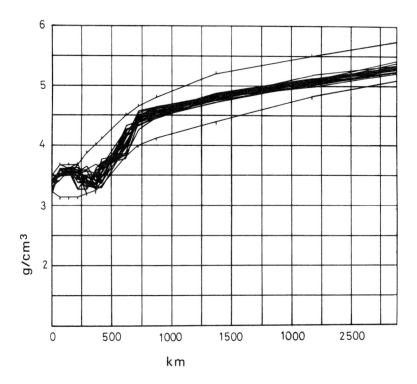

Fig. 3.19   Successful density distributions in the mantle (after Press (1970a))

Fig. 3.20   Successful density distributions in the core (after Press (1970a))

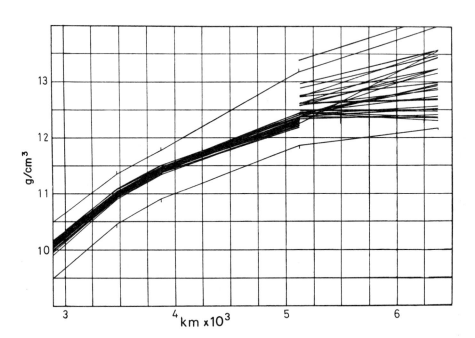

chemically and mineralogically zoned, and if high thermal gradients and partial melting take place. Using Kanamori's (1970) Rayleigh and Love wave phase velocity dispersion data for oceanic, continental shield and tectonic regions together with other more recent data, Press (1970b) constructed a number of regionalized earth models. He also confirmed that the earth's core is inhomogeneous. The density at the top of the core is constrained to the narrow range $9.9-10.2$ g/cm^3, a value appropriate for a mixture of iron with about 15 wt per cent silicon. Changes in the currently accepted value of the core radius range from -3 to $+10$ km. Wiggins (1969) has also used Monte Carlo techniques to investigate the nature of the non-uniqueness inherent in the interpretation of body wave data.

Haddon and Bullen (1969) have constructed a series of earth models (HB) using free oscillation data consisting of the observed periods of fundamental spheroidal and toroidal oscillations for $0 \leqslant n \leqslant 48$ and $2 \leqslant n \leqslant 44$ respectively, and certain spheroidal overtones. Data from the records of both the Chilean (May 1960) and Alaskan (March 1964) earthquakes were used. Their procedure was to start from models derived independently of the oscillation data and to produce a sequence of models showing improved agreement with all the available data. In passing from one model to the next, a guiding principle was to introduce and vary one or more of the parameters in the model description at any stage in order to satisfy the oscillation data. They thus tried to establish models described in terms of the minimum number of parameters demanded by the data. A major difference in principle between their method and the Monte Carlo procedure of Press is the comparatively large number of parameters that Press permits to be randomly varied. Haddon and Bullen point out that the predominance of complex models found by Press is inherent in his method, since a simple random walk would automatically have a low probability. Haddon and Bullen also stress that the 'average' earth to which average periods of free oscillation modes relate is not necessarily the same as an earth model to which the currently available average seismic body wave travel times apply, since earthquake epicentres and recording stations are not randomly distributed over the earth's surface.

One of the aims of Haddon and Bullen was to examine the conclusion of Landisman et al. (1965) that the oscillation data require $d\rho/dr$ to be abnormally low ($\simeq 0$) throughout much of the lower mantle. This result can only be accepted if the core radius is unchanged. However, one of the main findings of Haddon and Bullen is that the oscillation data required a significant increase in the core radius (about 15–20 km). Hales and Roberts (1970) also inferred an increase of about 15 km in the core radius from their analysis of travel-times of S waves in the range $30°-126°$. There is some evidence that the effect of damping on the oscillation periods may be significant—if so the increase in core radius may be somewhat less than 15 km. This would allow some reduction in the density gradient in the lower mantle but not nearly so much as that demanded by Landisman et al. Another result of Haddon and Bullen's work is a reduction in the thickness of the crustal layer A from 33 to 15 km. This is a result of the changes made in V_S and ρ in region B in the course of fitting the oscillation data.

Such a reduction is reasonable since it involves a change to a crust that may be more representative of average conditions in the earth as a whole.

In 1970, Bullen and Haddon used additional evidence to derive an improved earth model (HB_2). Model HB_2 incorporates the newer P travel-time data of Herrin (1968) and takes into account the abnormalities in the body wave observations in the lower 200 km of the mantle (region D″) and detailed structure of the lower core (Bullen (1965); see also § 10.5). In their earlier model HB_1 a simplified core structure was assumed (the whole core being considered fluid) since free earth oscillation data are incapable at present of resolving fine detail in the lower core.

An upper bound to the density increase at the boundary of the inner core has been obtained by Bolt and Qamar (1970) from measured amplitudes of PKiKP (i.e. reflections from the inner core boundary) and PcP phases recorded at the Large Aperture Seismic Array (LASA) in Montana. The PKiKP waves were reflected at steep angles from the inner core boundary (epicentral distance $\Delta < 40°$) and demonstrate the sharpness of that boundary (Engdahl *et al.* (1970)). Bolt and Qamar's analysis indicates a minimum value of 0·875 for the ratio of the densities at the inner core boundary. This would give a maximum density jump $\Delta\rho_{ic}$ at the inner core boundary of 1·8 g/cm³ if the density in the outer core at the boundary is 12·35 g/cm³ (see also Bolt (1972)). This result has been obtained on seismological evidence alone and does not depend on any assumptions concerning the chemical composition of the core. It is also in agreement with recent core models which indicate that the density in the inner core is much lower than was at one time supposed. Using his 1971 velocity model of the core, Buchbinder (1972b) estimated $\Delta\rho_{ic}$ to be 1·2 g/cm³, assuming a maximum central density ρ_0 of 13·6 g/cm³ (appropriate for an iron–nickel core), and 0·6–0·7 g/cm³, assuming ρ_0 to be 13·0 g/cm³ (appropriate for an iron core).

Derr (1969a) has developed a series of earth models using weighted observations of free oscillation data up to 1968. Weighting is necessary because of differences in record length, sharpness of spectral peaks and basic resolving power of the instruments. Some 1500 measurements from 46 different combinations of investigators and instruments were considered by Derr (1969b). Derr's philosophy of approach is similar to that of Bullen and Haddon; he first uses geophysical arguments to obtain an approximate earth model independent of any inferences from free oscillation data. Inversion is then used to derive from it another model which satisfies the observed free periods within prescribed limits. His final model shows a significant low velocity zone for S waves in the upper mantle, a rapid increase in density at depths of 500–800 km, an average shear velocity of 2·18 km/s in the inner core and $\Delta\rho_{ic} = 2$ g/cm³. Exact values of these last two parameters remain very uncertain until more reliable observations of low-order spheroidal overtones can be obtained. The model also indicates a jump of 26 per cent in k across the inner core boundary, in disagreement with Bullen's compressibility-pressure hypothesis (see § 10.5).

Dziewonski (1971) has stressed that only observations of overtones will allow the density distribution to be determined with sufficient detail to be able to make

94

meaningful estimates of the structure and composition of the earth's deep interior. The diversity of models which satisfy travel times and fundamental mode data indicates the insufficiency of the constraints provided by the more limited set of data. Dziewonski was able to identify overtones with periods greater than 250–300 s on records of the March 1964 Alaska earthquake from stations belonging to the Worldwide Standard Seismograph Network. Using this additional information, he constructed a number of earth models. His final models show a change in composition in the bottom 500 km of the mantle. His density distribution shows systematic differences from the HB_1 model and Press' models in the lower mantle, although all models lie within Press' band of solutions—except that lower shear velocities are found in the depth range 480–650 km. Although agreement with these other models is good for the fundamental modes and the first two spheroidal overtones, it is poor for the third and fourth spheroidal and torsional overtones. A solid inner core is also demanded by Dziewonski's data—if the shear velocity in the inner core is 3·5 km/s, $\Delta\rho_{ic} = 0·81$ g/cm³.

Anderssen and Seneta (1971, 1972) have examined in detail the Monte Carlo method of inversion of geophysical data; in particular they developed a statistical procedure for estimating the reliability of non-uniqueness bounds defined by a family of randomly generated models. Anderssen et al. (1972) have applied these techniques to the problem of obtaining the density distribution within the earth. In a later paper (Worthington et al. (1972)) they attempted to resolve the major discrepancies between the density models of Bullen and Haddon and those of Press. The successful models of Press (1970a) demand that the density of the mantle between depths of 70 and 150 km cannot be less than 3·40 g/cm³, and is more likely to be nearer 3·50 g/cm³. On the other hand the models of Haddon and Bullen (1969) and those of Wang (1970, 1972) yield lithospheric densities around 3·3 g/cm³. Worthington et al. showed that these differences cannot be due to the different techniques employed to derive the models. They conclude that the differences are predominantly due to differences in assumptions about the permissible range of values of the shear velocity in the upper mantle. Dziewonski (1970) had already found a high correlation between shear wave velocity in the depth range 250–800 km and density between 33 and 200 km. In particular, a change in velocity of as little as 0·05 km/s around 400 km may result in a density perturbation of almost 0·2 g/cm³ at a depth of 100 km. The Haddon and Bullen model fails because their upper mantle shear wave velocity profile does not agree with the results of surface wave group velocity observations, which in turn casts doubt on their shear wave velocity distribution in the deeper regions of the upper mantle. On the other hand, Worthington et al. do not accept Press' model, believing that his lower bound on shear velocity in the upper mantle is too constrained. It may well be that our present knowledge of shear wave velocities in the upper mantle is not precise enough to derive a more definitive estimate of density in the lithosphere.

The 'inverse' problem, i.e. the problem of determining the distribution of some physical parameter from a set of observations, has received a lot of attention during the last few years—especially in the USA and the USSR. Backus and

Gilbert (1967) outlined a general procedure for obtaining solutions, and in a later paper (Gilbert and Backus (1968)) presented approximate solutions to the inverse normal mode problem. Backus and Gilbert (1968) define a gross earth datum as 'a single measurable number describing some property of the whole earth, such as mass, moment of inertia, or the frequency of oscillation of some identified elastic-gravitational normal mode'. They show how to determine whether a given finite set of gross earth data can be used to specify an earth structure uniquely (except for fine-scale detail); and how to determine the shortest length scale which the given data can resolve at any particular depth. The principal result of their work is that it is possible to draw rigorous conclusions about the internal structure of the earth from a finite set of gross earth data. Infinite resolution can never be achieved but rigorous answers can be given to a number of qualitative questions, such as whether there are low velocity zones or density inversions in the mantle. The question of uniqueness has been further discussed by Dziewonski (1970).

In a later paper, Backus and Gilbert (1970) discuss the inversion of a finite set G of inaccurate gross earth data. They show that from some sets G it is possible to determine the structure of the earth (except for fine-scale detail) within certain limits of error. They also show how to determine whether a given set G will permit the construction of such localized averages of earth structure, and how to find the shortest length scale over which G gives a local average structure at a given depth if the variance of the error is to be less than a given amount. An excellent outline of the theory developed in the above three very important papers by Backus and Gilbert has been given by Parker (1970) who applies their technique to the inverse problem of determining the electrical conductivity in the mantle (see § 7.6). Parker (1972) has also examined the inverse problem with grossly inadequate data. This is the case, not only when the number of observations is small, but also if the inverse problem is intrinsically non-unique, as is the case of attempting to determine the density inside a body from gravity observations outside the body. Parker shows that although inadequate data cannot yield detailed structure, nevertheless they can be used to rule out certain classes of structures and provide bounds on acceptable models. The general inverse problem has also been recently reviewed by Wiggins (1972).

References

Aggarwal, Y. P., L. R. Sykes, J. Armbruster and M. L. Sbar (1973) Premonitory changes in seismic velocities and prediction of earthquakes, *Nature*, **241**, 101.

Aki, K., T. De Fazio, P. Reasenberg and A. Nur (1970) An active experiment with earthquake fault for an estimation of the *in situ* stress, *Bull. seism. Soc. Amer.*, **60**, 1315.

Alterman, Z., H. Jarosch and C. L. Pekeris (1959) Oscillations of the earth, *Proc. roy. Soc.*, **A 252**, 80.

Anderson, D. L. (1965) Recent evidence concerning the structure and composition of the earth's mantle, in: *Physics and Chemistry of the Earth*, Vol. 6 (Pergamon Press).

Anderssen, R. S. and E. Seneta (1971) A simple statistical estimation procedure for Monte Carlo inversion in geophysics, *Pure appl. Geophys.*, **91**, 5014.

Anderssen, R. S. and E. Seneta (1972) A simple statistical estimation for procedure for Monte Carlo inversion in geophysics, II. Efficiency and Hempel's paradox, *Pure appl. Geophys.*, **96**, 5.

Anderssen, R. S., M. H. Worthington and J. R. Cleary (1972) Density modelling by Monte Carlo inversion—I Methodology, *Geophys. J.*, **29**, 433.

Backus, G. E. and J. F. Gilbert (1967) Numerical applications of a formalism for geophysical inversion problems, *Geophys. J.*, **13**, 247.

Backus, G. E. and J. F. Gilbert (1968) The resolving power of gross earth data, *Geophys. J.*, **16**, 169.

Backus, G. and F. Gilbert (1970) Uniqueness in the inversion of inaccurate gross earth data, *Phil. Trans. roy. Soc.*, **266**, 123.

Barazangi, M. and J. Dorman (1969) World seismicity maps compiled from ESSA Coast and Geodetic Survey epicentre data, 1961–1967, *Bull. seism. Soc. Amer.*, **59**, 369.

Bath, M. (1966a) Earthquake energy and magnitude, in: *Physics and Chemistry of the Earth*, Vol. 7 (Pergamon Press).

Bath, M. (1966b) Earthquake seismology, *Earth Sci. Rev.*, **1**, 69.

Benioff, H. (1964) Earthquake source mechanisms, *Science*, **143**, 1399.

Ben-Menahem, A. and M. N. Toksöz (1963) Source mechanisms from spectrums of long period surface waves, 2. The Kamchatka earthquake of November 4, 1952, *J. geophys. Res.*, **68**, 5207.

Birch, F. (1952) Elasticity and constitution of the earth's interior, *J. geophys. Res.*, **57**, 227.

Birch, F. (1964) Density and composition of mantle and core, *J. geophys. Res.*, **69**, 4377.

Bolt, B. A. (1962) Gutenberg's early PKP observations, *Nature*, **196**, 122.

Bolt, B. A. (1968) Estimation of PKP travel times, *Bull. seism. Soc. Amer.*, **58**, 1305.

Bolt, B. A. (1972) The density distribution near the base of the mantle and near the earth's centre, *Phys. Earth Planet. Int.*, **5**, 301.

Bolt, B. A. and A. Qamar (1970) Upper bound to the density jump at the boundary of the earth's inner core, *Nature*, **228**, 148.

Bridgman, P. W. (1945) Polymorphic transitions and geological phenomena, *Amer. J. Sci.*, **243A**, 90.

Buchbinder, G. G. R. (1968) Properties of the core-mantle boundary and observations of PcP, *J. geophys. Res.*, **73**, 5901.

Buchbinder, G. G. R. (1971) A velocity structure of the earth's core, *Bull. seism. Soc. Amer.*, **61**, 429.

Buchbinder, G. G. R. (1972a) Travel times and velocities in the outer core, *Earth Planet. Sci. Letters*, **14**, 161.

Buchbinder, G. G. R. (1972b) An estimate of the inner core density, *Phys. Earth Planet. Int.* **5**, 123.

Bullen, K. E. (1954) *Seismology* (Methuen and Co., Ltd., London).

Bullen, K. E. (1962) Earth's central density, *Nature*, **196**, 973.

Bullen, K. E. (1963a) *An Introduction to the Theory of Seismology* (Camb. Univ. Press).

Bullen, K. E. (1963b) An index of chemical inhomogeneity in the earth, *Geophys. J.*, **7**, 584.

Bullen, K. E. (1965) Models for the density and elasticity of the earth's lower core, *Geophys. J.*, **9**, 233.

Bullen, K. E. and R. A. W. Haddon (1970) Evidence from seismology and related sources on the earth's present internal structure, *Phys. Earth Planet. Int.*, **2**, 342.

Clark, S. P. Jr. and A. E. Ringwood (1964) Density distribution and constitution of the mantle, *Rev. Geophys.*, **2**, 35.

Cook, A. H. (1963) The contribution of observations of satellites to the determination of the earth's gravitational potential, *Space Sci. Rev.*, **2**, 355.

Davies, D. (1973) Monitoring underground explosions, *Nature*, **241**, 19.

Dennis, J. G. and C. Walker (1965) Earthquakes resulting from metastable phase transitions, *Tectonophysics*, **2**, 401.

Derr, J. S. (1969a) Internal structure of the earth inferred from free oscillations, *J. geophys. Res.*, **74**, 5202.

Derr, J. S. (1969b) Free oscillation observations through 1968, *Bull. seism. Soc. Amer.*, **59**, 2079.

Dziewonski, A. M. (1970) Correlation properties of free period partial derivatives and their relation to the resolution of gross earth data, *Bull. seism. Soc. Amer.*, **60**, 741.

Dziewonski, A. M. (1971) Overtones of free oscillations and the structure of the earth's interior, *Science*, **172**, 1336.

Engdahl, E. R., E. A. Flinn and C. F. Romney (1970) Seismic waves reflected from the earth's inner core, *Nature*, **228**, 852.

Evans, D. (1966) Man-made earthquakes in Denver, *Geotimes*, **11**, 11.

Evison, F. F. (1967) On the occurrence of volume change at the earthquake source, *Bull. seism. Soc. Amer.*, **57**, 9.

Ewing, W. M., W. S. Jardetsky and F. Press (1957) *Elastic Waves in Layered Media* (McGraw-Hill).

Gilbert, J. F. and G. E. Backus (1968) Approximate solutions to the inverse normal mode problem, *Bull. seism. Soc. Amer.*, **58**, 103.

Gutenberg, B. and C. F. Richter (1954) *Seismicity of the Earth* (2nd ed.) (Princeton Univ. Press).

Haddon, R. A. W. and K. E. Bullen (1969) An earth model incorporating free earth oscillation data, *Phys. Earth Planet. Int.*, **2**, 35.

Hales, A. L. and J. L. Roberts (1970) Shear velocities in the lower mantle and the radius of the core, *Bull. seism. Soc. Amer.*, **60**, 1427.

Haskell, N. A. (1953) Dispersion of surface waves in multi-layered media, *Bull. seism. Soc. Amer.*, **43**, 17.

Healy, J., W. Rubey, D. Griggs and C. Raleigh (1968) The Denver earthquakes, *Science*, **161**, 1301.

Herrin, E. (1968) Introduction to the 1968 Seismological Tables for P phases, *Bull. seism. Soc. Amer.*, **58**, 1193.

Honda, H. (1962) Earthquake mechanism and seismic waves, *J. Phys. Earth, Tokyo*, **10**, 1.

Jeffreys, H. (1936) On travel times in seismology, *Bur. Cent. Seism. Inter. A., Fasc.*, **14**, 1.

Jeffreys, H. (1939a) The times of PcP and ScS, *Mon. Not. Roy. Astr. Soc. geophys. Suppl.*, **4**, 537.

Jeffreys, H. (1939b) The times of core waves, *Mon. Not. Roy. Astr. Soc. geophys. Suppl.*, **4**, 548.

Jeffreys, H. (1961) *Theory of Probability* (Clarendon Press, Oxford).

Jeffreys, H. and K. E. Bullen (1935) Times of transmission of earthquake waves, *Bur. Cent. Seism. Inter. A., Fasc.*, **11**.

Jeffreys, H. and K. E. Bullen (1940) *Seismological Tables* (Brit. Assoc. Gray-Milne Trust).

Johnson, A. G., R. L. Kovach, A. Nur and J. R. Booker (1973) Pore pressure changes during creep events on the San Andreas fault. *J. geophys. Res.*, **78**, 851.

Julian, B. R., D. Davies and R. M. Sheppard (1972) PKJKP, *Nature*, **235**, 317.

Kanamori, H. (1970) Velocity and Q of mantle waves, *Phys. Earth Planet. Int.*, **2**, 259.

Knopoff, L. (1961) Green's function for eigenvalue problems and the inversion of Love wave dispersion data, *Geophys. J.*, **4**, 161.

Knopoff, L. and M. Randall (1970) The compensated linear-vector dipole: a possible mechanism for deep earthquakes, *J. geophys. Res.*, **75**, 4957.

Landisman, M., Y. Sato and J. Nafe (1965) Free vibrations of the earth and the properties of its deep interior regions, Part I Density, *Geophys. J.*, **9**, 439.

Lehmann, I. (1935) P', *Pub. Bur. Cent. Seism. Inter.*, **A 14**, 3.

McKenzie, D. and J. N. Brune (1972) Melting on fault planes during large earthquakes, *Geophys. J.*, **29**, 65.

McQueen, R. G., J. N. Fritz and S. P. Marsh (1964) On the composition of the earth's interior, *J. geophys. Res.*, **72**, 2947.

Mizutani, H. and K. Abe (1972) An earth model consistent with free oscillation and surface wave data, *Phys. Earth Planet. Int.*, **5**, 345.

Parker, R. L. (1970) The inverse problem of electrical conductivity in the mantle, *Geophys. J.*, **22**, 121.

Parker, R. L. (1972) Inverse theory with grossly inadequate data, *Geophys. J.*, **29**, 123.

Pekeris, C. L. (1966) The internal constitution of the earth, *Geophys. J.*, **11**, 85.

Pekeris, C. L., Z. Alterman and H. Jarosch (1961) Rotational multiplets in the spectrum of the earth, *Phys. Rev.*, **122**, 1692.

Press, F. (1956) Determination of crustal structure from phase velocity of Rayleigh waves, Part I Southern California, *Bull. geol. Soc. Amer.*, **67**, 1647.

Press, F. (1968a) Density distribution in the earth, *Science*, **160**, 1218.

Press, F. (1968b) Earth models obtained by Monte Carlo inversion, *J. geophys. Res.*, **73**, 5223.

Press, F. (1970a) Earth models consistent with geophysical data, *Phys. Earth Planet. Int.*, **3**, 3.

Press, F. (1970b) Regionalized earth models, *J. geophys. Res.*, **75**, 6575.

Randall, M. and L. Knopoff (1970) The mechanism at the focus of deep earthquakes, *J. geophys. Res.*, **75**, 4965.

Reid, H. F. (1911) The elastic rebound theory of earthquakes, *Bull. Dept. Geol. Univ. Calif.*, **6**, 413.

Richter, C. F. (1935) An instrumental earthquake magnitude scale, *Bull. seism. Soc. Amer.*, **25**, 1.

Richter, C. F. (1958) *Elementary Seismology* (W. H. Freeman and Co., San Francisco).

Ringwood, A. E. (1967) The pyroxene–garnet transformation in the earth's mantle, *Earth Planet. Sci. Letters*, **2**, 255.

Ringwood, A. E. (1972) Phase transformations and mantle dynamics, *Earth Planet. Sci. Letters*, **14**, 233.

Ryall, A. and G. Boucher (1970) Earthquakes and nuclear detonation, *Science*, **167**, 1013.

Savarensky, E. F. (1968) On the prediction of earthquakes, *Tectonophysics*, **6**, 17.

Semonov, A. N. (1969) Change in the travel-time ratio of transverse and longitudinal waves before strong earthquakes, *Izv. Earth Phys.* (English Transl.), No. 4, 72.

Simon, R. (1970) Denver earthquake, *Science*, **167**, 1519.

Smith, S. W. (1966) Free oscillations excited by the Alaska earthquake, *J. geophys. Res.*, **71**, 1183.

Stauder, W. (1962) The focal mechanism of earthquakes, in: *Advances in Geophysics*, Vol. 9 (Acad. Press).

Stonely, R. (1961) The oscillations of the earth, in: *Physics and Chemistry of the Earth*, Vol. 4 (Pergamon Press).

Wang, C.-Y. (1968) Constitution of the lower mantle as evidenced from shock wave data for some rocks, *J. geophys. Res.*, **73**, 6459.

Wang, C.-Y. (1969) Equation of state of periclase and some of its geophysical implications, *J. geophys. Res.*, **74**, 1451.

Wang, C.-Y. (1970) Density and constitution of the mantle, *J. geophys. Res.*, **75**, 3264.

Wang, C.-Y. (1972) A simple earth model, *J. geophys. Res.*, **77**, 4318.

Whitcomb, J. H., J. D. Garmany and D. L. Anderson (1973) Earthquake prediction: variation of seismic velocities before the San Fernando earthquake, *Science*, **180**, 632.

Wiggins, R. A. (1969) Monte Carlo inversion of body wave observations, *J. geophys. Res.*, **74**, 3171.

Wiggins, R. A. (1972) The general linear inverse problem: implications of surface waves and free oscillations for earth structure, *Rev. Geophys. Space Phys.*, **10**, 251.

Williamson, E. D. and L. H. Adams (1923) Density distribution in the earth, *J. Wash. Acad. Sci.*, **13**, 413.

Worthington, M. H., J. R. Cleary and R. S. Anderssen (1972) Density modelling by Monte Carlo inversion—II Comparison of recent earth models, *Geophys. J.*, **29**, 445.

4

The Figure of the Earth and Gravity

4.1 The rotation of the earth

As a result of the attraction of the sun and moon on the earth's equatorial bulge and of movements of mass within the earth, the angular velocity of the earth is not constant; there are fluctuations not only in the rate of spin (i.e. changes in the length of the day) but also in the direction of the axis of rotation, i.e. the earth 'wobbles'. Munk and MacDonald (1960) have written a comprehensive monograph on the subject, and Fig. 4.1, which is taken from their book, shows for different parts of the frequency spectrum of the earth's rotation the methods of observing the fluctuations and what are believed to be the principal causes. It is not possible here to discuss the subject in any detail.

The annual wobble is principally due to seasonal shifts in air masses. In 1891, Chandler isolated a component with a 14-month period (since called the Chandler wobble). For a rigid earth, it can be shown that the period of its free nutation is about 10 months: the effect of the earth's deformation is thus to increase the period approximately 40 per cent. The Chandler wobble results in a periodic variation in latitude of a point on the earth's surface (see Fig. 4.2). An account of the methods of observation and reduction by the International Polar Motion Service (which succeeded the International Latitude Service in 1962) has been given by Yumi (1968). The annual wobble must be removed by careful spectral analysis of the latitude observations before the Chandler wobble can be investigated. The mean amplitude of the Chandler wobble is about 0·15″ and broadening of the spectral peak indicates a $Q*$ of about 60. It has been suggested that the Chandler peak is split into two narrowly separated resonances, and that

---

*Departure from ideal elasticity can be expressed in terms of the reciprocal of the dimensionless parameter Q (Knopoff (1964)). Q is defined by the equation

$$\frac{2\pi}{Q} = \frac{\Delta E}{E}$$

where ΔE is the energy dissipated per cycle and E is the peak energy stored. The mechanical Q defined above is mathematically equivalent to the Q of an oscillatory electrical circuit.

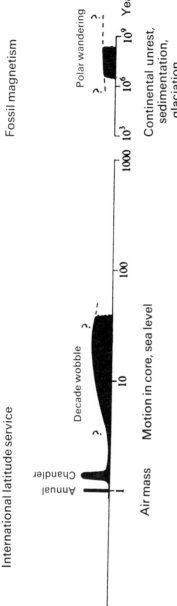

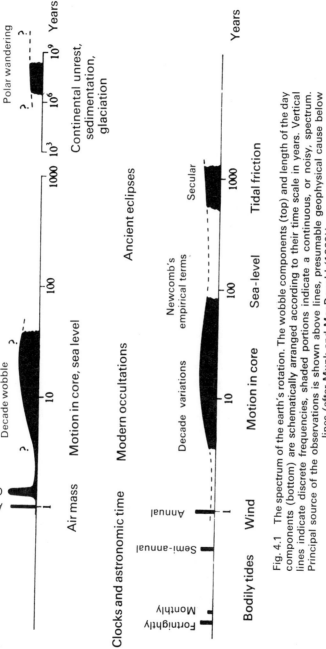

Fig. 4.1 The spectrum of the earth's rotation. The wobble components (top) and length of the day components (bottom) are schematically arranged according to their time scale in years. Vertical lines indicate discrete frequencies, shaded portions indicate a continuous, or noisy, spectrum. Principal source of the observations is shown above lines, presumable geophysical cause below lines (after Munk and MacDonald (1960))

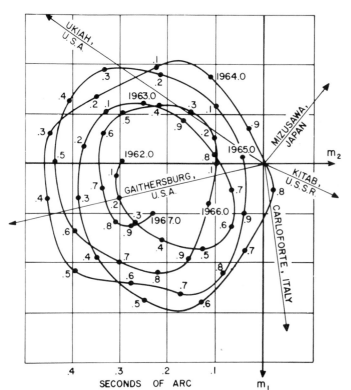

Fig. 4.2 The motion of the pole 1962·0–1967·0 (International Polar Motion Service stations)

this would explain the relatively low Q of the total peak compared with seismic values. However, the existing 80 year records would be barely capable of resolving them, even if there were no doubts about the level of systematic errors. Observations also show that the Chandler wobble is damped, so that it must be constantly re-excited. The excitation mechanism of the Chandler wobble has been the subject of much debate. Changes in the mass distribution of the atmosphere fail by at least one order of magnitude. Electro-magnetic core-mantle coupling fails by three orders of magnitude on a linearized model, although a much stronger, non-linear interaction has recently been suggested.

A possible connection with seismic activity has often been proposed. Estimates of the contribution of earthquakes to the excitation of the Chandler wobble have until recently been several orders of magnitude too small, mainly because the displacement fields of even the largest earthquakes were thought to extend no more than a few hundred kilometres from the focus. Following the work of Press (1965), which indicated that for great earthquakes a measurable displacement field may extend several thousand kilometres from the epicentre, Mansinha and Smylie (1967, 1968) estimated, using dislocation theory, the changes in the products of inertia of the earth arising from several large faults associated with major earthquakes. They found that the cumulative effect (based on earthquake statistics) could account for both the excitation of the Chandler wobble and a slow secular shift of the mean pole of rotation. A symposium on earthquake displacement fields and the rotation of the earth was held in 1969 (see Mansinha *et al.* (1970)). The damping mechanism of the Chandler wobble is also unknown. The general opinion is that damping occurs in the oceans: electromagnetic core-mantle coupling cannot make a significant contribution.

Other mechanisms are needed to explain the rather broad spectrum of 'wobble' variations with periods of the order of a decade. Motions in the earth's core and changes in sea level have been suggested as possible causes. Finally on a geologic time-scale, paleomagnetic and other indirect evidence indicates very large displacements of the pole. These are probably associated with movements of the continents and possible convective motions in the earth's mantle. Such questions are discussed in more detail in Chapter 8; they are intimately connected with the anelasticity of the earth.

The annual variation in the length of the day is caused by winds and shorter period terms by bodily tides. However, the rapid irregular variations in the length of the day over a decade cannot be explained by surface phenomena. No transport of mass at the surface could alter the earth's moment of inertia sufficiently to account for such large changes. It has been suggested that these 'decade' fluctuations are caused by the transfer of angular momentum between the mantle and liquid core. This in turn implies some form of core-mantle coupling. It is difficult, however, to make quantitative estimates of the horizontal stresses at the core-mantle boundary. Calculations indicate that neither viscous coupling nor electromagnetic coupling is really adequate to account for the decade fluctuations (see e.g. Rochester (1970)). More recently Hide (1969) has suggested the possibility of topographic coupling as a result of irregular features (bumps) at the core-mantle boundary. These questions will be discussed in more detail in §5.4.

Fluctuations with a time-scale of about 100 years (Newcomb's empirical terms) may be due to changes in the earth's moment of inertia. Changes over geologic time are predominantly a constant deceleration as a result of tidal friction. It is difficult to estimate the rate at which this deceleration has taken place, since our knowledge of palaeogeography is scant and the result depends critically on a few shallow seas. Urey (1952) has suggested on the other hand that, because of differentiation of the materials of the earth and the growth of the core, the moment of inertia of the earth about its axis of rotation may have been reduced and as a result the length of the day decreased from about 30 to 24 h. The changes caused by a growing core are considerably smaller (and of opposite sign) than those due to tidal friction. A method of determining the rate of the earth's rotation in the geologic past has recently been developed by the use of fossil 'clocks'. Some corals of Palaeozoic age show a characteristic banding on the outer skin (the epitheca). Wells (1963) was able to identify both a daily and an annual banding in corals of Middle Devonian age (about 370 m yr ago) and calculated that at that time there were about 400 days in the year. Later Scrutton (1964) found apparent monthly bands in corals of the same age and suggested that the Devonian year consisted of 13 lunar months of 30·5 days each. Using estimates of the length of the day and lunar month in Devonian times, Runcorn (1964) showed that it is possible to separate the slowing down of the earth by tidal friction from other processes which affect its rate of rotation. He found that nearly all the change in the length of the day is caused by tidal friction and that any residual effects (e.g. a change in the moment of inertia of the earth due to the growth of the core) are negligible.

Pannella *et al.* (1968) have obtained values of the length of the synodic month during other geologic periods using tidally controlled periodical growth patterns in molluscs and stromatolites (Fig. 4.3). They found that the slowing down of the earth's rotation has not taken place at a uniform rate. There appear to be two

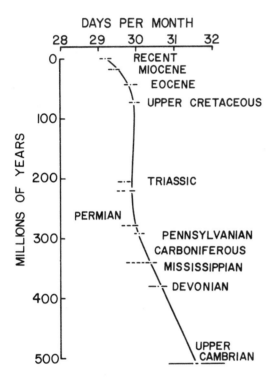

Fig. 4.3 Variations in the length of the synodic month through geologic time. The error bars show the standard error for each point (after Pannella *et al.* (1968))

major breaks in the slope of the curve, the slowing down being negligible over a 200 m yr period from the Pennsylvanian to the Upper Cretaceous. Pannella *et al.* suggest that these changes in slope may be related to particular events in the earth's history. Thus if the slowing down since the late Cambrian is attributed mainly to the loss of energy due to tidal torques in shallow seas, then a redistribution of continents, oceans and shallow seas should affect the amount of energy dissipated and the rotation of the earth. Again the high slope between the Cretaceous and the present could be attributed to the rapid drift period of the Upper Cretaceous and to the rise of the Alpine orogenic belt which created widespread shallow seas (see Chapter 8).

4.2 The figure of the earth

The gravitational field of a non-rotating uniform sphere is constant. The gravitational field of the earth, however, is distorted by a number of factors: the earth's rotation, departures from a true sphere and internal lateral variations in density (although a radial variation in density would not have any effect). The gravitational attraction of the earth can be derived from a potential V. The effects of its rotation can also be derived from a potential $\frac{1}{2}r^2\omega^2 \sin^2 \theta$ where ω is the angular velocity of the earth about its polar axis and (r, θ, ϕ) are spherical polar coordinates. Thus the combined potential (called the *geopotential*) is given by

$$U = V - \tfrac{1}{2}r^2\omega^2 \sin^2 \theta \tag{4.1}$$

Surfaces on which U is a constant are called equipotential surfaces. The surface of a liquid at rest is a surface of constant potential, for otherwise work would be done by the forces derived from the potential and the liquid would flow until its surface was one of constant potential. The surface of the sea is not one of constant potential because it is disturbed by wave action, tides and currents. If these effects are averaged out over a long time period, an equipotential surface (mean sea level) may be defined. This surface is called the *geoid*. For a physical picture of the geoid in continental areas we may imagine canals cut through the continents and connected with the oceans so that the water level is the geoidal surface. The gravitational acceleration at the surface is normal to the geoid and is given by

$$\mathbf{g} = -\operatorname{grad} U \tag{4.2}$$

The problem of calculating the form of the geoid reduces to that of obtaining an expression for V. If the distribution of mass within the earth were known exactly, V could be obtained by integration. In practice, we try and solve the inverse problem of obtaining information about the earth's interior from the shape of the geoid.

Consider first an ideal earth with no lateral variations in density. The surface corresponding to the geoid in this case is called the spheroid. The potential V satisfies Laplace's equation at all points outside the earth (and in the limit on the surface itself). Since we are assuming rotational symmetry about the z-axis (i.e. we are neglecting any dependence on longitude), V can be expressed as a series of zonal harmonics (see Appendix B),

$$V = -\frac{GM}{r}\left\{J_0 P_0(\cos\theta) - J_1\left(\frac{a}{r}\right)P_1(\cos\theta) - J_2\left(\frac{a}{r}\right)^2 P_2(\cos\theta)\ldots\right\} \tag{4.3}$$

where G is the gravitational constant, $J_0, J_1, J_2, \ldots,$ are dimensionless coefficients and M and a are the mass and equatorial radius of the earth. There are no terms involving powers of r/a since there is no matter outside the earth.

Since the potential at a great distance from the earth is to a first approximation

$-GM/r$, $J_0 = 1$. If the origin is taken as the centre of mass of the earth, $J_1 \equiv 0$, and to a first approximation

$$V = -\frac{GM}{r} + \frac{GMa^2}{2r^3} J_2 (3\cos^2\theta - 1) \qquad (4.4)$$

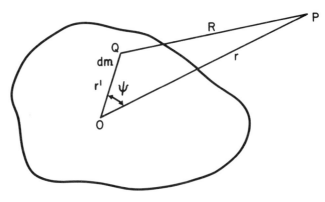

Fig. 4.4

It is easy to show that J_2 can be expressed in terms of the principal moments of inertia of the earth. The gravitational potential at P due to a mass dm at Q is (see Fig. 4.4)

$$dV = -\frac{G\,dm}{R}$$

The quantity

$$\frac{1}{R} = \frac{1}{r}\left(1 + \frac{r'^2}{r^2} - \frac{2r'}{r}\cos\psi\right)^{-1/2}$$

may be expanded in power of $1/r$,

$$\frac{1}{R} = \frac{1}{r}\left\{1 + \frac{r'}{r}\cos\psi - \frac{1}{2}\frac{r'^2}{r^2} + \frac{3}{2}\frac{r'^2}{r^2}\cos^2\psi \ldots\right\}$$

$$= \frac{1}{r}\left\{1 + \frac{r'}{r}\cos\psi + \frac{r'^2}{r^2} - \frac{3}{2}\frac{r'^2}{r^2}\sin^2\psi \ldots\right\}$$

Thus

$$V = -\frac{G}{r}\int dm - \frac{G}{r^2}\int r' \cos\psi\, dm - \frac{G}{r^3}\int r'^2\, dm + \frac{3G}{2r^3}\int r'^2 \sin^2\psi\, dm \qquad (4.5)$$

The first integral is $-(GM/r)$, and the second vanishes since the origin has been taken as the centre of mass. If (A, B, C) are the moments of inertia about the axes $O(x, y, z)$, which are taken as the principal axes, the third integral is

$-(G/2r^3)(A+B+C)$. The fourth integral is just the moment of inertia I about OP, so that equation (4.5) becomes

$$V = -\frac{GM}{r} - \frac{G}{2r^3}(A+B+C-3I) \tag{4.6}$$

which is MacCullagh's formula.

If (l, m, n) are the direction cosines of OP,

$$I = Al^2 + Bm^2 + Cn^2$$

Also for rotational symmetry about Oz

$$A \equiv B \quad \text{and} \quad n^2 = 1 - l^2 - m^2 = \cos^2\theta$$

Substituting these values in equation (4.6) yields

$$V = -\frac{GM}{r} + \frac{G}{2r^3}(C-A)(3\cos^2\theta - 1) \tag{4.7}$$

Comparing equations (4.4) and (4.7), the coefficient J_2 in the gravitational potential is given by

$$J_2 = \frac{C-A}{Ma^2} \tag{4.8}$$

The total geopotential U is given by equations (4.1) and (4.7)

$$U = -\frac{GM}{r} + \frac{G}{2r^3}(C-A)(3\cos^2\theta - 1) - \frac{1}{2}r^2\omega^2\sin^2\theta \tag{4.9}$$

The surface of the earth is an equipotential $U=U_0$. At the equator $(r=a, \theta=\pi/2)$ and at the poles $(r=c, \theta=0)$ so that

$$U_0 = -\frac{GM}{a} - \frac{G}{2a^3}(C-A) - \frac{1}{2}a^2\omega^2 \tag{4.10}$$

and

$$U_0 = -\frac{GM}{c} + \frac{G}{c^3}(C-A) \tag{4.11}$$

Hence by subtraction the ellipticity

$$f = \frac{a-c}{a} = \frac{C-A}{M}\left(\frac{1}{c^2} + \frac{c}{2a^3}\right) + \frac{1}{2}\frac{\omega^2 ca^2}{GM}$$

$$\simeq \frac{3}{2}\frac{C-A}{Ma^2} + \frac{1}{2}\frac{\omega^2 a^3}{GM} \tag{4.12}$$

to the first order in f, since $(\omega^2 a^3)/(GM)$ is also of order f. The equation of the surface $U=U_0$ follows from equation (4.9) by substituting a for r in the (small) second and third terms,

$$r = -\frac{GM}{U_0}\left\{1 - \frac{C-A}{2Ma^2}(3\cos^2\theta - 1) + \frac{\omega^2 a^3 \sin^2\theta}{2GM}\right\}$$

$$= -\frac{GM}{U_0}\left\{1 + \frac{C-A}{2Ma^2} + \frac{\omega^2 a^3}{2GM} - \left(\frac{3}{2}\frac{C-A}{Ma^2} + \frac{\omega^2 a^3}{2GM}\right)\cos^2\theta\right\}$$

$$= a + \frac{GMf}{U_0}\cos^2\theta$$

on using equations (4.10) and (4.12). Also from equation (4.10), $GM/U_0 = -a + O(f)$ so that the equation of the spheroid finally becomes

$$r = a(1 - f\cos^2\theta) \tag{4.13}$$

The difference $(C-A)$ between the polar and equatorial moments of inertia of the earth can be calculated from equation (4.12) if the ellipticity f is known from geodetic measurements. This was the method adopted in the past. However, the quantity $J_2 = (C-A)/(Ma^2)$ has been determined from satellite orbits with a precision that is more than an order of magnitude better than that obtained indirectly from geodetic data. Thus the ellipticity is now estimated from the value of J_2 determined from satellite data.

Writing

$$m = \frac{\omega^2 a^3}{GM}$$

$$= \frac{\omega^2 a}{GM/a^2} \tag{4.14}$$

$$= \frac{\text{centripetal acceleration at equator}}{\text{gravitational acceleration at equator}}$$

equation (4.12) may be written

$$f = \tfrac{3}{2}J_2 + \tfrac{1}{2}m \tag{4.15}$$

The value of $m = 3.4678 \times 10^{-3}$ so that using the satellite value of $J_2 = 1.08270 \times 10^{-3}$, obtained by Kaula (1965) and now adopted by the International Union of Geodesy and Geophysics as the standard for geodetic reference, the ellipticity $f = 3.3579 \times 10^{-3}$. If second order terms are retained,

$$f = 3.35280 \times 10^{-3}$$

$$= \frac{1}{298.26}$$

The normal to the spheroid differs from the radial direction by a small angle of

108

order f. Thus using equations (4.9) and (4.8), the value of g to the first order of small quantities is given by

$$-g = \frac{\partial U}{\partial r}$$

$$= \frac{GM}{r^2} - \frac{3G}{2r^4} Ma^2 J_2(3\cos^2\theta - 1) - r\omega^2 \sin^2\theta \qquad (4.16)$$

The value of r on the spheroid is given by equation (4.13). $r = a$ is a sufficient approximation in the second and third terms in equation (4.16) since they are of the order f times the first term. Hence on the spheroid

$$-g = \frac{GM}{a^2}(1 + 2f\cos^2\theta) - \frac{3}{2}\frac{GM}{a^2} J_2(3\cos^2\theta - 1) - a\omega^2(1 - \cos^2\theta)$$

which, on using equation (4.14) becomes

$$-g = \frac{GM}{a^2}\left(1 + \frac{3}{2}J_2 - m\right) + \frac{GM}{a^2}\left(2f - \frac{9}{2}J_2 + m\right)\cos^2\theta \qquad (4.17)$$

The value of g at the equator $(\theta = \pi/2)$ is thus

$$g_e = \frac{GM}{a^2}\left(m - \frac{3}{2}J_2 - 1\right) \qquad (4.18)$$

and the value of g at any point (to the first order in f) is

$$g = g_e\left\{1 + \left(2f + m - \frac{9}{2}J_2\right)\cos^2\theta\right\} \qquad (4.19)$$

Using equations (4.14) and (4.15) this may be written in the alternative forms

$$g = g_e\{1 + (2m - \tfrac{3}{2}J_2)\cos^2\theta\} \qquad (4.20)$$

or

$$g = g_e\{1 + (\tfrac{5}{2}m - f)\cos^2\theta\} \qquad (4.21)$$

This last equation is known as Clairaut's theorem (1743). It has been extended, taking into account terms of higher order, by Stokes (1849) and Helmert (1884)— see *Die mathematischen und physikalischen Theorien der höheren Geodäsie*, vol. II, Leipzig (1884). The resulting equation is

$$g = g_e\left\{1 + \left(\frac{5}{2}m - f - \frac{17}{14}mf + \frac{15}{4}m^2\right)\cos^2\theta\right.$$

$$\left. + \left(\frac{f^2}{8} - \frac{5}{8}mf\right)\sin^2 2\theta + \cdots\right\} \qquad (4.22)$$

The values of the constants in the international gravity formula adopted in 1930 have changed significantly as a result of satellite orbital studies and new astro-geodetic determinations. However, the older values have continued to be used for convenience as a standard reference. With these older values of the constants, and using latitude ϕ, instead of the colatitude θ, equation (4.22) gives

$$g = 978 \cdot 0490 \, (1 + 0 \cdot 005 \, 288 \, 4 \sin^2 \phi - 0 \cdot 000 \, 005 \, 9 \sin^2 2\phi) \qquad (4.23)$$

The units of g here are cm/s². An acceleration of 1 cm/s² has been called a gal in honour of Galileo—a mgal is defined as 0·001 gal.

An equipotential ellipsoid is determined by four constants. The International Union of Geodesy and Geophysics has chosen a, GM, J_2 and ω, and at the General Assembly in Switzerland, 1967 the following values were adopted

$$a = 6378 \, 160 \text{ m}$$
$$GM = 398 \, 603 \times 10^9 \text{ m}^3/\text{s}^2$$
$$J_2 = 0 \cdot 001 \, 0827$$
$$\omega = 7 \cdot 292 \, 115 \, 1467 \times 10^{-5} \text{ rad/s}.$$

ω is known to greater accuracy than the other three parameters. Further parameters may be calculated by standard formulae (see e.g. Heiskanen and Moritz (1967)). The gravity formula with the newer values of the constants is

$$g = 978 \cdot 031 \, 85(1 + 0 \cdot 005 \, 278 \, 895 \sin^2 \phi + 0 \cdot 000 \, 023 \, 462 \sin^4 \phi) \qquad (4.24)$$

to an accuracy of 0·004 mgal. To an accuracy of 0·1 mgal this may be written,

$$g = 978 \cdot 031 \, 8(1 + 0 \cdot 005 \, 302 \, 4 \sin^2 \phi - 0 \cdot 000 \, 005 \, 9 \sin^2 2\phi) \qquad (4.25)$$

The difference in g (in mgal) as given by the international gravity formula (1930) and the more recent gravity formula (1967) follows from equations (4.23) and (4.25)

$$g_{1967} - g_{1930} = (-17 \cdot 2 + 13 \cdot 6 \sin^2 \phi) \qquad (4.26)$$

4.3 Satellite observations

Geometrical (surveying) methods enable the form of the external equipotential surfaces to be found; in addition, gravity may be measured on the surface of the earth. Neither method, however, is very reliable when applied to the earth as a whole. The advent of artificial satellites has had a tremendous impact on geodesy, since it has provided a means by which the external potential can be found relatively easily with high accuracy on a worldwide scale, so that it has become possible to obtain reliable ideas of large scale features of the geoid.

Consider initially an axially symmetric earth whose external potential is given by equation (4.4). To the first order, a satellite's motion is controlled by the first term in this equation and its orbit is thus an ellipse with the centre of the earth at one focus. Because of the second term in the potential (which is due to the earth's equatorial bulge) and because the satellite is also affected by the drag of the atmosphere, its orbit shows small departures from an ellipse. Fortunately it is possible to distinguish between the effects of these two types of perturbations. Fig. 4.5 is a diagram defining some of the elements of the elliptical orbit, i.e. the

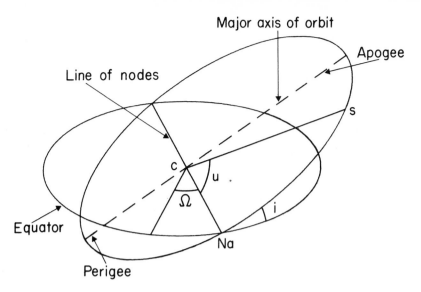

Fig. 4.5   Geometry of satellite orbit. Plane of orbit relative to the equator. c is the centre of mass of the earth (after Cook (1963))

parameters by which the orbit is defined. The time rates of change of these elements are given by Lagrange's equations of planetary motion (see e.g. Smart (1953); Cook (1963)). The only one of these equations which we shall use here is that giving the rate of change of longitude of the ascending node, namely,

$$\dot{\Omega} = \frac{rF_n \sin u}{na_s^2 \eta \sin i} \qquad (4.27)$$

In this equation F_n is the force acting on the satellite at right angles to the plane of the orbit, n is the mean angular velocity of the satellite in its orbit, a_s the semi-major axis, $\eta^2 = 1 - e^2$ (where e is the eccentricity) and the angles Ω, u and i are defined in Fig. 4.5. To a first approximation the force due to air drag is tangential to the orbit—the only forces with components perpendicular to the plane of the orbit are gravitational. We will calculate the motion of the node of a circular orbit due to the perturbing force $\mathbf{F}$ arising from the second term V' in equation (4.4). The three orthogonal components of $\mathbf{F}$ in spherical polar coordinates are given by

$$F_r = \frac{\partial V'}{\partial r} = \frac{-3GM}{2a_s^4} a^2 J_2 (3 \cos^2 \theta - 1)$$

$$F_\theta = \frac{1}{r} \frac{\partial V'}{\partial \theta} = \frac{-3GM}{a_s^4} a^2 J_2 \cos \theta \sin \theta \qquad (4.28)$$

$$F_\phi = 0$$

From Fig. 4.6, it follows from spherical trigonometry that $\cos \theta = \sin u \sin i$, so that

$$F_\theta = \frac{-3GM}{a_s^4} a^2 J_2 \sin u \sin i \sin \theta$$

111

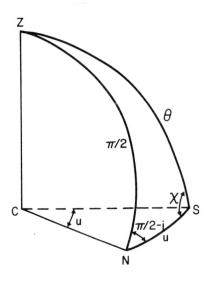

Fig. 4.6

Since $F_\phi = 0$, $F_n = F_\theta \sin \chi$ (see Fig. 4.6) where

$$\sin \chi = \frac{\cos i}{\sin \theta}$$

Hence

$$F_n = \frac{-3GM}{a_s^4} a^2 J_2 \sin u \sin i \cos i \tag{4.29}$$

This force varies periodically round the orbit and is greatest for orbits inclined at 45° to the equator. Substituting equation (4.29) in equation (4.27), the rate of change of the node in a circular orbit ($\eta = 1$), is given by

$$\dot{\Omega} = \frac{-3GMa^2}{na_s^5} J_2 \cos i \sin^2 u \tag{4.30}$$

$\dot{\Omega}$ varies round the orbit. It is zero when $u = 0$ and π, i.e. at the nodes, and obtains its maxima when $u = \pi/2$ and $3\pi/2$. The average rate of change of $\dot{\Omega}$ is

$$\bar{\dot{\Omega}} = \frac{-3GMa^2}{2na^5} J_2 \cos i$$

$$= -\frac{3}{2}\left(\frac{a}{a_s}\right)^2 nJ_2 \cos i \tag{4.31}$$

since, from Kepler's third law, $n^2 r^3 = GM$. For a satellite close to the earth, $\dot{\Omega}$ is nearly 6° per day for an inclination of 45°. Since $\dot{\Omega}$ can be measured to better than 0·1 per cent, equation (4.31) provides a very sensitive method for determining J_2. It can be shown that in the case of an elliptical orbit, the ellipse revolves in its plane at a rate which is also dependent upon J_2.

If the earth is regarded as axially symmetric, the gravitational potential can be represented as a series of zonal harmonics. All the even harmonics give rise to steady changes in the longitudes of the node and perigee, as well as to small periodic terms in these and other elements which are negligible in nearly circular orbits. Most orbits close to the earth, which are those that are most effective for the determination of zonal harmonics of higher order, have small eccentricities, and as the coefficients in the expression for the change of node, namely

$$\delta\Omega = a_2 J_2 + a_4 J_4 + \cdots + a_{2n} J_{2n} + \cdots$$

do not depend strongly on e, orbits that differ only in eccentricity are for this purpose identical. The coefficients a_n are proportional to $(a/a_s)^n$, and because a and a_s are very nearly the same for close satellites, most orbits that differ only in the semi-major axis are equivalent. The only orbits that lead to appreciable differences between the coefficients a_n are those with different inclinations. The number of distinct inclinations is limited for practical reasons—the highest even zonal harmonic that has been estimated is J_{20} (Kozai (1969)). Estimates of these higher harmonics may be considerably in error.

The odd zonal harmonics produce a long period oscillation in perigee height. For a typical orbit inclined at $45°$ to the equator, this oscillation has an amplitude of about 5 km and a period of about 3 months. The odd zonal harmonics also give rise to variations of eccentricity and inclination.

Periodic variations in the eccentricity of the orbit of the Vanguard satellite 1959 β_2 were found which can be explained by the presence of a third zonal harmonic in the earth's gravitational field. This modifies the geoid, making it pear-shaped (see Fig. 4.7). This figure greatly exaggerates the shape, the changes being only of the order of 30 m at the south pole, 10 m at the north pole and 7·5 m in middle latitudes. However, the presence of a third harmonic of this magnitude indicates a very substantial load on the surface of the earth. The stresses involved must be supported either by a mechanical strength larger than

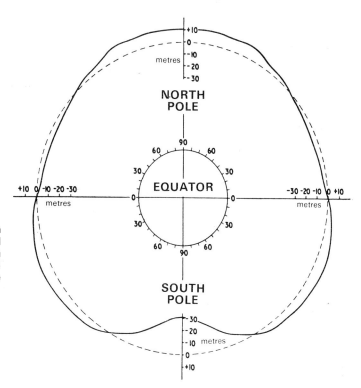

Fig. 4.7 Height of the mean meridional section of the geoid (solid curve) relative to a spheroid of flattening 1/298·25 (dotted curve) (after King-Hele *et al.* (1969))

Table 4.1 Values of the zonal harmonics, obtained from four recent papers, given as conventional harmonics multiplied by 10^6

	King-Hele et al. [1969]	King-Hele et al. [1966]	Kozai [1966]	Kozai [1969]
2		1082·68	1082·645	1082·628
3	−2·54		−2·546	−2·538
4		−1·61	−1·649	−1·593
5	−0·021		−0·210	−0·230
6		0·71	0·646	0·502
7	−0·40		−0·333	−0·361
8		0·13	−0·270	−0·118
9			−0·053	−0·100
10		0·09	−0·054	−0·354
11			0·302	0·202
12		−0·31	−0·357	−0·042
13			−0·114	−0·123
14			0·179	−0·073
15	−0·20			−0·174
16				0·187
17				0·085
18				−0·231
19				−0·216
20				−0·005
21	0·26			0·145
22				
23				
24				
25				
26				
27	−0·15			

that usually assumed for the interior of the earth or by large scale convection currents in the mantle. The depressed areas of the geoid are those which have recently been covered with ice and are rapidly rising; Wilson (1960) has pointed out that there is thus no evidence that the earth will remain pear-shaped.

The odd zonal harmonics up to J_{31} have been evaluated by King-Hele et al. (1969). They found that J_9 may be taken as zero and that for $9 < n < 33$, the odd J_n do not differ significantly from zero unless n is a multiple of 3, i.e. only J_{15}, J_{21} and J_{27} need be considered. As with the even harmonics, the values of the higher harmonics may be considerably in error. Table 4.1 gives the most recent estimates of the zonal harmonics. The values of J_2, J_3, J_4, J_5 and J_6 are thought to be reliable, but the values of the higher harmonics are probably little more than a guide to their order of magnitude.

The geoid shows in addition small departures from the simple spheroidal shape. Although these are smaller by a factor of 1000 or more than the ellipticity, they are nevertheless observable in an analysis of satellite orbits. The methods of finding tesseral harmonics are, however, considerably more complex and will not be discussed here (see e.g. Cook (1963); Kaula (1966); Caputo (1967); Lundquist and Veis (1966)). Tesseral harmonic coefficients have been evaluated with increasing success over the last few years, combining both satellite orbital data and geometrical triangulation using photographic observations, radio

Doppler measurements and, more recently, laser measurements. Gaposchkin and Lambeck (1970, 1971) have given 296 tesseral harmonic coefficients, complete up to the 16th order with additional terms of higher degree.

To improve the accuracy of the geoid we need greater accuracy of the higher order harmonics. One method is the use of resonant orbits, i.e. orbits which follow the same set of ground tracks day after day. For example, 14th order resonance occurs, for polar orbits, if the satellite makes exactly 14 revolutions per sidereal day. The along-track perturbations caused by tesseral harmonics of the 14th order then build up and can be accurately measured. Stable resonant orbits of order 9, 12, 13 and 14 were used by Gaposchkin and Lambeck (1970) in their Smithsonian Standard earth. Resonant orbits of order 15 have not been used because the satellites are then so low (height 500 km or less) that they are seriously affected by air drag which quickly destroys the resonance. However Gooding (1971) has shown that it is possible to analyse the change in the inclination of the orbit as a satellite passes through the 15th order resonance under the action of air drag. The magnitude of the change depends on the values of 15th order harmonics. By analysing the orbit of Ariel 3 as it passed through resonance, Gooding obtained what are probably the best values for a set of 'lumped' 15th order harmonics. By using other satellites at different inclinations, it may become possible to determine sets of harmonics of degree 15 and higher, free of the influence of other harmonics.

The results of the determination of the harmonics may be represented as a map of the geoid; the corresponding terms in the radius vector of the geoid may be found and the departure of the geoid from a spheroid having the flattening corresponding to the value of J_2 may be shown as a contour map such as Fig. 4.8. The broad features of such maps, those up to harmonics of order 5 or 6, are probably reliable but no great weight can be attached to details of higher order. The geoid map of Fig. 4.8 is probably accurate to 5 m. It is significant

Fig. 4.8   Map of the geoid referred to a flattening of 1/298·258 based on satellite and surface gravity observations. The contour interval is 10 m (after Gaposchkin and Lambeck (1971))

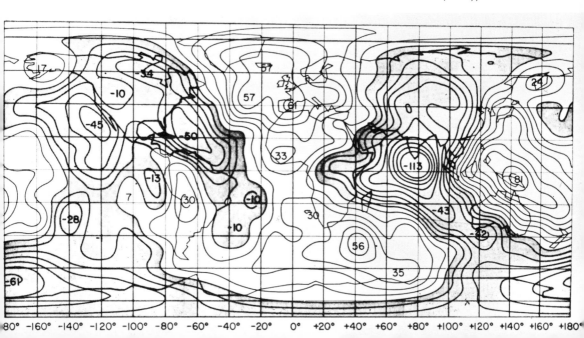

that the highs and lows of the geoid show no correlation with the crustal features of the earth (the oceans and continents). If the continents were simply super-imposed upon a perfectly layered ellipsoidal earth, there would be an exact correlation of the geoid with continental shapes, and the differences between highs and lows would be about 10 times greater than they are. This independence of geoidal and continental features shows that the features of the geoid are due either to density differences deep within the mantle or to density differences maintained by convection, in which case they are more likely to be in the upper mantle. The inequalities of density that give rise to the lowest harmonics in the potential could be anywhere in the earth, but the existence of harmonics of order greater than 12 shows that there must be inequalities in the mantle, since, if supposed to be at greater depths, they could not give rise to detectable harmonics of this order unless the variations of density were improbably large.

For the past decade, the coefficients appearing in the expression for the geopotential have been derived by analysing satellite observations under the assumption that the geopotential is time independent. Only combinations of observations made for different periods from different satellites have provided sufficient information to determine the geopotential. There are too many un-knowns to be determined, and not enough observations with sufficient accuracy and distribution, both in time and space, to allow a search for time dependence in the geopotential. However, the geopotential does vary because of tides, atmo-spheric motion, earthquakes, and other mass motions inside and on the earth; whenever mass displacements occur on or inside the earth, the geopotential changes.

It has been possible from satellite observations to identify tidal effects (i.e. perturbations in the orbit due to the tidal deformation of the earth), and to identify an annual variation in J_2, the coefficient of the most dominant term in the geopotential. More accurate orbit determination resulting from better tracking data, better coordinates of observing stations, and related improvements are needed to further refine and identify temporal variations in the geopotential. The benefits of such a programme are considerable and include improved nutation constants, the ability to correct refined surface gravity measurements, and, hopefully, added insight into core-mantle coupling phenomena by finding whether there are correlations between temporal variations of the geopotential and other phenomena such as temporal geomagnetic variations. Not only periodic variations, but also secular and irregular variations in J_2 and other coefficients, as well as in G (the gravitation constant), may be detected if accurate satellite tracking is continued for many years.

4.4 Gravity anomalies—isostasy

No account will be given of the experimental methods of measuring gravity. The absolute determination of g at any point on the earth's surface to an accuracy of 1 p.p.m. is extremely difficult. However, g occurs in many physical formulae and a more precise determination of its value is of interest in other branches of science besides geophysics. The earlier determinations of the absolute value of g were made using some form of reversible pendulum. In 1946 Volet suggested that the timing of a rod in free fall could give results that could compete in accuracy with those obtained from pendulum measurements. In the following years many free fall experiments were carried out—see e.g. Preston-Thomas et al. (1960); Cook (1967) and Faller (1967) for an account of the experimental techniques. The precision was increased by Faller (1965) who used one element of an optical interferometer as a freely falling body. Greater accuracy still has been obtained using a stabilized He–Ne laser as the light source (Hammond and Faller (1971)). The r.m.s. accuracy of these latest measurements is approximately 5 parts in 10^8.

Gravity differences are far easier to determine than absolute values. The ratio of the values of gravity at two stations is inversely proportional to the squares of the observed periods of a pendulum. The demands of exploration geophysics have led during the last 30 years to the development of precise, portable gravimeters which are capable of measuring gravity differences to better than 1 in 10^8. Such gravimeters essentially balance the force on a fixed mass against the elastic stresses in a system of springs or torsion fibres.

The large scale variations of g over the earth's surface are essentially an expression of the figure of the earth. Departures from the international gravity formula (4.23) provide evidence of lateral density variations in the earth—these variations are only a small fraction of the total gravitational field. Before comparing any measured value of g with equation (4.23) it must be corrected for elevation above sea level. Since

$$g \simeq -\frac{GM}{r^2}$$

$$\frac{\partial g}{\partial r} \simeq \frac{2GM}{r^3} = -\frac{2g}{r} \tag{4.32}$$

At sea level this gradient is -0.3086 mgal/m. This correction does not take into account the additional mass involved through change in elevation and is called the *free air* correction. If the extra mass is approximated by an infinite horizontal slab of thickness h and mean density ρ, the additional attraction is $2\pi G\rho h$. Assuming normal crustal rocks with a density of 2.67 g/cm^3, its value is 0.1119 mgal/m. When this is combined with the free air correction it is called the *Bouguer anomaly*. The net effect for an increase in elevation of 1 m is -0.1967 mgal. Woollard (1969) has pointed out that a better value for ρ in calculating the Bouguer anomaly is 2.93 g/cm^3 rather than 2.67 g/cm^3. Since a change of 0.1 g/cm^3 in ρ results in a change of about 4.2 mgal for the mass correction for each 1000 m change in elevation, this would significantly alter the Bouger anomaly values. It might appear that the topography surrounding a station would make the slab approximation unrealistic. However, if the angle between the horizontal and the top or bottom of the adjacent topographical feature is less than 10°, the effect does not exceed 1 mgal. Additional corrections for observations near a precipitous slope on the other hand could amount to 20 mgal or more.

In 1749 Bouguer estimated the deflection of a plumb bob due to the mountain Chimborazo in the Andes and found that it was less than the calculated value. A similar experiment by Petit in 1849 in the Pyrenees showed that the plumb bob was actually deflected away from the mountains. More quantitative work by Pratt (1855) and Airy (1855) on the attraction of the Himalayas indicated that elevated regions of the earth's surface are 'compensated' by mass deficiencies beneath the surface, although they gave different physical models to explain their results (see Fig. 4.9). Airy proposed a low density crust of variable thickness,

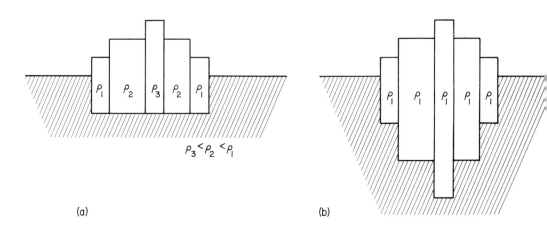

Fig. 4.9 Schematic representation of isostatic equilibrium according to (a) Pratt and (b) Airy

differences in elevation being accompanied by differences in thickness of the outer crust. This is essentially a statement of Archimedes' principle, the outer crust being in hydrostatic equilibrium, 'floating' in a dense subcrustal layer. Pratt on the other hand suggested that all crustal columns have the same mass above some level, called by Hayford (1910) the *depth of compensation*. Thus any column higher than sea level has a lower density. Dutton (1889) introduced the term *isostasy* to describe such 'compensation' of the topography and the state of hydrostatic stress below a certain depth. Large negative values of the Bouguer anomaly in mountainous regions is further proof of mass deficiencies at depth. The effect of the 'roots' of mountains may be calculated using either the Pratt or Airy hypothesis and when added to the Bouguer anomaly leaves the isostatic anomaly. The isostatic anomaly is, however, not unique, depending on the density and depth of compensation of the model chosen.

The postglacial uplift of many continental areas (such as Fennoscandia) provides some of the most convincing evidence for the reality of isostasy. Fenno-scandia sank under the heavy load of ice during the Pleistocene ice age. Following the melting of the ice, it has been rising again to achieve isostatic equilibrium under its present loading. Tilted beaches around the Great Lakes and raised features throughout Arctic Canada which have been dated by radio carbon analyses have enabled curves to be drawn of postglacial uplift against time for a number of areas (see e.g. Farrand (1962)). The curves show a strong exponential decrease from the time of removal of the ice to the present day (see Fig. 4.10).

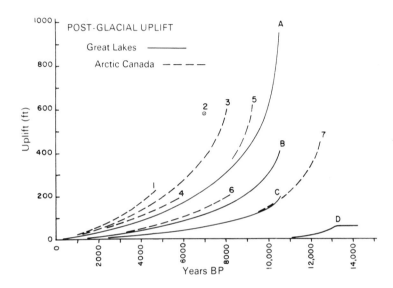

Fig. 4.10 Postglacial uplift curves for North America. Solid curves are from Lake Huron; dashed curves are from Arctic Canada. A, North Bay; B, Sault Ste. Marie; C, Cape Rich; D, Port Huron. 1, Igloolik; 2, Carr Lake; 3, James Bay; 4, Southampton Island; 5, Coronation Gulf; 6, northern Ellesmere Island; 7, northwest Victoria Island (after Farrand (1962))

Such vertical motions of the earth's crust yield valuable information on our knowledge of the rheology of the earth (see § 9.2). More recent work by Andrews (1968, 1970) indicates that the data are better explained by a logarithmic, rather than an exponential, decay. A review of the structure of the earth deduced from a study of glacio-isostatic rebound has recently been given by Walcott (1973) (see also § 9.3).

Kaula (1969a) attempted a tectonic classification of the earth's gravitational field based on a combination of satellite data and terrestrial gravimetry. He examined 43 areas and classified 19 of them as markedly positive, 14 as markedly negative and 10 as exceptionally 'mild'. The major features of the gravity field, particularly the dominant positive features, appear to be associated with relatively recent geologic activity. Thus Canada is negative because the recent glaciation dominates the character of the ancient shield. Kaula suggested that the major tectonic disturbances are associated with positive gravity anomalies, whereas a greater part of the negative anomalies are the consequence of passive isostatic response. The major exceptions are the more rapidly spreading ocean rises, glaciated areas and recent orogenies without extrusives (such as the Himalayas). The strongest correlation was found between positive gravity anomalies and Quaternary volcanism.

Kaula extended his work in a later paper (1970) using the more recent determination of the gravity field by Gaposchkin and Lambeck (1971). Fig 4.11 shows a world map of the free air anomalies. The improved resolution over Kaula's earlier (1969a) map results in significant changes in the relationship between gravity and tectonics in the southern oceans. Large positive anomalies are located along the ocean rises, the only exception being the positive anomaly lying over Hawaii. The trench and island arc belts are also predominantly positive. The commonest negative anomaly features are the ocean basins, located on the flanks of the rises. Two negative features (among the largest in the world) are not, however, obviously related to the global tectonic system: the Antarctic negative anomaly which is at least three times too large to be explained by glacial melting, and is probably caused by flow in the asthenosphere, and the negative anomaly over Turkestan, Siberia and Mongolia, the only negative area close to a compressive belt. Most features of the gravity field appear to be interpretable as a consequence of the varying behaviour of the lithosphere in response to flow in the asthenosphere.

Correlations have been made from time to time between features of the earth's gravitational field and other geophysical measurements—the question of a possible relationship between gravity and magnetic anomalies is discussed in § 5.4. Lee and MacDonald (1963) carried out a spherical harmonic analysis (to the third order) of the terrestrial heat flow and showed that anomalies in the geoid correlated roughly with the main features of the heat flow field. Where the heat flow is lower than average, the geoid is raised and vice versa. Such a correlation is in the sense to be expected if the anomalies in both fields are associated with convection in the upper mantle. Lee and MacDonald also showed, by an order of magnitude calculation, that such anomalies are consistent with convection currents having velocities of a few centimetres per year. Such a correlation has received support from Wang (1965) and Girdler (1967). Girdler found that the heat flow over all regions of negative gravity is significantly higher than that over all regions of positive gravity with one notable exception; the region of negative gravity to the west of the Atlantic ridge has lower heat flow than the world mean. However, the expansion of heat flow data into spherical harmonics can hardly be meaningful outside the restricted areas from which the bulk of the heat flow data have been obtained. Studies by Strange and Khan (1965) and Kaula (1967) indicate no significant correlation on a global scale between heat flow and gravity variations, although there appears to be some evidence of an inverse correlation on a local scale. A more detailed spherical harmonic analysis (up to the seventh order) of the terrestrial heat flow field by Horai and Simmons (1969)

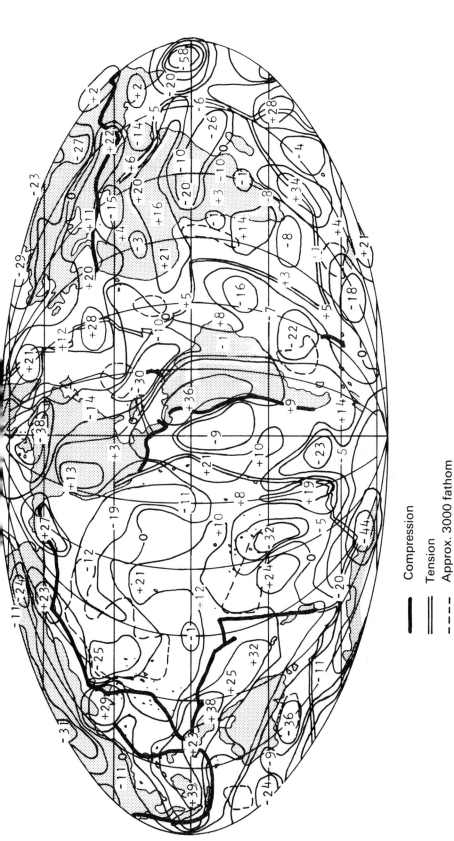

Fig. 4.11 Free air anomalies in mgal referred to an ellipsoid of flattening 1/299·8. They are calculated from the spherical harmonic coefficients of the gravitational field of degrees 2 through 16 of Gaposhkin and Lambeck (1971). Non zero contours enclosing only one value have been omitted. Global tectonic lines of compression and tension are taken from Isacks et al. (1968); major basins are indicated by the approximate 3000-fathom line (after Kaula (1970))

—— Compression

══ Tension

---- Approx. 3000 fathom

also shows no correlation with the earth's gravitational field. This is perhaps not too surprising since the basic nature of these two quantities in the earth is quite different. A large part of the mass anomaly that produces undulations in the geoid probably lies in the mantle; on the other hand, crustal and upper mantle inhomogeneities are more likely to be the major cause of spatial variations in the observed heat flow. This is confirmed by Toksöz *et al.* (1969) who analysed the global variations of a number of different geophysical measurements in an investigation of the broad lateral heterogeneities in the earth's mantle. They found that surface heat flow variations are controlled primarily by the shallow structure and tectonic features of the earth, and are uncorrelated with geopotential variations; also at long wave lengths, gravitational, heat flow, and seismic travel time variations show no correlation with topography.

4.5 Mascons

Muller and Sjogren (1968) have used tracking data of Orbiter 5 to derive a gravimetric map of the nearside of the moon (see Fig. 4.12). They found several large positive gravity anomalies centred on the lunar ringed maria: Imbrium,

Fig. 4.12   Variations in the moon's gravitational field based upon Muller and Sjogren's (1968) analysis of Orbiter 5 results (after Kaula (1969c))

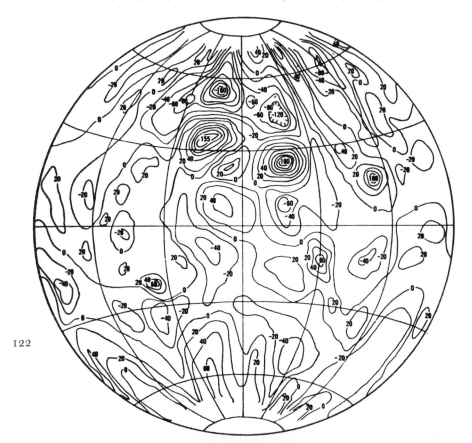

Serenitatis, Crisium, Nectaris and Humorum. Their analysis also showed that the variation of gravity for the remainder of the front side of the moon is quite mild. Subsequent analyses of Lunar Orbiter and Apollo data have largely confirmed these initial results and yielded a number of smaller positive anomalies and a few negative features. The areas of high gravity indicate the presence of high density mass concentrations near the surface and have been called *mascons*. It has been estimated that they extend laterally from 50 to 200 km. In spite of their high density, they have not sunk into the moon to reach isostatic equilibrium. An obvious explanation for the existence of mascons is that large objects (iron meteorites?) of high density material collided with the moon and flattened out beneath the surface where they have remained ever since. Urey (1968) and Stipe (1968) have suggested that mascons are dense meteoritic material, remnants of the same bodies that excavated the mare basins on which the mascons are centred.

A different point of view put forward by Conel and Holstrom (1968) is that mascons are flat plates of dense rock, probably lava, filling crater basins in less dense crustal material, rather than buried spherical masses. One possibility is that mascons were formed as the mare lavas cooled, contracted and became more dense, but did not sink because of the strength of the underlying crust. Improved measurements of the lunar gravitational field obtained by Doppler radio tracking of Apollo 15 support the seismic evidence that a rigid crust existed at the time the maria were formed. A slightly different interpretation has been given by Kaula (1969b) and Wise and Yates (1970) who proposed that plugs of denser mantle material were pushed up into the mare basins. Kaula suggested that the plug was driven beyond the position of isostatic equilibrium by excess lithostatic pressures generated in the moon by more rapid cooling and consequent shrinkage of the outer layers. Wise and Yates stop the mantle plug at the isostatically compensated level, the remainder of the crater being filled by lava. All mascon interpretations require a volume of additional material or over-dense material that is supported non-isostatically by the strength and rigidity of the lunar rocks beneath it. It appears that the moon has been strong enough at a depth of ~ 100 km to sustain stresses of ~ 50 bars throughout most of geologic time. This has led Urey (1969) and others to conclude that the lunar interior is relatively cool and has been so throughout geologic history. Different workers on the other hand have argued, from the fact that igneous rocks continued to appear at the lunar surface for more than a thousand m yr after the formation of the moon, that the lunar interior has been, and must still be, quite hot. Fresh volcanic features on the lunar surface also tend to indicate that volcanism has continued into very recent times (see also § 2.4.4).

In the later Apollo flights, the command service module and the lunar module (LM) were placed at a periapsis altitude of 16 km for 12 revolutions of the moon prior to the LM's descent. This enabled much higher resolution Doppler gravity measurements to be obtained (the data from the earlier Orbiter and Apollo missions were obtained at an altitude of 100 km). Gravity measurements over a 100 km band from $+70°$ to $-70°$ longitude were obtained from Apollo 14;

direct correlations were obtained between gravity variations and surface features (Sjogren *et al.* (1972)). Figure 4.13 shows three acceleration profiles over the mascon in Mare Nectaris. The consistency in shape and amplitude of these three profiles is striking—the amplitude should increase with later orbits since the moon is rotating the centre of Nectaris closer to the orbit plane. Figure 4.13 also

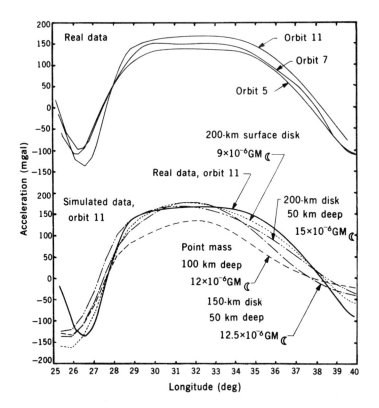

Fig. 4.13 Apollo 14 gravity profiles over Mare Nectaris (after Sjogren *et al.* (1972))

shows simulated accelerations for the 11th orbit for a 100 km deep spherical mass and for disks of 150 and 200 km radius. The point-mass curve is low in amplitude in spite of the fact that it was made 50 per cent larger than required to match earlier Apollo 12 LM data. Its shape is also a poor match. On the other hand a broad surface disk matches the data fairly well; the fit could be improved by removing some mass from the centre and adding it to the eastern edge.

The existence of mascons places an additional constraint on possible evolutions of the moon and on its temperature and rheological properties. There is still much controversy over the interpretation of mascons, and their origin, like that of the maria themselves, has not been finally settled.

4.6 Palaeogravity

One of the main difficulties in any geophysical investigation is that observations of natural phenomena cover only a very small fraction of the earth's lifetime—less than about one part in 10^6. With such an extremely small sample, it is not to be wondered at that our knowledge of many geophysical processes is so rudimentary. Two fields of research, however, have yielded valuable information about the early history of the earth: isotopic studies and palaeomagnetism. Isotopic studies were discussed briefly in § 1.2 and Chapter 6 is devoted to an account of palaeomagnetic investigations. This section discusses the possibility of obtaining information about the strength of the earth's gravitational field in the past. A determination of the gravitational field throughout time is of great importance to cosmological theories and planetary development. The possibility of a variation with time of the gravitational 'constant' G has been discussed in § 1.4; no experimental confirmation of any decrease in G has yet been obtained.

Faytel'son (1969) has reported g varying throughout large areas of the Soviet Union with mean rates over a decade of roughly ± 1 part in 10^7 per year. If such a rapid fluctuation is confirmed, it would rule out the possibility of direct measurements of any steady decrease; to obtain mean rates of change over millions of years a gravity sensitive geological system is required. There is no shortage of gravity controlled phenomena, but gravity seldom leaves a permanent record in the rocks and the effects of gravity are usually small compared with those of other often unpredictable variables. This is in sharp contrast to palaeomagnetism.

Stewart (1970) has suggested a number of phenomena, both geological and biological, which could possibly be used for this purpose. These include the gravitational compaction of clays, the compaction of clays beneath glacial ice, palaeobarometry, the size of flying animals, the depth of animals' footprints, the dimensions of the frames of land animals, and the growth of diapirs. Stewart later (1972) developed a method which, while not determining accurate values of palaeogravity, has been able to define a limit to the decrease of g with time which, though broad, is not inconsistent with theoretical predictions. Stewart argued that if gravity in the past had been higher than today it is conceivable that some fine-grained sedimentary deposits would be over consolidated, i.e. compacted more by the smaller sedimentary column above them in the past than by the larger one existing today. Since compaction is relatively rapid and largely irreversible the requisite evidence could be preserved. Stewart measured the degree of over consolidation in sediments in the London basin. He found that the London clay now exposed has been consolidated by higher pressures than would have been produced by what is now the greatest thickness of sediment overburden to be found anywhere in the London basin. It is of course possible that this extra pressure was derived from sediment younger than any now observed, additional overburden which has since been eroded away. But if the additional pressure was caused solely by a higher value of g in the past (about 26 m yr ago), Stewart's results indicate that gravity at that time could not have been more than twice its present value. This implies that the maximum possible decrease in g over the past 26 m yr is 4 parts in 10^8 per yr.

References

Andrews, J. T. (1968) Postglacial rebound in Arctic Canada: similarity and prediction of uplift curves, *Can. J. Earth Sci.*, **5**, 39.

Andrews, J. T. (1970) Present and postglacial rates of uplift for glaciated northern and eastern North America derived from postglacial uplift curves, *Can. J. Earth Sci.*, **7**, 703.

Caputo, M. (1967) *The Gravity Field of the Earth from Classical and Modern Methods* (Acad. Press).

Conel, J. E. and G. B. Holstrom (1968) Lunar mascons—a near surface interpretation, *Science*, **162**, 1403.

Cook, A. H. (1963) The contribution of observations of satellites to the determination of the earth's gravitational potential, *Space Sci. Rev.*, **2**, 355.

Cook, A. H. (1967) A new absolute determination of the acceleration due to gravity at the National Physical Laboratory, England, *Phil. Trans. Roy. Soc.*, **A 261**, 211.

Faller, J. E. (1965) Results of an absolute determination of the acceleration of gravity, *J. geophys. Res.*, **70**, 4035.

Faller, J. E. (1967) Precision measurements of the acceleration of gravity, *Science*, **158**, 60.

Farrand, W. R. (1962) Postglacial uplift in North America, *Amer. J. Sci.*, **260**, 181.

Faytel'son, A. S. (1969) Secular variations in the gravity on the Russian platform, *Dokl. Acad. Nauk SSR*, **188**, 579.

Faytel'son, A. S. (1969) Secular changes in the force of gravity in the Aral-Caspian region, *Dokl. Acad. Nauk SSR*, **189**, 1240.

Gaposchkin, E. M. and K. Lambeck (1970) 1969 Smithsonian Standard Earth (II), *Smithsonian Astrophys. Obs. Spec. Rept.* **315**.

Gaposchkin, E. M. and K. Lambeck (1971) Earth's gravity field to the sixteenth degree and station coordinates from satellite and terrestrial data, *J. geophys. Res.*, **76**, 4855.

Girdler, R. W. (1967) A review of terrestrial heat flow, in: *Mantles of the Earth and Terrestrial Planets* (ed. S. K. Runcorn) (Interscience Publ.).

Gooding, R. H. (1971) Lumped fifteenth-order harmonics in the geopotential, *Nature Phys. Sci.*, **231**, 168.

Hammond, J. A. and J. E. Faller (1971) Results of absolute gravity determinations at a number of different sites, *J. geophys. Res.*, **76**, 7850.

Heiskanen, W. A. and H. Moritz (1967) *Physical Geodesy* (Freeman and Co., San Francisco).

Hide, R. (1969) Interaction between the earth's liquid core and solid mantle, *Nature*, **222**, 1055.

Horai, K. and G. Simmons (1969) Spherical harmonic analysis of terrestrial heat flow, *Earth Planet. Sci. Letters*, **6**, 386.

Isacks, B., J. Oliver and L. R. Sykes (1968) Seismology and the new global tectonics, *J. geophys. Res.*, **73**, 5855.

Kaula, W. M. (1966) Tests and combinations of satellite determinations of the gravity field with gravimetry, *J. geophys. Res.*, **71**, 5303.

Kaula, W. M. (1967) Geophysical implications of satellite determinations of the earth's gravitational field, *Space Sci. Rev.*, **7**, 769.

Kaula, W. M. (1969a) A tectonic classification of the main features of the earth's gravitational field, *J. geophys. Res.*, **74**, 4807.

Kaula, W. M. (1969b) Interpretation of lunar mass concentrations, *Phys. Earth Planet. Int.*, **2**, 123.

Kaula, W. M. (1969c) The gravitation field of the moon, *Science*, **166**, 1581.

Kaula, W. M. (1970) Earth's gravity field: relation to global tectonics, *Science*, **169**, 982.

King-Hele, D. G., G. E. Cook and D. W. Scott (1966) Even zonal harmonics in the earth's gravitational potential: a comparison of recent determinations, *Planet. Space Sci.*, **14**, 49.

King-Hele, D. G., G. E. Cook and D. W. Scott (1969) Evaluation of odd zonal harmonics in the geopotential of degree less than 33, from the analysis of 22 satellite orbits, *Planet. Space Sci.*, **17**, 629.

Knopoff, L. (1964) *Q*, *Rev. Geophys.*, **2**, 625.

Kozai, Y. (1966) The zonal harmonic coefficients, *Smithsonian Astrophys. Obs. Spec. Rept. 200*, Vol. 2, 67.

Kozai, Y. (1969) Revised values for coefficients of zonal spherical harmonics in the geopotential, *Smithsonian Astrophys. Obs. Spec. Rept. 295*.

Lee, W. H. K. and G. J. F. MacDonald (1963) The global variation of terrestrial heat flow, *J. geophys. Res.*, **68**, 6481.

Lundquist, C. A. and G. Veis (ed) (1966) *Smithsonian Astrophys. Obs. Spec. Rept. 200*.

Mansinha, L. and D. E. Smylie (1967) Effect of earthquakes on the Chandler wobble and the secular polar shift, *J. geophys. Res.*, **72**, 4731.

Mansinha, L. and D. E. Smylie (1968) Earthquakes and the earth's wobble, *Science*, **161**, 1127.

Mansinha, L., D. E. Smylie and A. E. Beck (ed.) (1970) *Earthquake Displacement Fields and the Rotation of the Earth* (D. Reidel Publ. Co., Holland).

Muller, P. M. and W. L. Sjogren (1968) Mascons: lunar mass concentrations, *Science*, **161** 680.

Munk, W. H. and G. J. F. MacDonald (1960) *The Rotation of the Earth* (Camb. Univ. Press).

Pannella, G., C. MacClintock and M. N. Thompson (1968) Palaeontological evidence of variations in length of synodic month since late Cambrian, *Science*, **196**, 792.

Press, F. (1965) Displacements, strains and tilts at teleseismic distances, *J. geophys. Res.*, **70**, 2395.

Preston-Thomas, H., L. G. Turnbull, E. Green, T. M. Dauphinee and S. N. Kalra (1960) An absolute measurement of the acceleration due to gravity at Ottawa, *Can. J. Phys.*, **38**, 824.

Rochester, M. G. (1970) Core-mantle interactions: geophysical and astronomical consequences, in: *Earthquake Displacement Fields and the Rotation of the Earth* (ed. L. Mansinha, D. E. Smylie and A. E. Beck) (D. Reidel Publ. Co., Holland).

Runcorn, S. K. (1964) Changes in the earth's moment of inertia, *Nature*, **204**, 823.

Scrutton, C. T. (1964) Periodicity in Devonian coral growth, *Palaeontology*, **7**, 552.

Sjogren, W. L., P. Gottlieb, P. M. Muller and W. Wollenhaupt (1972) Lunar gravity via Apollo 14 Doppler radio tracking, *Science*, **175**, 165.

Smart, W. M. (1953) *Celestial Mechanics* (Longmans, Green and Co., London).

Stewart, A. D. (1970) Palaeogravity, in: *Palaeogeophysics* (ed. S. K. Runcorn) (Acad. Press).

Stewart, A. D. (1972) Palaeogravity from the compaction of fine-grained sediments, *Nature*, **235**, 322.

Stipe, J. G. (1968) Iron meteorites as mascons, *Science*, **162**, 1402.

Strange, W. E. and M. A. Khan (1965) On the relation between satellite gravity results, heat flow and convection currents, *Trans. Amer. Geophys. Union.*, **46**, 544.

Toksöz, M. N., J. Arkani-Hamed and C. A. Knight (1969) Geophysical data and long wave heterogeneities of the earth's mantle, *J. geophys. Res.*, **74**, 3751.

Urey, H. C., (1952) *The Planets, their Origin and Development* (Yale Univ. Press).

Urey, H. C. (1968) Mascons and the history of the moon, *Science*, **162**, 1408.

Urey, H. C. (1969) Early temperature history of the moon, *Science*, **165**, 1275.

Volet, C. (1946) Sur la mesure absolue de la gravité, *Compt. Rend. Acad. Sci.*, **222**, 373.

Walcott, R. I. (1973) Structure of the earth from glacio-isostatic rebound, *Ann. Rev. Earth Sci.*, **1**, 15.

Wang, C.-Y. (1965) Some geophysical implications from gravity and heat flow data, *J. geophys. Res.*, **70**, 5629.

Wells, J. W. (1963) Coral growth and geochronometry, *Nature*, **197**, 948.

Wilson, J. T. (1960) Some consequences of expansion of the earth, *Nature*, **185**, 880.

Wise, D. U. and M. T. Yates (1970) Mascons as structural relief on a lunar 'Moho', *J. geophys. Res.*, **75**, 261.

Woollard, G. P. (1969) Standardization of gravity measurements, in: *The Earth's Crust and Upper Mantle* (AGU Geophysical Monograph 13).

Yumi, S. (1968) *Annual Report of the International Polar Motion Service for 1966* (Mizusawa, Japan) (See also previous annual reports).

5

Geomagnetism and Aeronomy

5.1 Introduction

Geomagnetism occupies a rather special place in geophysics since it has applications in other disciplines as well. It cannot be divorced from space physics and the physics of the upper atmosphere (aeronomy). Again, since rocks may become permanently magnetized at the time of their formation, palaeomagnetic measurements may yield information on the past history of the field (see Chapter 6). Magnetic measurements have also played a key role in the development of the new concepts of plate tectonics—these will be discussed in some detail in § 8.4.

At its strongest near the poles, the earth's magnetic field is several hundred times weaker than that between the poles of a toy horseshoe magnet, being less than a gauss (Γ).* Thus in geomagnetism we are measuring extremely small magnetic fields and a more convenient unit is the gamma (γ), defined as $10^{-5}\ \Gamma$. Although the earth's field is weak, it occupies a very large volume, and since the energy of a magnetic field is proportional to the volume, the earth's field plays an important role in extra-terrestrial relationships. It screens the earth's equatorial regions from cosmic rays with energies of a few tens of billions electron volts, traps charged particles in the Van Allen radiation belts, and is distinguishable from the background field of interplanetary space out to distances of 10–13 earth radii.

In a magnetic compass, the needle is weighted so that it will swing in a horizontal plane and its deviation from geographical north is called the declination, D. A non-magnetic needle which is balanced horizontally on a pivot, becomes inclined to the vertical when magnetized. Over most of the northern hemisphere the north-seeking end of the needle will dip downwards, the angle it makes with the horizontal being called the magnetic dip or inclination, I. The total intensity

---

*Strictly speaking the unit of magnetic field strength is the oersted, the gauss being the unit of magnetic induction. The distinction is somewhat pedantic in geophysical applications since the permeability of air is virtually unity in c.g.s. units. The traditional unit used in geomagnetism, the gauss, has been retained in this book.

F, the declination D and the inclination I completely define the magnetic field at any point, although other components are often used. The horizontal and vertical components of **F** are denoted by H and Z. H may be further resolved into two components X and Y, X being the component along the geographical meridian and Y the orthogonal component. Figure 5.1 illustrates these different

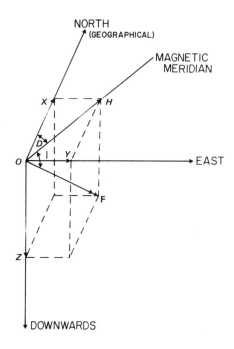

Fig. 5.1

magnetic elements. They are simply related to one another by the following equations,

$$H = F \cos I, \qquad Z = F \sin I, \qquad \tan I = Z/H \tag{5.1}$$

$$X = H \cos D, \qquad Y = H \sin D, \qquad \tan D = Y/X \tag{5.2}$$

$$F^2 = H^2 + Z^2 = X^2 + Y^2 + Z^2 \tag{5.3}$$

The variation of the magnetic field over the earth's surface is best illustrated by isomagnetic charts, i.e. maps on which lines are drawn through points at which a given magnetic element has the same value. Contours of equal intensity in any of the elements X, Y, Z, H or F are called isodynamics. Figures 5.2 and 5.3 are world maps showing contours of equal declination (isogonics) and equal inclination (isoclinics) for the year 1965. It is remarkable that a phenomenon (the earth's magnetic field) whose origin, as we shall see later, lies within the earth should show so little relation to the broad features of geography and geology. The isomagnetics cross from continents to oceans without disturbance and show

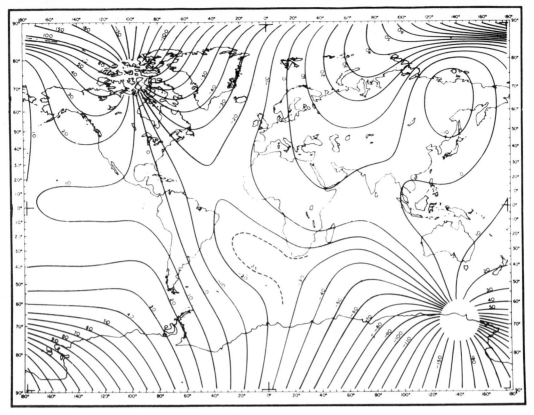

Fig. 5.2 World map showing contours of equal declination for 1965, International Geomagnetic
Reference Field (after Zmuda (1971))

Fig. 5.3 World map showing contours of equal inclination for 1965, International Geomagnetic
Reference Field (after Zmuda (1971))

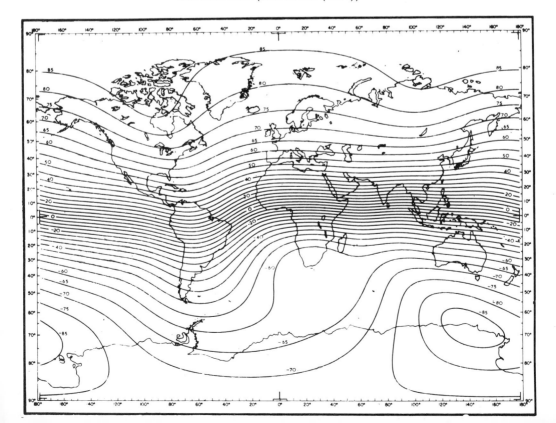

no obvious relation to the great belts of folding or to the pattern of submarine ridges. In this respect the magnetic field is in striking contrast to the earth's gravitational field and to the distribution of earthquake epicentres, both of which are closely related to the major features of the earth's surface.

Not only do the intensity and direction of magnetization vary from place to place across the earth, but they also show a time variation. There are two distinct types of temporal changes: transient fluctuations and long term secular changes. Transient variations produce no large or enduring changes in the earth's field; they arise from causes outside the earth, and are discussed in § 5.7. Secular changes, on the other hand are due to causes within the earth, and over a long period of time the net effect may be considerable. If successive annual mean values of a magnetic element are obtained from a particular station, it is found that the changes are in the same sense over a long period of time, although the rate of change does not usually remain constant. Figure 5.4 shows the changes in declination and inclination at London, Boston and Baltimore. The declination

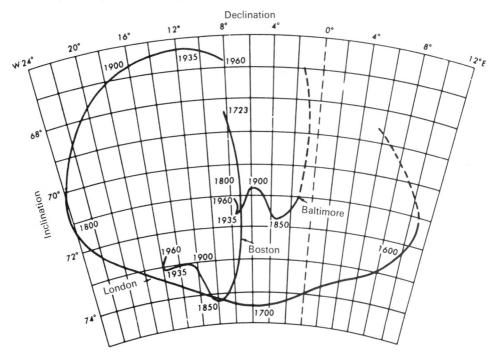

Fig. 5.4   Secular change of declination and inclination at London, Boston and Baltimore
(after Nelson *et al.* (1962))

at London was $11\frac{1}{2}°$E in 1580 and $24\frac{1}{2}°$W in 1819, a change of almost $36°$ in 240 yr. Lines of equal secular change (isopors) in an element form sets of ovals centring on points of local maximum change (isoporic foci). Figures 5.5 and 5.6 show the secular change in Z for the years 1922·5 and 1942·5. It can be seen

132

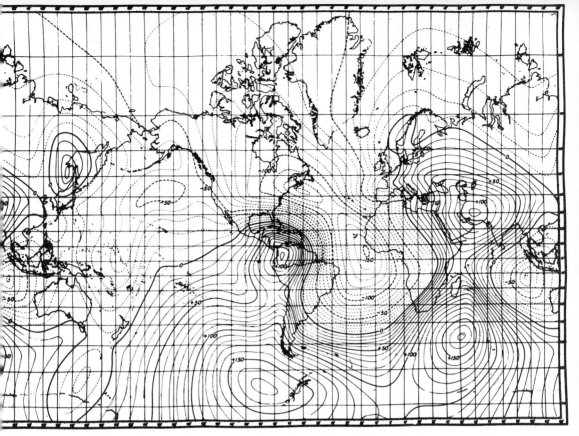

Fig. 5.5 World map showing the geomagnetic secular variation of the vertical component Z.
Epoch 1922·5 (after Vestine *et al.* (1947))

Fig. 5.6 World map showing the geomagnetic secular variation of the vertical component Z.
Epoch 1942·5 (after Vestine *et al.* (1947))

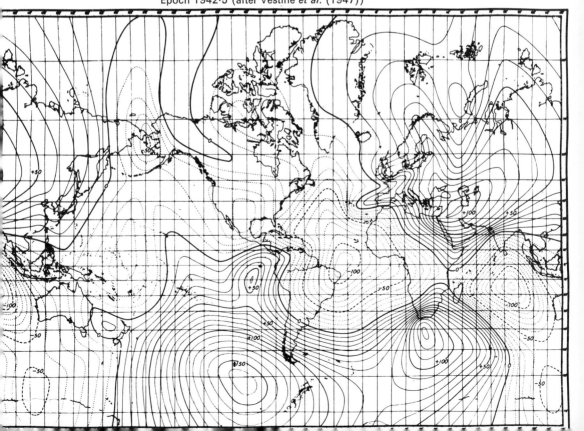

that the secular variation is a regional rather than a planetary phenomenon and that considerable changes can take place in the general distribution of isopors even within 20 years.

5.2 The field of a uniformly magnetized sphere

Before 1600, William Gilbert had investigated the variation in direction of the magnetic force over the surface of a piece of the naturally magnetized mineral lode-stone which he had cut in the shape of a sphere. He found that the variation of the inclination was in agreement with what was then known about the earth's magnetic field, and came to the conclusion that the earth behaved substantially as a uniformly magnetized sphere, its magnetic field being due to causes within the earth and not from any external influence as was supposed at that time. Since 1600 the direction and intensity of the earth's magnetic field has been measured at many widely scattered points over the earth's surface, although no attempt was made to represent the field mathematically before 1839. In that year Gauss showed by a spherical harmonic analysis that the field of a uniformly magnetized sphere is an excellent first approximation to the earth's magnetic field. Gauss further analysed the irregular part of the earth's field, i.e. the difference between the actual observed field and that due to a uniformly magnetized sphere, and showed that both the regular and irregular components of the earth's field are of internal origin.

Since the north-seeking end of a compass needle is attracted towards the northern regions of the earth, those regions must have opposite polarity. Consider therefore the field of a uniformly magnetized sphere whose magnetic axis runs north–south, and let P be any external point distant r from the centre O and θ the angle NOP, i.e. θ is the magnetic co-latitude (see Fig. 5.7). It can be shown

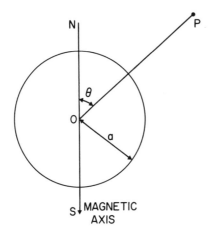

Fig. 5.7

that the field of a uniformly magnetized sphere is the same as that of a dipole at its centre. If M is the magnetic moment of the dipole, the potential at P is

$$V = -\frac{M \cos \theta}{r^2} \tag{5.4}$$

The inward radial component of force corresponding to the magnetic component Z is given by

$$Z = \frac{\partial V}{\partial r} = \frac{2M \cos \theta}{r^3} \tag{5.5}$$

and the component at right angles to OP in the direction of decreasing θ, corresponding to the magnetic component H by

$$H = \frac{1}{r}\frac{\partial V}{\partial \theta} = \frac{M \sin \theta}{r^3} \tag{5.6}$$

The inclination I is then given by

$$\tan I = \frac{Z}{H} = 2 \cot \theta \tag{5.7}$$

and the magnitude of the total force $\mathbf{F}$ by

$$F = (H^2 + Z^2)^{1/2} = \frac{M}{r^3}(1 + 3 \cos^2 \theta)^{1/2} \tag{5.8}$$

The equations of the lines of force can readily be obtained from equations (5.5) and (5.6). By symmetry they lie on surfaces of revolution about the axis, and in any axial plane are given by

$$\frac{dr}{Z} = \frac{r\,d\theta}{H}$$

i.e.

$$\frac{dr}{2M \cos \theta/r^3} = \frac{r\,d\theta}{M \sin \theta/r^3}$$

or

$$\frac{dr}{r} = \frac{2 \cos \theta}{\sin \theta}\,d\theta$$

which integrates to give

$$\ln r = 2 \ln \sin \theta + \text{constant}$$

or

$$r = C \sin^2 \theta \tag{5.9}$$

where different values of C correspond to different lines of force.

Equation (5.9) may be written in the form

$$\frac{r}{a} = \frac{\sin^2 \theta}{\sin^2 \theta_0} \tag{5.10}$$

where θ_0 is the value of θ at the point where the line of force meets the sphere $r=a$. The maximum distance of a line of force from the sphere is obtained when $\theta = 90°$, i.e. above the equator, where $r = r_{\max} = a/\sin^2 \theta_0$. The geomagnetic poles, i.e. the points where the axis of the geocentric dipole which best approximates the earth's field meets the surface of the earth, are situated at approximately $79°N$, $70°W$ and $79°S$, $110°E$. The geomagnetic axis is thus inclined at about $11°$ to the earth's geographical axis. If greater accuracy is needed, it is usually more convenient to expand the potential of the earth's magnetic field in a series of spherical harmonics (see § 5.3). It can be shown that a better approximation to the earth's field can be obtained by displacing the centre of the equivalent dipole by about 300 km in the direction of Indonesia.

5.3 Spherical harmonic analysis of the earth's magnetic field

Assuming that there is no magnetic material near the ground, the earth's magnetic field can be derived from a potential function V which satisfies Laplace's equation and can thus be represented as a series of spherical harmonics in the form (cf equation (B.15))

$$V = a \sum_{n=1}^{\infty} \sum_{m=0}^{n} P_n^m (\cos \theta) \left[\{ c_n^m (r/a)^n + (1 - c_n^m)(a/r)^{n+1} \} A_n^m \cos m\phi \right.$$
$$\left. + \{ s_n^m (r/a)^n + (1 - s_n^m)(a/r)^{n+1} \} B_n^m \sin m\phi \right] \tag{5.11}$$

c_n^m and s_n^m are numbers lying between 0 and 1, and represent the fractions of the harmonic terms $P_n^m (\cos \theta) \cos m\phi$ and $P_n^m (\cos \theta) \sin m\phi$ in the expansion of V which, on the surface of the sphere $(r=a)$, are due to matter outside the sphere. In geomagnetism it is more convenient to use coefficients c_n^m and s_n^m which have the dimensions of magnetic field. Thus the factor $1/a$ in the expression for V in equation (B.15) has been replaced by a in equation (5.11). There is also no term with $n=0$, which would correspond to a magnetic monopole within the earth. It is also assumed that there are no electric currents flowing across the surface of the earth—if there were they would set up a non-potential field and thus contribute a part of the earth's magnetic field which could not be represented by equation (5.11).

The potential V cannot of course be measured directly; what can be determined are the three components of force $X = (1/r)(\partial V/\partial\theta)$ (horizontal, northward), $Y = (-1/r \sin \theta)/(\partial V/\partial\phi)$ (horizontal, eastward) and $Z = \partial V/\partial r$ (vertical, downward) at the earth's surface $r=a$.

136

Z (at $r=a$) may be expanded as a series of spherical harmonics

$$Z = (\partial V/\partial r) = \sum_{n=1}^{\infty} \sum_{m=0}^{n} P_n^m(\cos\theta)(\alpha_n^m \cos m\phi + \beta_n^m \sin m\phi) \qquad (5.12)$$

and the coefficients α_n^m, β_n^m may be determined from the observed values of Z.

By differentiating equation (5.11) with respect to r and then writing $r=a$, we have

$$(\partial V/\partial r) = \sum_{n=1}^{\infty} \sum_{m=0}^{n} P_n^m(\cos\theta)[\{nc_n^m - (n+1)(1-c_n^m)\}A_n^m \cos m\phi$$
$$+ \{ns_n^m - (n+1)(1-s_n^m)\}B_n^m \sin m\phi] \qquad (5.13)$$

The coefficients of each separate harmonic term for each n and m must be equal in the two expansions of $\partial V/\partial r$ given by equations (5.12) and (5.13).

Hence

$$\alpha_n^m = \{nc_n^m - (n+1)(1-c_n^m)\}A_n^m \qquad (5.14)$$

and

$$\beta_n^m = \{ns_n^m - (n+1)(1-s_n^m)\}B_n^m$$

Again from an analysis of the observed values of X and Y, the coefficients in the following two expansions derived from equation (5.11) may be obtained

$$Y_{r=a} = \left(\frac{-1}{r \sin\theta} \frac{\partial V}{\partial\phi}\right)_{r=a}$$

$$= \frac{1}{\sin\theta} \sum_{n=1}^{\infty} \sum_{m=0}^{n} P_n^m(\cos\theta)(mA_n^m \sin m\phi - mB_n^m \cos m\phi) \qquad (5.15)$$

$$X_{r=a} = \left(\frac{1}{r} \frac{\partial V}{\partial\theta}\right)_{r=a}$$

$$= \sum_{n=1}^{\infty} \sum_{m=0}^{n} \frac{d}{d\theta} P_n^m(\cos\theta)(A_n^m \cos m\phi + B_n^m \sin m\phi) \qquad (5.16)$$

Both these equations contain A_n^m and B_n^m and thus if values of X are known all over the world, values of Y can be deduced. If there is disagreement between observed and calculated values of Y, it would imply that the field was not completely derivable from a potential V and hence that earth–air currents do exist. When Gauss first carried out such calculations in 1839 he found no discrepancy. From a knowledge of the coefficients A_n^m, B_n^m, α_n^m and β_n^m, equations (5.14) determine c_n^m and s_n^m, Gauss found from the data available at that time that $c_n^m = s_n^m = 0$, i.e. the source of the earth's magnetic field is entirely internal.

The cofficients of the field of internal origin are

$$g_n^m = (1-c_n^m)A_n^m, \qquad h_n^m = (1-s_n^m)B_n^m \qquad (5.17)$$

and are known as Gauss coefficients. If the external field is negligible, equations (5.17) reduce to $g_n^m = A_n^m$, and $h_n^m = B_n^m$. Values of these coefficients as obtained by different investigators since the time of Gauss are given in Table 5.1. It is clear that by far the most important contribution to V comes from the term containing g_1^0, which is proportional to $P_1(\cos \theta)/r^2$, i.e. $\cos \theta/r^2$, and corresponds to the field of a geocentric dipole. Harmonic analysis of most force fields yields a first degree term, but the formal calculation of the term provides little insight into whether the term describes a distinct physical entity or process. For this some other observations or some theoretical basis are needed. The first order terms in the potential at the surface $r=a$ are, from equation (5.11), given by

$$V/a = g_1^0 P_1^0(\cos \theta) + (g_1^1 \cos \phi + h_1^1 \sin \phi) P_1^1(\cos \theta)$$
$$= g_1^0 \cos \theta + (g_1^1 \cos \phi + h_1^1 \sin \phi) \sin \theta$$

Let

$$H_0^2 = (g_1^0)^2 + (g_1^1)^2 + (h_1^1)^2, \quad \cos \theta_0 = g_1^0/H_0, \quad \text{and}$$
$$\tan \phi_0 = h_1^1/g_1^1 \quad \text{where} \quad H_0 > 0 \tag{5.18}$$

Then

$$g_1^0 = H_0 \cos \theta_0, \qquad g_1^1 = H_0 \sin \theta_0 \cos \phi_0, \qquad h_1^1 = H_0 \sin \theta_0 \sin \phi_0$$

and

$$V/a = H_0 \cos \theta_0 \cos \theta + H_0 \sin \theta_0 \sin \theta (\cos \phi \cos \phi_0 + \sin \phi \sin \phi_0)$$
$$= H_0 \cos \theta_0 \cos \theta + H_0 \sin \theta_0 \sin \theta \cos (\phi - \phi_0) \tag{5.19}$$
$$= H_0 \cos \Theta$$

where Θ is the angle between the general direction (θ, ϕ) and the particular direction (θ_0, ϕ_0). Equation (5.19) is the same as the potential (at $r=a$) of the field of a sphere uniformly magnetized along the direction $(-\theta_0, -\phi_0)$.

An outstanding feature of Table 5.1 is the secular variation of the individual coefficients which is clearly apparent in spite of individual scatter. There appears to have been an overall decrease in the dipole moment of the earth's field of about 6 per cent during the past century. Over the same time interval, the inclination of the dipole axis appears to have remained sensibly constant. The non-dipole components of the field are subject to strong and comparatively rapid secular variations with apparently no constant components.

Determinations of the main geomagnetic field have been made for more than a century. One of the most detailed is that of Vestine et al. (1947) who carried out an exhaustive analysis of both the main field and its secular variation between 1905 and 1945. There is still a poor distribution and uneven quality of the raw data from ground stations, although the situation has improved considerably through the use of aircraft and, later, satellites. Recent analyses show that secular change cannot be extrapolated accurately over intervals longer than 4 or 5 years. Thus only space craft can perform surveys quickly enough to maintain accurate models of the field. It now appears possible to be able to construct very accurate models of the main field and its secular change using only low altitude total-field

138

Table 5.1 Dipole and quadrupole Gauss-Schmidt coefficients (gauss), dipole moment (gauss cm³), and polar and azimuth angles (in degrees), at various epochs

Author	Epoch	g_1^0	g_1^1	h_1^1	g_2^0	g_2^1	g_2^2	h_2^1	h_2^2	$M \times 10^{25}$ emu	θ polar	ϕ azimuth
Erman-Petersen	1829	−·3201	−·0284	·0601	−·0008	·0257	−·0014	−·0004	·0146	8·454	11·7°	−64·7°
Gauss	1835	−·3235	−·0311	·0625	·0051	·0292	−·0002	·0012	·0157	8·558	12·2	−63·5
Adams	1845	−·3219	−·0278	·0578	·0009	·0284	−·0004	−·0010	·0135	8·488	11·3	−64·3
Adams	1880	−·3168	−·0243	·0603	−·0049	·0297	−·0006	−·0075	·0149	8·363	11·6	−68·1
Neumayer	1880	−·3157	−·0248	·0603	−·0053	·0288	·0065	−·0075	·0146	8·336	11·7	−67·6
Fritsche	1885	−·3164	−·0241	·0591	−·0035	·0286	·0068	−·0075	·0142	8·347	11·4	−67·8
Schmidt	1885	−·3174	−·0236	·0598	−·0050	·0278	·0065	−·0071	·0149	8·375	11·5	−68·5
Dyson-Furner	1922	−·3095	−·0226	·0592	−·00887	·02991	·01443	−·01241	−·00843	8·165	11·4	−68·8
Jones-Melotte	1942·5	−·3039	−·0218	·0555	−·0117	·02940	·0156	−·0150	·0051	8·009	11·1	−68·5
Vestine et al.	1905	−·31423	−·02270	·05981	−·00773	·02952	·01107	−·01051	·01156	8·291	11·4	−70·0
Vestine et al.	1915	−·31176	−·02176	·05912	−·00842	·02940	·01345	−·01144	·00986	8·225	11·4	−69·9
Vestine et al.	1925	−·30892	−·02166	·05839	−·00946	·02946	·01510	−·01284	·00814	8·149	11·4	−70·0
Vestine et al.	1935	−·30662	−·02129	·05792	−·01086	·02959	·01608	−·01460	·00676	8·088	11·4	−69·9
Vestine et al.	1945	−·30570	−·02116	·05805	−·01265	·02960	·01632	−·01658	·00535	8·065	11·4	−70·0
Afanasieva	1945	−·3032	−·0229	·0590	−·0125	·0288	·0150	−·0146	·0048	8·010	11·8	−68·8
U.S.C. & G.S.	1945	−·3057	−·0219	·0579						8·066	11·4	−69·3
Franselau-Kautzleben	1945	−·30668	−·02160	·0577	−·01279	·029596	·01547	−·01673	·005811	8·090	11·4	−69·5
U.S.C. & G.S.	1955	−·3046	−·0212	·0576						8·035	11·4	−69·8
Finch-Leaton	1955	−·3055	−·0227	·0590	−·0152	·0303	·0158	−·1090	·0024	8·067	11·7	−69·0
Nagata-Oguti	1958·5	−·3045	−·0222	·0584	−·0151	·0295	·0149	−·0194	·0021	8·038	11·6	−69·2
Cain et al.	1959	−·30674	−·01923	·05762	−·02055	·0344	·0124	−·01977	·00671	8·086	11·2	−71·5
Fougere	1960	−·30509	−·02181	·05841	−·01464	·02971	·01673	−·01988	·00198	8·053	11·6	−69·5
Adam et al.	1960	−·3046	−·0214	·0580	−·0150	·0299	·0164	−·0194	·0027	8·037	11·5	−69·7
Jensen-Cain	1960	−·30411	−·02147	·05799	−·01602	·02959	·01545	−·01912	·00812	8·025	11·5	−69·7
Leaton et al.	1965	−·30375	−·02087	·05769	−·01648	·02954	·01579	−·01995	·00116	8·013	11·4	−70·1
Hurwitz et al.	1965	−·30388	−·02117	·05760	−·01640	·02983	·01583	−·02004	·00125	8·017	11·4	−69·8

mean 11·49°

(After McDonald and Gunst (1967))

139

Table 5.2 IGRF 1965.0 coefficients

n	m	Main field, gammas		Secular change, gammas/year	
		g_n^m	h_n^m	$\dot{g}_n^m$	$\dot{h}_n^m$
1	0	-30339		15·3	
1	1	-2123	5758	8·7	$-2·3$
2	0	-1654		$-24·4$	
2	1	2994	-2006	0·3	$-11·8$
2	2	1567	130	$-1·6$	$-16·7$
3	0	1297		0·2	
3	1	-2036	-403	$-10·8$	4·2
3	2	1289	242	0·7	0·7
3	3	843	-176	$-3·8$	$-7·7$
4	0	958		$-0·7$	
4	1	805	149	0·2	$-0·1$
4	2	492	-280	$-3·0$	1·6
4	3	-392	8	$-0·1$	2·9
4	4	256	-265	$-2·1$	$-4·2$
5	0	-223		1·9	
5	1	357	16	1·1	2·3
5	2	246	125	2·9	1·7
5	3	-26	-123	0·6	$-2·4$
5	4	-161	-107	0·0	0·8
5	5	-51	77	1·3	$-0·3$
6	0	47		$-0·1$	
6	1	60	-14	$-0·3$	$-0·9$
6	2	4	106	1·1	$-0·4$
6	3	-229	68	1·9	2·0
6	4	3	-32	$-0·4$	$-1·1$
6	5	-4	-10	$-0·4$	0·1
6	6	-112	-13	$-0·2$	0·9
7	0	71		$-0·5$	
7	1	-54	-57	$-0·3$	$-1·1$
7	2	0	-27	$-0·7$	0·3
7	3	12	-8	$-0·5$	0·4
7	4	-25	9	0·3	0·2
7	5	-9	23	0·0	0·4
7	6	13	-19	$-0·2$	0·2
7	7	-2	-17	$-0·6$	0·3
8	0	10		0·1	
8	1	9	3	0·4	0·1
8	2	-3	-13	0·6	$-0·2$
8	3	-12	5	0·0	$-0·3$
8	4	-4	-17	0·0	$-0·2$
8	5	7	4	$-0·1$	$-0·3$
8	6	-5	22	0·3	$-0·4$
8	7	12	-3	$-0·3$	$-0·3$
8	8	6	-16	$-0·5$	$-0·3$

(After IAGA Commission 2, Working Group 4 (1969))

observations from satellites. Systematic global coverage with a precision of $\pm 0·1\ \gamma$ by a low altitude (< 350 km) polar orbiting satellite (which would require an accuracy in altitude of ± 10 m) would help resolve the higher harmonics of the field.

There have been a number of analyses of the magnetic field in recent years. A

World Magnetic Survey (WMS) was proposed at the Toronto meeting of the International Union of Geodesy and Geophysics in 1957. As a result, an international geomagnetic reference field (IGRF) for the epoch 1965·0 has now been established (IAGA, 1969; IAGA, 1971). Table 5.2 gives the IGRF coefficients for the period 1955–1972. Eighty spherical harmonic coefficients were used both for the main field and for the secular variation. Mead (1970) transformed IGRF 1965·0, which is referenced to a geocentric spherical polar coordinate system, into a model in geomagnetic dipole coordinates. This latter model is of more use for studies of external field sources (e.g. ring currents, magnetopause, tail fields, etc.; see § 5.7) which depend on the direction of the magnetic dipole axis rather than on the direction of the rotation axis.

Benkova *et al.* (1973) have carried out spherical harmonic analyses of the palaeomagnetic fields of the earth for a number of geological periods. In spite of the low accuracy, small number, and uneven distribution of palaeomagnetic data, the geomagnetic field appears to have maintained its global structure throughout the past 300 m yr. The best fit is that of a dipole displaced some 300–400 km from the earth's centre in the direction of the western Pacific ocean.

5.4   The secular variation and westward drift

It has been known for over 400 years that the earth's main magnetic field is not steady but shows a secular change; this has already been discussed briefly in § 5.1. The features of the secular variation as shown by observatory data have been summarized by Hide (1966). The data cover a period less than 10^{-7} that of the age of the earth and it is not possible to make dogmatic statements about the behaviour of the field in the geologic past. The non-dipole component of the earth's field, though much weaker than the dipole component, shows more rapid changes. The time scale of the non-dipole changes is measured in decades and that of the dipole in centuries. The isoporic foci drift westward at a fraction of a degree per year. The drift of the dipole field is slower than that of the non-dipole field, at least by a factor of three. In addition to the westward drift, the pattern of the secular variation field may alter appreciably in a few decades. In some areas, however, the amplitude of the secular change is anomalously large or small. At present it is larger than average in the Antarctic and smaller in the Pacific hemisphere (see Fig. 5.8). Individual features of the non-dipole field are not associated with individual continents, but, on a larger scale, the hemisphere containing most of the continents also contains most of the major features of the non-dipole field. Conversely, it is in the centre of the oceanic hemisphere that these are absent.

Cox and Doell (1964) pointed out that there are difficulties in attempting to explain the absence of the non-dipole field in the central Pacific by shielding.

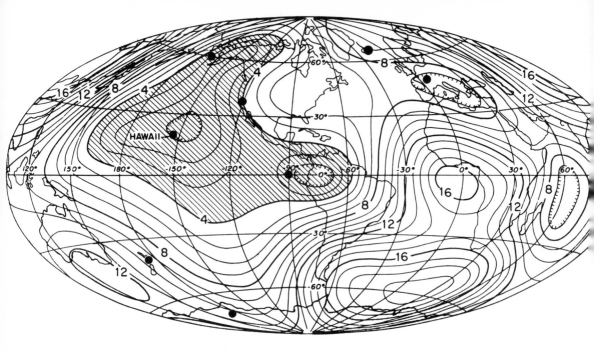

Fig. 5.8 Intensity of the non-dipole portion of the 1965 International Geomagnetic Reference Field. Contour interval, 1000γ. Shaded area, intensity less than 4000γ. The map is equal area with the pole of projection on the equator at 80°W (after Doell and Cox (1971))

Conductivities sufficiently high to attenuate the secular variation with periods in the range 10^2–10^3 yr would reduce variations with periods of 1–10 yr below the noise level of magnetic measurements, yet variations with periods in this shorter time range appear in the records of observatories in the central Pacific area. A possible alternative explanation is that some process in the core selects against formation of non-dipole sources beneath the central Pacific region (Cox (1962)). These non-dipole sources may be associated with thermal convection in the core, in which case their pattern is probably sensitive to any lateral differences in temperature at the base of the mantle.

Doell and Cox (1971) analysed lateral variations in the amplitude of the geomagnetic secular variation which may reflect lateral variations in the properties of the lower mantle or undulations of the core-mantle boundary. They considered several different types of records: direct measurements from geomagnetic observatories; palaeomagnetic measurements on Hawaiian lava flows with accurately known ages in the time interval 0–200 yr ago and on other flows with ages between 200–10 000 yr ago; and worldwide palaeomagnetic measurements of the average geomagnetic angular dispersion recorded in lava flows during the past 0·7 m yr. All these magnetic records indicate that during this time the non-dipole component of the earth's field was lower in the central Pacific than elsewhere (as is the case today). This implies that there must be some type of inhomogeneity in the lower mantle which is coupled to the core in such a way as to suppress the generation of the non-dipole field beneath the central Pacific. In this connection Alexander and Phinney (1966) and Phinney and Alexander (1966, 1969) have found evidence that seismic parameters, sensitive to physical properties in the lower 200 km of the mantle, are different beneath the Pacific and Atlantic regions.

The most detailed recent work on the secular variation is that of McDonald and Gunst (1967, 1968). They have examined several features of the secular variation by analysing the spherical harmonic coefficients obtained by numerous authors from 1835–1965 (see Table 5.1). They found that the magnetic energy density in the outer core has been increasing at a small rate while the field has become hemispherically more asymmetric. Vestine and his colleagues in a series of papers (Kahle *et al.* (1967a, b); Vestine *et al.* (1967)) have inferred fluid motions at the surface of the earth's core from a knowledge of the magnetic field and secular change observed at the earth's surface. There has been some controversy over the reliability of extrapolating higher order harmonics to the core-mantle boundary and on the importance of magnetic diffusion. The above papers by Kahle *et al.* assume that the effects of magnetic diffusion are negligible, although in a later paper (Ball *et al.* (1969)) some allowance is made for this effect. Similar computations by Booker (1969), following a detailed discussion on the kinematics of the secular variation in a perfectly conducting core by Backus (1968), disagree with the above papers both in the results obtained and in some aspects of the theory.

Munk and Revelle (1952) suggested that the westward drift may be related to irregular fluctuations in the length of the day, electromagnetic torques adjusting the balance of angular momentum between the earth's core and mantle. Consider a model of the earth in which the mantle and core rotate as two solid bodies (Runcorn (1954)). The principal moments of inertia of the mantle and core are $I_M = 7 \cdot 2 \times 10^{44}$ g cm^2 and $I_C = 0 \cdot 85 \times 10^{44}$ g cm^2, respectively. If their angular velocities are Ω_M and Ω_C, then conservation of angular momentum gives

$$I_M \Omega_M + I_C \Omega_C = \text{constant}$$

i.e.

$$I_M \, \delta\Omega_M + I_C \, \delta\Omega_C = 0 \tag{5.20}$$

The rate of change of westward drift is given by

$$-\delta\dot{\phi} = -(\delta\Omega_C - \delta\Omega_M) \tag{5.21}$$

From equations (5.20) and (5.21) it follows that

$$\delta\Omega_M = -\frac{\delta\dot{\phi}}{1 + (I_M/I_C)} = -0 \cdot 105 \, \delta\dot{\phi} \tag{5.22}$$

The reported increase in the rate of westward drift between the epochs 1905–1925 and 1925–1945 is $\delta\dot{\phi} = -0 \cdot 87 \times 10^{-10}$ rad/s. Equation (5.22) then gives $\delta\Omega_M/\Omega_M =$

143

$12 \cdot 5 \times 10^{-8}$. On the other hand, astronomical observations give $\delta \Omega_M / \Omega_M = 6 \times 10^{-8}$ for the period 1910–1930. Thus

$$\frac{(\delta \Omega_M)_{\text{observed}}}{(\delta \Omega_M)_{\text{computed}}} = 0 \cdot 48$$

and the changes in the length of the day are thus not inconsistent with the reported fluctuations in the rate of westward drift, provided a substantial part of the outer core is involved in the fluctuations. Vestine and Kahle (1968) have further investigated the possibility of a correlation between changes in the length of the day and the secular variation, and Ball *et al.* (1968) have found a correlation with a few years phase lag between changes in the rate of the earth's rotation and the westward drift of the eccentric dipole. The latest decrease in the rotation rate appears to be reflected in a recent slowing of the westward drift of the eccentric dipole, as shown by both satellite results (Kahle *et al.* (1969)) and analyses of observatory values. All the above calculations imply that there exists an adequate couple to produce the observed change in angular velocity. As already pointed out in § 4.1 neither viscous coupling nor electromagnetic coupling seems adequate to account for the decade fluctuations in the length of the day (Rochester (1970)).

Hide (1969) has suggested the possibility of a different kind of coupling: topographic coupling, as a result of irregular features at the core-mantle boundary. Analysis of travel times of compressional waves reflected at this boundary shows that any such topographic features cannot exceed a few kilometres in height, this being the limit of resolution of present day seismic techniques. Topographic 'bumps' on the core-mantle boundary may also be the cause of horizontal density variations responsible for regional gravity anomalies—it can readily be shown (e.g. Hide and Horai (1968)) that because of the density contrast at the core-mantle boundary, bumps with horizontal dimensions up to thousands of kilometres and a kilometre or so in height would make a significant (although not dominant) contribution to the observed distortion of the gravitational field at the earth's surface. Hide (1967, 1969, 1970) has also suggested that bumps on the core-mantle boundary might affect the flow pattern in the core and thus influence the detailed configuration of the geomagnetic field and its time variations. While the liquid core of the earth is the only likely location of electric currents responsible for the main geomagnetic field, it is the most unlikely place to find density variations of sufficient magnitude to cause the observed distortions of the gravitational field—these must arise largely in the mantle. Thus any correlation between gravity and magnetic anomalies would reflect processes at the core-mantle boundary. Hide and Malin (1970) thus argued that if both gravity and magnetic anomalies are the result of the same topographic features, it should be possible to find a statistically significant correlation between them; in fact they found, for spherical harmonic coefficients up to degree 4, a correlation coefficient of $0 \cdot 84$ between large scale features of the earth's non-dipole magnetic field (for epoch 1965) and the gravitational field, provided the magnetic field is displaced $160°$ eastward in longitude λ. Hide and Malin also

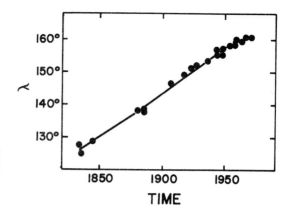

Fig. 5.9 The variation with time of the eastward displacement in longitude λ between the earth's magnetic and gravitational fields (after Hide and Malin (1970))

showed that λ has increased linearly with time since 1835 (see Fig. 5.9), the date of the earliest reliable spherical harmonic analysis of the geomagnetic field (by Gauss). Their result is

$$\lambda = (126 \cdot 2 \pm 0 \cdot 2)^\circ + (0 \cdot 273 \pm 0 \cdot 005)(t - 1835 \pm 10)^\circ \qquad (5.23)$$

where t is the epoch (year A.D.). This dependence of λ on t is associated with the westward drift of the geomagnetic field. There has been some controversy over the above correlation, particularly concerning the statistical procedures used (see e.g. Khan (1971); Lowes (1971) and reply by Hide and Malin (1971)). It must be stressed that even if the correlation does exist (as seems most probable) it does not by itself prove the existence of bumps on the core-mantle boundary. It is possible that quite small temperature variations over the core-mantle boundary could, through their effects on core motions, produce measurable distortions of the geomagnetic field. If these temperature variations reflect the density structure of the lower mantle, then there would be a correlation between gravity and geomagnetic anomalies. An international conference on the core-mantle interface was held recently in 1972 (see Cox and Cain (1972)).

5.5 The origin of the earth's magnetic field

There has been much speculation on the problem of the origin of the earth's magnetic field (and secular variation). Although it is now generally believed that the cause lies in motions of the fluid, electrically-conducting core, there are still many points that remain unresolved and no completely satisfactory theory has yet been given.

Other sources for the field can easily be shown to be inadequate. Consider first the possibility of permanent magnetization. The temperature gradient in the crust is approximately 30°C/km, so that at a depth of about 25 km a temperature of the order of the Curie point for iron is reached, and all ferromagnetic substances will have lost their magnetic properties at greater depths. Hence in order to account for the earth's magnetic moment, an intensity of magnetization in the earth's crust of about 6 Γ is necessary, which is impossible. The magnetization of rocks depends on the amount and nature of the iron oxide minerals contained in them, and for most rocks the intensity is less than 10^{-2} Γ. Permanent magnetization also fails to account for other features of the earth's magnetic field, such as the close proximity of the magnetic and geographical poles, the secular variation, and reversals of polarity.

A number of theories have been proposed that invoke the rotation of the earth. These usually involve the gyromagnetic effect (magnetic polarization) of rotating ferromagnetic bodies or the rotation of separate electric charges in the earth. The first effect can be observed in the laboratory at high speeds, but the angular velocity of the earth is so small that the magnetization arising from such causes is several orders of magnitude too low. In the second case, the associated electric field that would be expected has not been observed.

Some workers have postulated changes in the fundamental laws of physics which would only become significant in rotating bodies of cosmic size, but such theories have not met with much success. In this regard, Blackett in 1947 suggested that the dipole moment of a massive rotating body is proportional to its angular momentum. This was prompted by the discovery by Babcock in 1947 that the star 78 Virginis possessed a magnetic field. When Blackett first proposed his theory it appeared to predict correctly the relative magnetic moments of the sun and earth in terms of their angular momenta. A revised value for the sun's magnetic moment later nullified this agreement. One consequence of Blackett's hypothesis is that a small magnetic field should be produced by a dense body rotating with the earth. In a long and detailed paper, Blackett (1952) described the results of a 'negative experiment' in which a sensitive astatic magnetometer failed to detect a field of the order of magnitude predicted near dense bodies at rest in the laboratory. However, in developing an astatic magnetometer of greatly increased sensitivity, Blackett paved the way for the renewed activity in rock magnetism. Again, any rotational theory can only explain the component of the magnetic moment along the axis of rotation; further arguments are necessary to account for the transverse component, which at present is about one-fifth of the axial. Reversals of the earth's magnetic field (§ 6.3) are also a difficulty with most rotational theories.

We are thus forced to the conclusion that electric currents flow in the earth's interior and set up a magnetic field by induction. Electric currents may have been initiated by chemical irregularities which separated charges and set up a battery action generating weak currents. Palaeomagnetic measurements have shown that the earth's main field has existed throughout geologic time and that its strength has never differed significantly from its present value. In a bounded, stationary, electrically conducting body, any system of electric currents will decay. The field or the current may be analysed into normal modes, each of which decays exponentially with its own time constant. The time constant is proportional to σl^2 where σ is the electrical conductivity and l a characteristic length representing the distance in which the field changes by an appreciable

146

amount. For a sphere the size of the earth, the most slowly decaying mode is reduced to $1/e$ of its initial strength in a time of the order of 100 000 yr. Since the age of the earth is more than 4000 m yr, the geomagnetic field cannot be a relic of the past, and a mechanism must be found for generating and maintaining electric currents to sustain the field.

Possible causes of electromotive forces deep within the earth are thermoelectric and chemical. The suggestion that thermoelectric currents circulate within the earth was first put forward by Elsasser in 1939. Thermoelectric electromotive forces are generated whenever two materials with different electrical properties are in contact at points which are at different temperatures. Elsasser proposed that thermoelectric electromotive forces are due to inhomogeneities in the core material, which are created and continuously regenerated by turbulent fluid motion. Only a small fraction of the current generated by such a mechanism could be responsible for the surface magnetic field, the greater part producing only a contained field, i.e. a field with lines of force not cutting the earth's surface. Runcorn (1954) suggested that thermoelectric currents may be generated at the core-mantle boundary where there is a contact between two materials with different electrical properties. Temperature differences between different parts of the core-mantle boundary could be due to the eccentricity of the earth (producing a temperature contrast between the poles and the equator) or thermal convection in the core. All thermoelectric currents generated by Runcorn's hypothesis would also produce contained fields, and further recourse has to be made to inductive interaction between these fields and the fluid flow in the core. It is extremely difficult to make a quantitative assessment of any thermoelectric theory, but it seems that rather implausible values of the physical properties of the material in the core and lower mantle would be necessary. It is also by no means certain that the required temperature differences demanded by the theory can be realized.

A more promising mechanism for generating and maintaining electric currents in the earth's core is the familiar action of the dynamo. The dynamo theory of the earth's magnetic field was due originally to Sir Joseph Larmor who in 1919 suggested that the magnetic field of the sun might be maintained by a mechanism analogous to that of a self-exciting dynamo. The pioneering work in dynamo theory was later carried out by Elsasser (1946a, b, 1947) and Bullard (1949a, b). The earth's core is a good conductor of electricity and a fluid in which motions can take place, i.e. it permits both mechanical motion and the flow of electric current, and the interaction of these could generate a self-sustaining magnetic field. It has not been possible to demonstrate in the laboratory the existence of such a dynamo action. If a bowl of mercury is heated from below, thermal convection will be set up—but no electric currents or magnetism will be detected in the bowl. Such a model experiment fails because electrical and mechanical processes do not scale down in the same way. An electric current in a bowl of mercury 30 cm in diameter would have a decay time of about one-hundredth of a second. The decay time, however, increases as the square of the diameter; an electric current in the earth's core would persist for about 10 000 yr before it decayed. This time is more than sufficient for the current and its associated magnetic field to be altered and amplified by motions in the fluid, however slow.

The energetics of the earth's core are discussed in § 5.6, but even if sufficient energy sources exist to maintain the field, there remains the critical problem of sign, i.e. it must also be shown that the inductive reaction to an initial field is regenerative. In an engineering dynamo, the coil has the symmetry of a clock face in which the two directions of rotation are not equivalent; it is this very feature which causes the current to flow in the coil in such a direction that it produces a field which reinforces the initial field. A sphere does not have this property; any asymmetry can exist only in the motions. This is the crux of the problem: whether asymmetry of motion is sufficient for dynamo action or whether asymmetry of structure is necessary as well.

Sir Joseph Larmor originally suggested (in the case of the sun) that, through inductive interaction with a small inducing field parallel to the magnetic axis, steady meridional circulation of matter might produce zonal electric currents which would amplify the inducing field. Cowling (1934) showed, however, that a steady motion confined to meridional planes cannot amplify a field of the required type. Elsasser (1947) further showed that purely zonal motion parallel to latitude circles cannot amplify an inducing dipole field. Cowling's result has been extended by Backus and Chandrasekhar (1956), and it now appears that homogeneous dynamos must possess a low degree of symmetry. A number of other 'non-dynamo' theorems that prohibit particular types of motion in a sphere from acting as dynamos have been proved. Most kinds of symmetry are excluded. Some of these theorems restrict the types of possible motions and others restrict the field. Braginskii (1965a, b) has shown that if the motion is nearly axially symmetric, the departure from symmetry must contain terms in both the sine and cosine of some multiple of the longitude relative to the axis of symmetry. Again, although an axially symmetric field cannot be produced by a dynamo, G. O. Roberts has shown (unpublished) that a symmetric motion can maintain an asymmetric field.

In 1958 Backus and Herzenberg showed independently that it was possible to postulate a pattern of motions in a sphere filled with a conducting fluid which would act as a dynamo producing a magnetic field outside the conductor. Although rigorous mathematical solutions were obtained, in each case the motions were physically very improbable. In order to ensure that other fields generated by induction will not develop in such a way that they eventually destroy the whole process, the motions obtained by Backus require periods when the fluid is at rest. Herzenberg's model consists of two spheres in the core each of which rotates as a rigid body at a constant angular velocity about a fixed axis. The axially symmetric component of the magnetic field of one of the spheres is twisted by rotation resulting in a toroidal field which is strong enough to give rise to a magnetic field in the other sphere. The axial component of this field is twisted as well and fed to the first sphere. If the rotation of the spheres is sufficiently rapid, a steady state may be reached.

Lowes and Wilkinson (1963) have built a working model of what is effectively

148

a homogeneous self-maintaining dynamo based on Herzenberg's theory. For mechanical convenience they used, instead of spheres, two cylinders placed side by side with their axes at right angles to one another so that the induced field of each is directed along the axis of the other. If the directions of rotation are chosen correctly, any applied field along one axis will lead, after two stages of induction, to a parallel induced field. If the velocities are large enough, the induced field will be larger than the applied field which is no longer needed, i.e. the system would be self-sustaining. Rikitake and Hagiwara (1966) have investigated the stability of a Herzenberg dynamo and concluded that it is unstable for small disturbances about its steady state. Their analysis is not applicable to the case of the earth however, since numerical integrations could only be performed for parameters very different from those in the earth's core. More recently Childress (1969) and G. O. Roberts (1970, 1972) have investigated dynamos in infinite bodies of fluids. In particular Roberts has shown that 'almost all' motions spatially periodic in three dimensions act as dynamos, and that these spatially periodic dynamos can be wrapped around an axis and enclosed in a sphere. These dynamos are remarkable in that, although the motion extends to infinity, the field does not. Also the field, unlike the motion, is not spatially periodic.

A different approach has been to consider small-scale turbulence as a means of generating large-scale magnetic fields. The main work in this direction has been carried out by Parker in the USA and by Krause, Rädler and Steenbeck in Germany. The scope of these investigations goes far beyond the terrestrial dynamo and is important in the general field of cosmical electrodynamics. The more recent (1970, 1971) papers of Parker relative to the origin of the earth's field are given in the list of references: P. H. Roberts and Stix (1971) have produced a translation of the relevant work of the German school on turbulent dynamos. No attempt will be made to review this field.

Hide (1966) has shown that 'magnetohydrodynamic' shear waves may play an important role in the earth's liquid core. In the absence of a magnetic field, an ordinary fluid can transmit only compressional waves. When a magnetic field is present, if the fluid is a good conductor of electricity (and not too viscous) it can then transmit shear waves which travel along the magnetic lines of force with a speed dependent on the strength of the magnetic field.

Hide investigated small amplitude hydromagnetic oscillations of a rotating spherical shell of incompressible fluid about a mean state characterized by a uniform (mainly toroidal) magnetic field. In the absence of rotation, the oscillations correspond to the superposition of ordinary non-dispersive hydromagnetic waves propagating at the Alfvén speed in both directions along the magnetic lines of force. The effect of rotation is to reduce the phase speed of waves propagating in one of these directions and to increase that of waves moving in the other direction. Both types of waves are highly dispersive. The periods of the slow waves are decades or even centuries if the strength of the toroidal magnetic field in the core lies between 50 and 200 Γ; those of the fast waves are of the order of days. Thus the slow waves have oscillation periods and dispersion times comparable with the time-scale of the secular variation. On the other hand, any magnetic signals generated inside the core with periods less than approximately 4 years are effectively cut off from observation by the conducting lower mantle, i.e. although the electrical conductivity of the mantle is weak, it is sufficient to suppress any observations at the earth's surface of magnetic variations in the core on the time-scale of the fast waves. The slow waves move westward relative to the core material, and Hide suggested that the westward drift of the geomagnetic field relative the earth's surface is a manifestation of free hydromagnetic oscillations. A review of the origin of the geomagnetic field and its secular change, with particular reference to the work in the USSR, has been given by Braginskii (1971).

5.6 The energetics of the earth's core

Elsasser (1954) has shown by a dimensional analysis that in geophysical and astrophysical problems the displacement current and all purely electrostatic effects are negligible, as are all relativistic effects of order higher than v/c where $\mathbf{v}$ is the fluid velocity. Thus the electromagnetic field equations (in emu) are the usual Maxwell equations

$$\text{curl } \mathbf{E} = -\mu \frac{\partial \mathbf{H}}{\partial t} \tag{5.24}$$

$$\text{curl } \mathbf{H} = 4\pi \mathbf{j} \tag{5.25}$$

$$\text{div } \mathbf{H} = 0 \tag{5.26}$$

where $\mathbf{H}$ and $\mathbf{E}$ are the magnetic and electric fields respectively, and $\mathbf{j}$ the electric current density. The magnetic permeability μ and electrical conductivity σ will be assumed constant. The electromotive forces which give rise to $\mathbf{j}$ are due both to electric charges and to motional induction so that the total current $\mathbf{j}$ is given by

$$\mathbf{j} = \sigma(\mathbf{E} + \mu \mathbf{v} \times \mathbf{H}) \tag{5.27}$$

Taking the curl of equation (5.25) and using equations (5.27) and (5.24), $\mathbf{E}$ can be eliminated, leading to the equation

$$\text{curl curl } \mathbf{H} = 4\pi\mu\sigma(-\partial \mathbf{H}/\partial t + \text{curl } \mathbf{v} \times \mathbf{H}) \tag{5.28}$$

Since curl curl $\mathbf{H} = \text{grad div } \mathbf{H} - \nabla^2 \mathbf{H} = -\nabla^2 \mathbf{H}$, on using (5.26) we finally obtain

$$\partial \mathbf{H}/\partial t = \text{curl } (\mathbf{v} \times \mathbf{H}) + \nu_m \nabla^2 \mathbf{H} \tag{5.29}$$

where

$$\nu_m = 1/4\pi\sigma\mu \tag{5.30}$$

is the 'magnetic diffusivity'.

Equations (5.26) and (5.29) give the relations between $\mathbf{H}$ and $\mathbf{v}$ which have to be satisfied from electromagnetic considerations. In the case of the earth, the only other condition is that at the surface all components of the field should be continuous with an external field that vanishes at infinity at least as rapidly as $1/r^3$, where r is distance from the centre. The term $\nabla^2 \mathbf{H}$ causes the field to decay and for a dynamo it is essential that this term should be balanced, and the decay prevented, by the term $\mathrm{curl}\,(\mathbf{v} \times \mathbf{H})$ which represents the interaction of the velocity and the field.

To the electromagnetic equations must be added the hydrodynamical equation of fluid motion in the earth's core (the Navier-Stokes equation) together with the equation of continuity, which, for an incompressible fluid (the speed of flow $\mathbf{v}$ is much less than the speed of sound in the earth's core) reduces to

$$\mathrm{div}\,\mathbf{v} = 0 \tag{5.31}$$

The Navier-Stokes equation is

$$\rho\left\{\frac{\partial \mathbf{v}}{\partial t} + (\mathbf{v} \cdot \nabla)\mathbf{v} + 2\mathbf{\Omega} \times \mathbf{v} - \nu\,\nabla^2\mathbf{v}\right\} - \frac{\mu}{4\pi}\,\mathrm{curl}\,\mathbf{H} \times \mathbf{H} = -\nabla p + \rho\,\nabla W \tag{5.32}$$

where $\mathbf{v}$ is the velocity relative to a system rotating with angular velocity $\mathbf{\Omega}$, p is the pressure, W is the gravitational potential (in which is absorbed the centrifugal force) and ρ and ν are the density and kinematic viscosity, respectively. Equations (5.29) and (5.32) contain only the vectors $\mathbf{v}$ and $\mathbf{H}$ and are the basic equations of field motion.

Because of the complexity of the equations describing the hydromagnetic conditions in the earth's core, most effort has been directed to seeking solutions of Maxwell's equations for a given velocity distribution. This approach, known as the kinematic dynamo problem, is linear and has been the subject of much discussion. Expansion in spherical harmonics reduces equation (5.29) to an infinite set of differential equations containing the radial functions for each harmonic, their first and second radial derivatives and their first time derivatives. For a given set of initial conditions these could in theory be integrated in time steps. Little progress has been made in this direction, however, and most workers have looked for steady state solutions, putting $\partial \mathbf{H}/\partial t = 0$. In 1954, Bullard and Gellman, after a considerable amount of computation, appeared to have found one particular set of motions in the core that could set up a self-exciting dynamo. There was some doubt cast, however, on the convergence of their solutions, which included harmonics up to degree and order four, and Gibson and P. H. Roberts showed later (1969) that the solution including harmonics up to degree and order five was very different. It may be that the velocity field chosen by Bullard and Gellman was approximately symmetric about an axis and had planes of symmetry, thus violating Braginskii's criterion. Lilley (1970) appears to have obtained better convergence by introducing some asymmetry.

Although it is now known that kinematic dynamos exist, it must be stressed that solutions in which Maxwell's equations are solved for specified velocities are of limited geophysical interest, since there is no guarantee that there exist

151

forces in the earth's core that can sustain them. In a dynamical theory the velocities would be calculated from assumed forces; almost no work has been done on this more difficult non-linear problem. From a consideration of the order of magnitude of the terms in the Navier-Stokes equation (5.32) it can be shown that the inertial terms may be neglected and most probably the viscous forces. We then must have a balance between the Coriolis forces, the pressure gradient, the electromagnetic forces and the applied force. It can easily be verified that if there is no applied force, the electromagnetic force cannot be balanced by the other two terms, since neither can supply energy. The pressure term can be removed by taking the curl of the resulting equation; this leads to an expression from which the curl of the force can be found for any dynamo for which the velocity and the field are known, i.e. if we had a solution to Maxwell's equations for a specified velocity field, it is possible, in theory, to calculate the forces needed to drive the system. To solve the inverse problem with specified forces is, as already mentioned, very much more difficult.

A number of mechanisms have been suggested which might maintain fluid motions in the earth's core, although only two, thermal convection and the precession of the earth, seem at all possible. It is extremely difficult to evaluate quantitatively the effects of these two mechanisms. Whether radioactivity in the core and the processes for removing heat from outside the core are sufficient to produce the temperature gradient necessary to initiate convection are unknown (see § 7.4 and § 10.5). To investigate the matter in any detail, the equations of thermodynamics must be considered in addition to those of electromagnetism and hydrodynamics. Apart from the difficulty of estimating the total heat generated in the core, our estimates of such physical parameters as the thermal conductivity may well be out by a factor of 5 or more. The inhibiting effects of viscosity, rotation and magnetic fields on convection are also difficult to assess.

It is also possible that the forces which drive the earth's dynamo may be connected with the formation and evolution of the earth's core. Urey (1952) suggested that the core has been growing at the expense of the mantle by iron in the mantle slowly and continuously 'seeping' into the core, the heat which is released in the core by this process causing convection. The basic question here is how long did it take the earth's core to form and when did this event take place? Evidence from many different fields (see § 10.5) indicates that in all probability the event was comparatively rapid and took place very early in, or simultaneous with, the formation of the earth itself. It would thus seem that the processes which led to core formation have not continued throughout geologic time and thus could not be a major factor in the maintenance of motions in the earth's core.

The earth's mantle undergoes complicated accelerations due to a number of causes such as the bodily tide of the earth, tidal friction in the oceans, sudden changes in the rate of rotation of the earth and precession and nutation. It can be shown (see e.g. Hide (1956)) that of these only precession could have any appreciable effect on motions in the earth's core, and experiments by Malkus (1968) have indicated that the core would not precess with the mantle like a rigid body, as had previously been thought. Thus it is possible that precession

152

may produce turbulent motion in the core and hence drive the dynamo. Malkus (1963) had earlier suggested that precessional torques drive the earth's dynamo. There are unfortunately some errors in this paper, and no detailed treatment of a dynamo in a precessing turbulent core has as yet been given.

There are really two problems in assessing the role that precession plays in the generation of the earth's magnetic field: the correct value of the electrical conductivity σ in the core and the functional dependence on σ of parameters in any model of the earth. Malkus (1968) assigned an uncertainty of a factor of 3 in either direction to the value of σ he used in his calculations (7×10^{-6} emu). Stacey (1967) suggested that allowance for the effect of impurities in the iron in the earth's core could increase its electrical resistivity by a factor of 10. However, in a later paper Gardiner and Stacey (1971) withdrew this suggestion in favour of Bullard's original figure for σ of 3×10^{-6}. Although the conclusions of Gardiner and Stacey may well be right, some of their arguments in rebuttal of Stacey's (1967) earlier suggestion of a large increase in resistivity in the core appear to be incorrect (see Jacobs et al. (1972)). Jain and Evans (1972) have carried out a calculation of the resistivity of the earth's core based on a model for the electrical transport properties of simple liquid metals proposed by Ziman (1961, 1971). Their estimate of the resistivity is between 1 and 2×10^5 emu which is in the range of plausible values (1 to 6×10^5 emu) according to Gardiner and Stacey. It is also interesting to note the results of the experimental work of Kawai and Mochizuki (1971) on the metallic states in three 3d transition metal oxides (including Fe_2O_3) under very high pressures. They found a very sudden drop in resistivity of from four to six orders of magnitude at a pressure corresponding to that in the liquid core of the earth. Whether the resistivity of the earth's core is of the same order of magnitude as that of pure iron at its melting temperature at atmospheric pressure will depend on the alloying material (see § 10.5). If it is mainly silicon, the answer is probably, yes, since liquid silicon is a normal metal. If it is sulphur, the answer would be very different, since liquid sulphur is almost a perfect insulator. However, if the core material is in the form of the compound FeS, the resistivity can still be quite similar to that of pure iron since liquid FeS is a semiconductor under normal temperature and pressure conditions and there may be a Mott transition to metal at very high pressures.

With regard to the second problem, different dynamo models yield a different functional dependence of various critical parameters on σ—in particular the dependence of the ohmic dissipation rate Q on σ has, according to different authors, ranged from $Q \propto \sigma^{-3}$ to $Q \propto \sigma^2$ (see Jacobs et al. (1972)). This wide difference between the functional dependence of Q on σ for different models and the uncertainty in the value of σ in the core make it impossible at the moment to decide whether precessional torques are sufficient to drive the earth's dynamo. A detailed review of the magnetohydrodynamics of the earth's core has been given by P. H. Roberts and Soward (1972).

5.7 Transient magnetic variations

The records from any magnetic observatory show that on some days the field exhibits smooth and regular variations, while on other days it is disturbed and shows irregular fluctuations. These transient variations, unlike the secular variation, have their origin outside the earth, in the sun, and are but one manifestation of a number of phenomena collectively called solar-terrestrial relationships.

In general, day-to-day changes in the intensity of any disturbance follow a similar pattern over a wide area; magnetically quiet conditions are also usually widespread. Any disturbance D is superposed on a regular daily variation. The mean daily variation obtained from the five quietest days of the month is denoted by S_q and each magnetic element behaves in a characteristic manner. An outstanding feature of S_q is the equatorial enhancement of the daily range of the horizontal intensity. At the centre of a narrow belt $2°-3°$ of latitude wide, the daily range is nearly three times as large as at points 500 km to the north or south. A spherical harmonic analysis of the S_q field shows that, unlike the earth's main field, it is not produced wholly inside the earth: in fact the major part S_q^e is of external origin, and it is generally believed that this external field gives rise to the internal component S_q^i by electromagnetic induction.

The air above the earth's surface is essentially non-magnetic and non-conducting up to a height of about 70 km, above which there are two main ionized layers: the E layer at about 100 km and the F layer at about 200–300 km. The ionization in these layers is renewed daily by the sun. From a spherical harmonic analysis of the S_q field it is possible to calculate the strength and direction of an ionospheric current system which could produce the S_q^e field (see Price (1969) for a review of such analyses). Balfour Stewart had suggested as long ago as 1882 that such electric currents existed and were caused by electromotive forces induced by the daily periodic motion of the air in the presence of the earth's main magnetic field. Much work was done on this atmospheric dynamo theory even before Appleton demonstrated in 1924 the existence of the highly conducting E layer in the upper atmosphere. Ionospheric currents have now been detected from magnetic field profiles obtained from magnetometers flown in rockets.

In the atmospheric-dynamo theory of the S_q variations, it is the tidal motion of the air in the ionosphere in a horizontal direction across the lines of force of the earth's magnetic field that is assumed to induce electric currents. The behaviour of ocean tides has been well understood since the nineteenth century—they are caused by gravity. The same forces act on the atmosphere and must also produce tides in it. However, the identification of the lunar atmospheric tide by recording the barometric pressure at ground level is very difficult. It is so small that it is overlain by other effects and highly sophisticated methods of data analysis accumulated over very long time intervals must be employed. Strangely enough the solar influence is much easier to distinguish, which is the exact opposite in the case of oceanic tides. Laplace suggested early in the nineteenth century that the solar atmospheric tide might be caused by thermal rather than gravitational action of the sun on the atmosphere. However, if this were the case the main heating effects should vary diurnally, not semi-diurnally, and it is necessary to explain why the amplitude of the 12 h pressure wave is greater than that of the 24 h wave. The problem of atmospheric tides has had a long and chequered history, and has attracted the attention of some of the greatest names in geophysics.

It is generally agreed now that the diurnal tide in the atmosphere is thermally driven, and that the main sources of the thermal drive are direct insolation absorption by water vapour in the troposphere and by ozone in the stratosphere–mesosphere. It was not until 1967 that Lindzen solved Laplace's tidal equations and obtained the theoretical response of the atmosphere's winds, temperatures, etc., to these known drives. Lindzen also showed that the small amplitude of the diurnal surface pressure oscillation results from most of the thermal drive being used to activate trapped modes which do not propagate to the ground. A comprehensive account of atmospheric tides has been given by Lindzen and Chapman (1969).

In addition to S_q, large disturbances called magnetic storms may occur. These are worldwide phenomena and usually commence suddenly at almost the same instant (within $\frac{1}{2}$ min) all over the earth. The records of individual storms differ greatly among themselves, but the changes in the field at the earth's surface may be analysed into three parts:

(a) a part depending on time measured from the commencement of the storm (known as the storm-time variation D_{st});
(b) a daily variation S_D which is much greater in intensity and markedly different in character to S_q;
(c) an irregular part which is most prominent in high latitudes.

Figure 5.10 shows the striking difference in the storm-time variations between low latitudes (Honolulu) and those in the auroral zone (College, Alaska). On crossing the auroral zone (approximately magnetic latitude 65°) towards the magnetic pole, the characteristics undergo a further change to another type peculiar to the polar caps (Godhavn, Greenland).

In addition to large-scale geomagnetic variations, there are disturbances of much shorter duration such as polar magnetic sub-storms (see Akasofu (1968); Rostoker (1972)). Abrupt impulsive changes (sudden impulses) may also occur as well as variations with periods from about 0·1 s to 10 min which are called geomagnetic micropulsations. Information on these rather specialized phenomena can be found in recent reviews (see e.g. Troitskaya (1967); Campbell (1967); Saito (1969); Jacobs (1970); Orr (1973)).

In recent years transient variations in the geomagnetic field have been considered from the viewpoint of hydromagnetic waves generated by interactions between the geomagnetic field and the solar wind (see § 2.3). Chapman and Ferraro in a long series of papers (1931, 1932, 1933) had already predicted the confinement of the earth's magnetic field inside an elongated cavity during

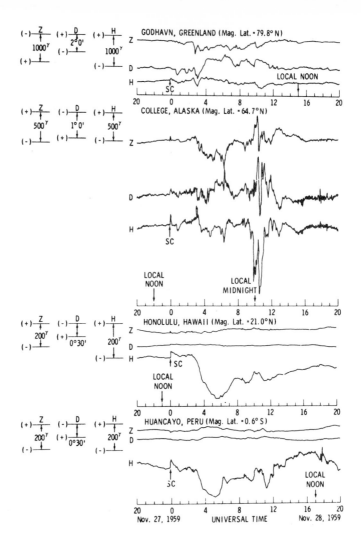

Fig. 5.10 Example of a magnetic storm of moderate intensity at different latitudes (after Sugiura and Heppner (1965))

magnetic storms. The region inside the cavity is called the *magnetosphere* and the boundary the *magnetopause*. The existence of the magnetopause and a bow shock wave have since been confirmed many times by satellites. The region between the magnetopause and the shock wave is referred to as the *magnetosheath* or *transition region*. Beyond the shock wave, conditions are characteristic of the interplanetary medium and the presence of the magnetized earth has little or no effect.

The distance from the centre of the earth to the magnetopause in the solar direction is around 10 earth radii (R_e) although distances less than $8R_e$ and greater than $13R_e$ have occasionally been observed. The shock wave is located several R_e beyond this distance. At 90° to the solar direction both the magnetopause and shock wave are observed to flare out to distances about 30–50 per cent greater than in the sub-solar direction. In the anti-solar direction the cavity extends out to very large distances; no definite closure of the magnetospheric tail has yet been observed by satellites. A schematic diagram of the geomagnetic field in the noon–midnight meridian plane is shown in Fig. 5.11.

When a transient plasma stream arrives at the boundary of the magnetosphere with a speed substantially greater than that of the steady solar wind, the outer

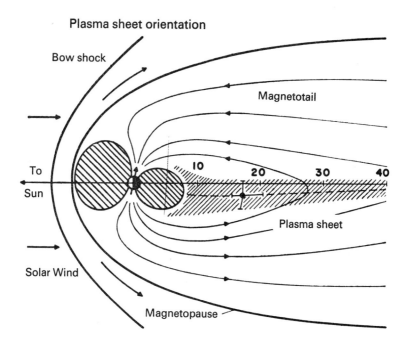

Fig. 5.11 Approximate configuration of the magnetosphere in the solar magnetospheric noon–midnight meridional plane for a 12° tilt of the magnetic dipole axis (after Bame *et al.* (1967))

region of the magnetosphere must experience a strong impact. A direct effect of the impact is the sudden compression of the magnetic field and plasma in that region. This compression is transmitted inward as a hydromagnetic perturbation which, on arrival at the earth, is observed as a sudden increase in the magnetic field (the sudden commencement of a magnetic storm). In low latitudes this increased level of H (the initial phase of the magnetic storm) lasts for a few hours. Analyses of records from many observatories around the world show that the worldwide decrease in H (the main phase of a magnetic storm) is due to a uniform magnetic field. Such a field could be produced by a ring-shaped westward current encircling the earth. Direct evidence for such a ring current at a distance of about $3 \cdot 5$ R_e has been obtained by Frank (1967) who measured, on the satellite OGO 3, the differential energy spectra of protons and electrons in the outer Van Allen radiation belt in the energy range from ~200 eV to 50 keV. The total energy of these low energy protons and electrons within the earth's magnetosphere is sufficient to account for the reduction of the geomagnetic field observed at the earth's surface at low and middle latitudes during the main phase of magnetic storms. The slow decay of the ring current (the recovery phase of the storm) is probably due to a charge exchange process between the protons in the ring current and the ambient slow neutral hydrogen in the magnetosphere.

References

Akasofu, S.-I. (1968) *Polar and Magnetospheric Substorms* (D. Reidel Publ. Co., Holland).

Alexander, S. S. and R. A. Phinney (1966) A study of the core-mantle boundary using P waves diffracted by the earth's core, *J. geophys. Res.*, **71**, 5943.

Babcock, H. W. (1947) Zeeman effect in stellar spectra, *Astrophys. J.*, **105**, 105.

Backus, G. E. (1958) A class of self-sustaining dissipative spherical dynamos, *Ann. Phys.*, **4**, 372.

Backus, G. E. (1968) Kinematics of geomagnetic secular variation in a perfectly conducting core, *Phil. Trans. Roy. Soc.*, **A 263**, 239.

Backus, G. E. and S. Chandrasekhar (1956) On Cowling's theorem on the impossibility of self-maintained axi-symmetric homogeneous dynamos, *Proc. Nat. Acad. Sci.*, **42**, 105.

Ball, R. H., A. B. Kahle and E. H. Vestine (1968) Variations in the geomagnetic field and in the rate of the earth's rotation, *Rand Corp. Rept. RM-5717-PR* (Santa Monica, Calif.).

Ball, R. H., A. B. Kahle and E. H. Vestine (1969) Determination of surface motions of the earth's core, *J. geophys. Res.*, **74**, 3659.

Bame, S. J., J. R. Ashbridge, H. E. Felthauser, E. W. Hones and I. B. Strong (1967) Characteristics of the plasma sheet in the earth's magnetotail, *J. geophys. Res.*, **72**, 113.

Benkova, N. P., A. N. Khramov, T. N. Cherevko and N. V. Adam (1973) Spherical harmonic analysis of the palaeomagnetic field, *Earth Planet. Sci. Letters*, **18**, 141.

Blackett, P. M. S. (1947) The magnetic field of massive rotating bodies, *Nature*, **159**, 658.

Blackett, P. M. S. (1952) A negative experiment relating to magnetism and the earth's rotation, *Phil. Trans. Roy. Soc.*, **A 245**, 309.

Booker, J. R., (1969) Geomagnetic data and core motions, *Proc. Roy. Soc.*, **A 309**, 27.

Braginskii, S. I. (1965a) Self-excitation of a magnetic field during the motion of a highly conducting fluid, Translated in *Sov. Phys. JETP*, **20**, 726.

Braginskii, S. I. (1965b) Theory of the hydromagnetic dynamo, Translated in *Sov. Phys. JETP*, **20**, 1462.

Braginskii, S. I. (1971) Origin of the geomagnetic field and its secular change, *IAGA Bull.*, **31**, 41.

Bullard, E. C. (1949a) The magnetic field within the earth, *Proc. Roy. Soc.*, **A 197**, 433.

Bullard, E. C. (1949b) Electromagnetic induction in a rotating sphere, *Proc. Roy. Soc.*, **A 199**, 413.

Bullard, E. C. and H. Gellman (1954) Homogeneous dynamos and terrestrial magnetism, *Phil. Trans. Roy. Soc.*, **A 247**, 213.

Campbell, W. H. (1967) Geomagnetic pulsations, in: *Physics of Geomagnetic Phenomena* (ed. S. Matsushita and W. H. Campbell) (Acad. Press).

Chapman, S. and V. C. A. Ferraro (1931, 1932, 1933) A new theory of magnetic storms, *Terr. Mag.*, **36**, 77 and 171; **37**, 147 and 421; **38**, 79.

Childress, S. (1969) A class of solutions of the magnetohydrodynamic problem, in: *The Application of Modern Physics to the Earth and Planetary Interiors* (ed. S. K. Runcorn) (Wiley, New York).

Cowling, T. G. (1934) The magnetic field of sunspots, *Mon. Not. Roy. Astr. Soc.*, **94**, 39.

Cox, A. (1962) Analysis of present geomagnetic field for comparison with palaeomagnetic results, *J. Geomag. Geoelec.*, **13**, 101.

Cox, A. and R. R. Doell (1964) Long-period variations of the geomagnetic field, *Bull. seism. Soc. Amer.*, **54**, 2243.

Cox, A. and J. C. Cain (1972) International conference on the core-mantle interface, *Trans. Amer. Geophys. Union, EθS*, **53**, 591.

Doell, R. R. and A. Cox (1971) Pacific geomagnetic secular variation, *Science*, **171**, 248.

Elsasser, W. M. (1939) On the origin of the earth's magnetic field, *Phys. Rev.*, **55**, 489.

Elsasser, W. M. (1946a) Induction effects in terrestrial magnetism Part I: Theory, *Phys. Rev.*, **69**, 106.

Elsasser, W. M. (1946b) Induction effects in terrestrial magnetism Part II: The secular variation, *Phys. Rev.*, **70**, 202.

Elsasser, E. M. (1947) Induction effects in terrestrial magnetism Part III: Electric modes, *Phys. Rev.*, **72**, 821.

Elsasser, W. M. (1954) Dimensional values in magnetohydrodynamics, *Phys. Rev.*, **95**, 1.

Frank, L. A. (1967) On the extra-terrestrial ring current during geomagnetic storms, *J. geophys. Res.*, **72**, 3753.

Gardiner, R. B. and F. D. Stacey (1971) Electrical resistivity of the core, *Phys. Earth Planet. Int.*, **4**, 406.

Gibson, R. D. and P. H. Roberts (1969) The Bullard-Gellman dynamo, in: *The Application of Modern Physics to the Earth and Planetary Interiors* (ed. S. K. Runcorn) (Wiley, New York).

Herzenberg, A. (1958) Geomagnetic dynamos, *Phil. Trans. Roy. Soc.*, **A 250**, 543.

Hide, R. (1956) The hydrodynamics of the earth's core, in: *Physics and Chemistry of the Earth*, **1**, 94.

Hide, R. (1966) Free hydromagnetic oscillations of the earth's core and the theory of the geomagnetic secular variation, *Phil. Trans. Roy. Soc.*, **A 259**, 615.

Hide, R. (1967) Motions of the earth's core and mantle and variations of the main geomagnetic field, *Science*, **157**, 55.

Hide, R. (1969) Interaction between the earth's liquid core and solid mantle, *Nature*, **222**, 1055.

Hide, R. (1970) On the earth's core-mantle interface, *Q. J. Roy. Met. Soc.*, **96**, 579.

Hide, R. and K.-I. Horai (1968) On the topography of the core-mantle interface, *Phys. Earth Planet. Int.*, **1**, 305.

Hide, R. and S. R. C. Malin (1970) Novel correlations between global features of the earth's gravitational and magnetic fields, *Nature*, **225**, 605.

Hide, R. and S. R. C. Malin (1971) Novel correlations between global features of the earth's gravitational and magnetic fields: further statistical considerations, *Nature Phys. Sci.*, **230**, 63.

IAGA (1969) Commission 2 Working Group 4 (Analysis of the Geomagnetic Field), International Geomagnetic Reference Field 1965.0, *J. geophys. Res.*, **74**, 4407.

IAGA (1971) Bulletin No. 28, *The World Magnetic Survey 1957–1969* (ed. A. J. Zmuda) (IUGG Publ. Off., 39 ter, Rue Gay-Lussac, Paris (v)).

Jacobs, J. A. (1970) Geomagnetic Micropulsations, *Physics and Chemistry in Space, Vol. I* (Springer-Verlag).

Jacobs, J. A., T. Chan and M. Frazer (1972) Precession and the earth's magnetic field, *Nature Phys. Sci.*, **236**, 24.

Jain, A. and R. Evans (1972) Calculation of the electrical resistivity of liquid iron in the earth's core, *Nature Phys. Sci.*, **235**, 165.

Kahle, A. B., R. H. Ball and E. H. Vestine (1967a) Comparison of estimates of fluid motions at the surface of the earth's core for various epochs, *J. geophys. Res.*, **72**, 4917.

Kahle, A. B., E. H. Vestine and R. H. Ball (1967b) Estimated surface motions of the earth's core, *J. geophys. Res.*, **72**, 1095.

Kahle, A. B., R. H. Ball and J. C. Cain (1969) Prediction of geomagnetic secular change confirmed, *Nature*, **223**, 165.

Kawai, N. and S. Mochizuki (1971) Metallic states in the three 3d transition metal oxides Fe_2O_3, Cr_2O_3 and TiO_2 under static high pressures, *Phys. Letters*, **36A**, 54.

Khan, M. A. (1971) Correlations between the earth's gravitational and magnetic fields, *Nature Phys. Sci.*, **230**, 57.

Larmor, J. (1919) How could a rotating body such as the sun become a magnet? *Rept. Brit. Assoc.*, **159**.

Lilley, F. E. M. (1970) On kinematic dynamos, *Proc. Roy. Soc.*, **A 316**, 153.

Lindzen, R. S. (1967) Thermally driven diurnal tide in the atmosphere, *Q. J. Roy. Met. Soc.*, **93**, 18.

Lindzen, R. S. and S. Chapman (1969) Atmospheric tides, *Space Sci. Rev.*, **10**, 3.

Lowes, F. J. (1971) Significance of the correlation between spherical harmonic fields, *Nature Phys. Sci.*, **230**, 61.

Lowes, F. J. and I. Wilkinson (1963) Geomagnetic dynamo: a laboratory model, *Nature*, **198**, 1158.

Malkus, W. V. R. (1963) Precessional torques as the cause of geomagnetism, *J. geophys. Res.*, **68**, 2871.

Malkus, W. V. R. (1968) Precession of the earth as the cause of geomagnetism, *Science*, **160**, 259.

McDonald, K. L. and R. H. Gunst (1967) An analysis of the earth's magnetic field from 1835 to 1965, *IERT Tech. Rept. 46-IESI* (Boulder, Colo).

McDonald, K. L. and R. H. Gunst (1968) Recent trends in the earth's magnetic field, *J. geophys. Res.*, **73**, 2057.

Mead, G. D. (1970) International Geomagnetic Reference Field 1965·0 in dipole coordinates, *J. geophys. Res.*, **75**, 4372.

Munk, W., and R. Revelle (1952) On the geophysical interpretation of irregularities in the rotation of the earth, *Mon. Not. Roy. Astr. Soc. Geophys. Suppl.*, **6**, 331.

Nelson, J. H., L. Hurwitz and D. G. Knapp (1962) Magnetism of the earth, *Publ. 40-1*, *U.S. Dept. Com. Coast Geod. Surv.*

Orr, D. (1973) Magnetic pulsations within the magnetosphere: a review, *J. Atmos. Terr. Phys.*, **35**, 1.

Parker, E. N. (1970) The generation of magnetic fields in astro-physical bodies I. The dynamo equations, *Astrophys. J.*, **162**, 665.

Parker, E. N. (1971) The generation of magnetic fields in astrophysical bodies IV. The solar and terrestrial dynamos, *Astrophys. J.*, **164**, 491.

Phinney, R. A. and S. S. Alexander (1966) P wave diffraction theory and the structure of the core-mantle boundary, *J. geophys. Res.*, **71**, 5959.

Phinney, R. A. and S. S. Alexander (1969) The effect of a velocity gradient at the base of the mantle on diffracted P waves in the shadow, *J. geophys. Res.*, **74**, 4967.

Price, A. T. (1969) Daily variations of the geomagnetic field, *Space Sci. Rev.*, **9**, 151.

Rikitake, T. and Y. Hagiwara (1966) Non-steady state of a Herzenberg dynamo, *J. Geomag. Geoelec.*, **18**, 393.

Roberts, G. O. (1970) Spatially periodic dynamos, *Phil. Trans. Roy. Soc.*, **A 266**, 535.

Roberts, G. O. (1972) Dynamo action of fluid motions with two-dimensional periodicity, *Phil. Trans. Roy. Soc.*, **A 271**, 411.

Roberts, P. H. and A. M. Soward (1972) Magnetohydrodynamics of the earth's core, *Ann. Rev. Fluid Mech.*, **4**, 17.

Roberts, P. H. and M. Stix (1971) The turbulent dynamo: a translation of a series of papers by F. Krause, K.-H. Rädler and M. Steenbeck, *NCAR Tech. Note 1A-60* (Boulder, Colo.).

Rochester, M. G. (1970) Core-mantle interactions: geophysical and astronomical consequences, in: *Earthquake Displacement Fields and the Rotation of the Earth* (ed. L. Mansinha, D. E. Smylie and A. E. Beck) (D. Reidel Publ. Co., Holland).

Rostoker, G. (1972) Polar magnetic sub-storms, *Rev. Geophys. Space Phys.*, **10**, 157.

Runcorn, S. K. (1954) The earth's core, *Trans. Amer. Geophys. Union*, **35**, 49.

Saito, T. (1969) Geomagnetic pulsations, *Space Sci. Rev.*, **10**, 319.

Stacey, F. D. (1967) Electrical resistivity of the earth's core, *Earth Planet. Sci. Letters*, **3**, 204.

Sugiura, M. and J. P. Heppner (1965) The earth's magnetic field, in: *Introduction to Space Science* (ed. W. N. Hess) (Gordon and Breach).

Troitskaya, V. A. (1967) Micropulsations and the state of the magnetosphere, in: *Solar-Terrestrial Physics* (ed. J. W. King and W. S. Newman) (Acad. Press).

Urey, H. C. (1952) *The Planets: their Origin and Development* (Yale Univ. Press).

Vestine, E. H., L. Laporte, C. Cooper, I. Lange and W. C. Hendrix (1947) Description of the earth's main magnetic field and its secular change, 1905–1945, *Carnegie Inst. Wash. Publ. No. 578*.

Vestine, E. H., R. H. Ball and A. B. Kahle (1967) Nature of surface flow in the earth's central core, *J. geophys. Res.*, **72**, 4927.

Vestine, E. H. and A. B. Kahle (1968) The westward drift and geomagnetic secular change, *Geophys. J.*, **15**, 29.

Ziman, J. M. (1961) A theory of the electrical properties of liquid metals I: The monovalent metals, *Phil. Mag.*, **6**, 1013.

Ziman, J. M. (1971) The calculation of Bloch functions, *Solid State Phys.*, **26**, 1 (Acad. Press).

Zmuda, A. J. (1971) (ed.) World Magnetic Survey 1957–1969, *IAGA Bulletin No. 28*.

6

Palaeomagnetism

6.1 Introduction

Most physical studies of the earth yield information about the earth only as it is at present; to obtain information about its past history we must use hypothetical models. Some disciplines, however, do give information about the earth's past. The radioactive decay of certain elements has enabled some rocks to be dated and an estimate made of the age of the earth (see § 1.2). Stable isotopes can also provide valuable information. Because of its wide occurrence and the relatively large fractional mass difference between its abundant isotopes, isotopic abundance measurements of oxygen have proved particularly useful. The isotopic composition of oxygen in glacier ice varies appreciably with temperature, altitude and season of precipitation. Thus O^{18}/O^{16} ratios can be used as natural tracers within a glacier to determine the site and season of accumulation and the subsequent history of the snow and ice. Oxygen isotope ratios have also been used to estimate palaeo-temperatures: equilibrium reactions in general result in a small but definite separation of isotopic molecules, and the equilibrium constant related to this separation is a function of the temperature at which the equilibrium is attained. As a further example recent studies of the banding shown in corals and fossil shells have given an estimate of the number of days in a year at various times in the past, and thus an indication of how the rate of rotation of the earth has varied (see § 4.1). However, it is in the field of geomagnetism that during the last few years some of the most significant and exciting advances in geophysics have been made. At the time a rock is formed it often acquires a very stable, permanent magnetization. If this magnetization does not change throughout geologic time, it gives an estimate of the position of the ancient magnetic pole and of the strength of the ancient magnetic field. This has direct application to such problems as polar wandering, continental drift and seafloor spreading, and will be discussed in Chapter 8.

The classic early work in palaeomagnetism is that of Chevallier (1925) who showed that the remanent magnetizations of several lava flows on Mount Etna were parallel to the earth's magnetic field measured at nearby observatories at the time the flow erupted. Later Johnson *et al.* (1948) measured changes in the direction and intensity of magnetization of a continuous vertical sequence of varved clays in New England, and were able to trace the history of the earth's magnetic field as far back as 15 000 B.C. Their results indicated that during this time the field underwent fluctuations in direction about geographic north (Fig. 6.1). Other investigations have confirmed this, and it seems that for at least several million years the earth's main field has corresponded on the average to that which would be produced by an axial dipole situated at the centre; in addition, there has been a continuous secular variation about the mean direction with a period of from 500 to 1000 years (*cf* Fig. 5.4).

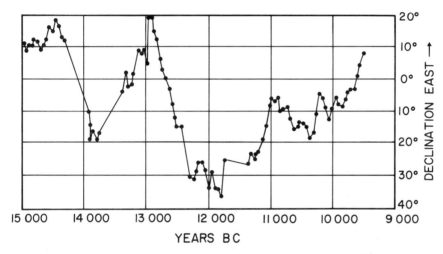

Fig. 6.1 Declination of the earth's magnetic field in New England, 15 000–9 000 B.C. (after Johnson *et al.* (1948))

Compared to the large number of ancient field directions which go back to Precambrian times, comparatively few accurate determinations have been made of the intensity of the geomagnetic field in the past. Smith (1967) has reviewed all palaeomagnetic intensity measurements, both from historic (archaeomagnetic) and geological specimens. The geomagnetic dipole moment does not remain constant within any given polarity but fluctuates, possibly with a period of the order of 10^4 years. Because of these fluctuations and non-dipole variations it is impossible to draw conclusions regarding the mean strength of the geomagnetic dipole at any particular time if only a few rock samples are measured. During the past 2000 years the dipole moment has decreased by about one-third of its peak value to its present value of 8.0×10^{25} emu. Prior to the last two thousand years it was increasing. Field intensities in the geologic past are given in Fig. 6.2, which shows that for at least the past 400 m yr the mean dipole moment was smaller than at present. However, in view of the scarcity of data before the Tertiary we cannot consider this increase with time as much more than a possibility.

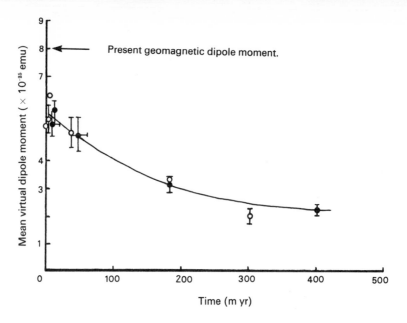

Fig. 6.2 Mean virtual dipole moments VDM plotted against time, i.e. equivalent dipole moments which would have produced the measured intensity at the magnetic palaeolatitudes of the samples. ● accurate values, ○ approximate values. Error bars in VDM represent standard errors of the means; error bars in time represent range of potassium–argon ages where appropriate (after Smith (1967))

6.2 The magnetization of rocks

Many minerals contain magnetic elements, particularly iron, but the majority of them have Curie temperatures less than 0°C and hence are of no direct interest in palaeomagnetic studies. In most rocks the remanence of importance to palaeomagnetism is carried by magnetite in igneous rocks and by hematite in sedimentary rocks. There are several different types of magnetism: diamagnetism, paramagnetism, ferromagnetism and anti-ferromagnetism. Diamagnetism and paramagnetism exist only in the presence of an external magnetic field; if the applied field is removed, the magnetism is lost.

In a typical substance there is no overall magnetism arising from the orbital motion of the electrons round the central nucleus. However, if it is placed in a magnetic field, a force is exerted on each of the orbital electrons tending to modify its orbit slightly. Since the resulting effect tends to oppose the applied field, the magnetization acquired in this way is negative (i.e. the susceptibility is negative) and is called *diamagnetism*. All substances possess diamagnetism but it is often obscured by other superimposed effects. Many common minerals such as quartz and feldspar are predominantly diamagnetic. The electrons also have a spin motion about their axes in addition to their orbital motion around the central nucleus. Each individual electron thus behaves like a small magnet. They are usually randomly oriented, but in the presence of a magnetic field they tend to line up in the direction of the field, giving an increase in the magnetization. This effect is called *paramagnetism*. In natural minerals, only a few important ions show significant paramagnetic properties, the commonest being Mn^{2+}, Fe^{3+} and Fe^{2+}. Since an applied magnetic field tends to orient the spins while thermal fluctuations tend to randomize them, paramagnetic effects are strongly dependent on temperature. (It can be shown that the paramagnetic susceptibility is inversely proportional to temperature, a fact first recognized by Curie.)

In palaeomagnetic studies we are more interested in ferromagnetic materials, i.e. those which exhibit a permanent (spontaneous) magnetization even in the absence of an external field. The phenomenon of *ferromagnetism* is due to the 'exchange' interaction between atoms. Some substances (e.g. Fe, Co, Ni) contain unpaired electrons which are magnetically coupled between neighbouring atoms. This interaction results in a strong spontaneous magnetization (i.e. without the application of an external field) and in the property of being able to retain the alignment imparted by an applied field after it has been removed. These properties are several orders of magnitude greater than those of diamagnetic atoms in the same substance. Below some critical temperature (the *Curie temperature*) the interaction dominates, but above this temperature, thermal disordering takes over and the behaviour is that of a simple paramagnetic material. Ferromagnetism is common in meteorites and lunar samples which contain large amounts of iron and nickel alloys.

A certain amount of energy (*magnetostatic energy*) is required just to maintain a permanent magnet. While the exchange energy tends to line up all the spins, the magnetostatic energy attempts to prevent this in order to minimize the total energy. The result is a balance of energy in which small zones, a few microns in size, are uniformly magnetized while adjacent zones may have their magnetization in some other direction. These zones are called *magnetic domains*, domain walls separating adjacent regions in which the spontaneous magnetization is in different directions. A sample of ferromagnetic material may thus have only a weak overall spontaneous magnetization, although in individual domains it may be quite large.

Within any crystal there are certain 'easy' directions of minimum energy along which the magnetic dipoles prefer to be aligned. However, in domain walls separating adjacent regions of uniform magnetization, some spins cannot be aligned in the preferred magnetocrystalline direction; these particular spins require extra energy to overcome the magnetocrystalline effect. In the absence of any external influences, domains form in such a manner as to reduce the total magnetostatic energy and wall energy to a minimum. When a body is placed in a magnetic field, the domain walls can move fairly easily allowing more of the grain to become magnetized in the direction of the applied field. Such magnetization is reversible when the applied field is small, and the domain walls move back again when the field is removed. If the field is increased, the domain walls are forced over small imperfections and impurities in the grain and cannot return to their original position when the field is removed. The process is then no longer reversible and a definite permanent magnetization is left in the body. If higher fields still are applied, all the spins line up in the direction of the applied field, overcoming both the magnetostatic and magnetocrystalline energies. At this point the body has spontaneous or saturation magnetization. Magnetization acquired in this way is referred to in palaeomagnetic studies as *isothermal remanent magnetism* (IRM) since no heating is involved. Other ways in which a rock can become magnetized are discussed later in this section.

The dominant minerals in rocks that preserve a permanent magnetization are not simple ferromagnetic substances. The internal ordering of the atomic spins can be quite complex with the result that both positive and negative exchange effects can exist in one material at the same time. If there are equal numbers of parallel and anti-parallel spins the material is called *anti-ferromagnetic*. Such a material behaves like a paramagnetic substance above a critical temperature (the *Néel temperature*) at which thermal disordering disrupts magnetic ordering. In some materials the number of ions in the parallel and antiparallel states is not equal with the result that it may be quite magnetic. Materials of this type are called *ferrimagnetic* or *ferrites*. Most naturally occurring magnetic minerals are either anti-ferromagnetic or ferrimagnetic. Ferrimagnetic minerals include magnetite, maghemite and some members of the ilmenite–hematite solid solution series; anti-ferromagnetic minerals include hematite, ilmenite, and ulvospinel. These last two minerals, however, have Curie or Néel temperatures well below room temperature.

We will now consider briefly the various ways in which rocks may become magnetized by natural processes.

(i) Thermoremanent magnetization (TRM) Igneous rocks cooling from above their Curie temperature in a magnetic field acquire a remanent magnetization, called thermoremanent magnetization (TRM). This magnetization is parallel to the applied field and for low field strengths is directly proportional to it. If a rock is cooled through various temperature intervals in the presence of a magnetic field, the TRM acquired in each interval is found to be independent of that acquired in each of the other intervals, a result known as the law of partial thermoremanent magnetism (PTRM). The sum of all the PTRM values is equal to the total TRM acquired by cooling from above the Curie temperature to room temperature (see Fig. 6.3).

(ii) Isothermal remanent magnetization (IRM) Magnetic materials may also acquire a remanent magnetization without heating if exposed to a magnetic field. This has already been discussed. In many natural materials, the IRM acquired in the earth's magnetic field is less than the TRM. This may not be true for magnetically soft materials however.

(iii) Viscous remanent magnetization (VRM) Viscous remanent magnetism (VRM) is essentially time-dependent IRM. If a magnetic material is exposed to a magnetic field it may slowly acquire a magnetization in the direction of that field. Rocks whose natural remanent magnetization (NRM) is subject to large viscous effects are unsuitable for palaeomagnetic studies.

166

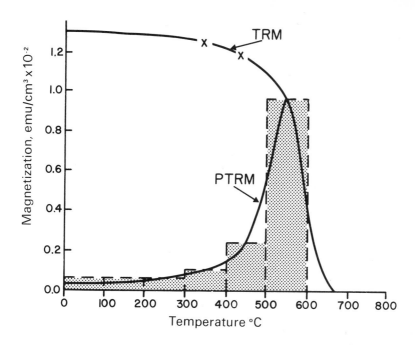

Fig. 6.3 The partial thermoremanence (PTRM) of the magnetization acquired in various temperature intervals. The sum of the individual PTRM's gives the TRM curve (after Strangway (1970).)

The rate at which VRM is acquired (or decays) depends on the temperature and particle size. For any one particle the relaxation time $\tau \propto \log v/T$ where v is the volume of the grain and T the temperature. Very fine ferromagnetic particles (less than domain size) have very short relaxation times even at normal temperatures and behave *superparamagnetically*. A small increase in volume can change a particle from superparamagnetic to single domain; a small decrease in temperature leads to the same effect. A material which contains many small particles will have some which are superparamagnetic and some which are single domain. As the temperature is lowered, more particles become single domain and are referred to as *blocked*. In natural substances such as rocks which contain particles of varying size, shape and composition there is a wide spectrum of relaxation times. The remanent magnetization observed under normal conditions is thus the resultant of all the blocked magnetizations.

(iv) Depositional remanent magnetization (DRM) Some sediments contain small magnetic particles which become aligned in the direction of the ambient magnetic field during settling. Fine-grained sediments which contain particles of magnetic material 10 μm or less in size tend to have a stable depositional remanent magnetization (DRM).

(v) Chemical remanent magnetization (CRM) Some sediments undergo post-depositional chemical changes in the iron minerals introducing a subsequent

magnetization. Such a process is referred to as chemical remanent magnetization (CRM). This may be the main cause of the magnetization of the well-known red sandstones.

More and more evidence (see, e.g. Strangway *et al.* (1968); Larson *et al.* (1969); Hargraves and Young (1969)) has been obtained which indicates that the stable component of NRM in rocks may reside only in the ultra fine single domain particles. These may constitute but a small fraction of the active magnetic materials, but, because of their uniform and stable magnetization, their contribution to NRM predominates in a rock that has been 'cleaned' by thermal or a.c. techniques. If this is generally true, the origin of the stable NRM in ultra fine particles can be attributed to the shape and isotropy of the individual single domain particles. This theory was originally suggested for non-interacting particles by Néel (1955) and was later modified for interacting particles by Dunlop (1968). In a later paper, Dunlop (1972) examined the behaviour of magnetite near the single domain threshold d_0. He found that particles larger than d_0 (570 ± 50 Å) but smaller than about 0.25 μm have size-dependent saturation remanences and coercive forces like those of multi-domain particles, but intense and stable TRM like that of single-domain particles. He suggested that the presence of magnetite grains in this size range could account for the essentially single-domain character of stable NRM in many volcanic and intrusive rocks.

Strangway *et al.* (1968) observed that oxidation-produced ilmenite lamellae in titanomagnetites can effectively produce the ultra fine domain particles necessary for stable NRM. This raises the possibility that the observed NRM consists of an original TRM (contemporaneous with the absolute age of the host material) together with a younger undesirable component of CRM of unknown age. Stable NRM has also been discovered recently in the so-called non-magnetic minerals, i.e. those that are diamagnetic, paramagnetic or anti-ferromagnetic. The NRM in these minerals has been shown to be due to extremely small (< 1 μm) ferrimagnetic regions caused by the precipitation of a second phase, or the presence of lattice imperfections. It is possible, however, that the NRM may be partly or wholly CRM.

6.3 Reversals of the earth's magnetic field

One of the most interesting results of palaeomagnetic studies is that many igneous rocks show a permanent magnetization approximately opposite in direction to the present field. Reverse magnetization was first discovered in 1906 by Brunhes in a lava from the Massif Central mountain range in France; since then examples have been found in almost every part of the world. About one-half of all rocks measured are found to be normally magnetized and one-half reversely. Dagley *et al.* (1967) recently carried out an extensive palaeomagnetic survey of Eastern Iceland sampling some 900 separate lava flows lying on top of each other. The direction of magnetization of more than 2000 samples representative of individual lava flows was determined covering a time interval of 20 m yr. At least 61 polarity zones or 60 complete changes of polarity were found giving an average rate of at least three inversions/m yr.

There is no *a priori* reason why the earth's field should have a particular polarity, and in the light of modern views on its origin there is no fundamental reason why its polarity should not change. It is easy to see that dynamos can produce a field in either direction. Equation (5.29) is linear and homogeneous in the field, and equation (5.32) inhomogeneous and quadratic. Thus if a given velocity field will support either a steady or a varying magnetic field, then it will also support the reversed field and the same forces will drive it. This, however, merely shows that the reverse field satisfies the equations, it does not prove that reversal will take place.

There have been many cases where reversely magnetized lava flows cross sedimentary layers. Where the sediments have been baked by the heat of the cooling lava flow, they are also found to be strongly magnetized in the same reverse direction as the flow. In fact, in about 95 per cent of all cases the direction of magnetization of the baked sediment was the same as that of the dike or lava which heated it, whether normal or reversed. It seems improbable that the adjacent rocks as well as the lavas themselves should possess a self reversal property, and such results seem difficult to interpret in any other way than by a reversal of the earth's field.

However, before such an explanation is accepted it must be asked whether there exist any physical or chemical processes whereby a material could acquire a magnetization opposite in direction to that of the ambient field. Néel (1951, 1955) suggested, on theoretical grounds, four possible mechanisms and within two years two of them had been verified, one by Nagata for a dacite pumice from Haruna in Japan, and one by Gorter for a synthetic substance in the laboratory.

The first and third of Néel's mechanisms only involve reversible physical changes while the second and fourth involve, in addition, irreversible physical and/or chemical changes. In his first mechanism, Néel imagined a crystalline substance with two sub-lattices A and B with the magnetic moments of all the magnetic atoms in lattice B oppositely directed to those of lattice A. If the spontaneous magnetizations of the two sets of atoms J_A and J_B vary differently with temperature, Néel suggested that the resultant magnetization of the whole, $J_A + J_B$, could reverse with change in temperature (see Fig. 6.4). Gorter and Schulkes two years later synthesized a range of substances with properties predicted by Néel, although no naturally occurring rock has been found which behaves in this manner.

In his third mechanism, Néel considered a substance containing a mixture of two different types of grains A and B, one with a high Curie point T_A and a low intensity of magnetization J_A and the other with a low Curie point T_B and a high

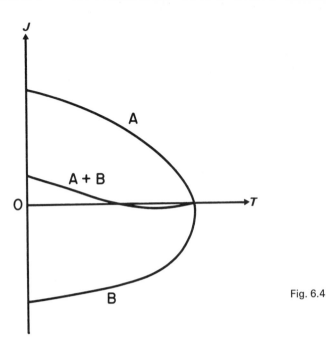

Fig. 6.4

magnetization J_B (see Fig. 6.5). When such a substance cools from a high temperature, substance A, because of its higher Curie point, becomes magnetized first in the direction of the ambient field. When the temperature falls below T_B, substance B becomes magnetic but will be subject to the dual influence of the ambient field and of the field due to the grains of substance A. Néel suggested that under suitable geometrical conditions, the resultant direction of magnetization of B could on the average be opposite to that of the ambient field. At room temperature the greater value of J_B causes the resultant magnetization of the whole

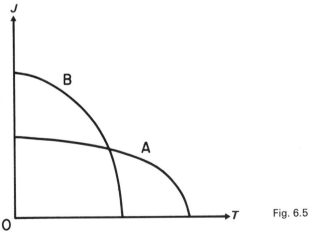

Fig. 6.5

to be in the opposite direction to the ambient field. In 1951 Nagata found a dacite pumice from Haruna in Japan which was reversely magnetized in the field and which behaved in the laboratory in the way which Néel had predicted. However, the great majority of igneous rocks which show reversal in the field do not show this property in the laboratory. Néel's second and fourth mechanisms depend on the possibility of subsequent demagnetization by physical or chemical changes. Thus reverse magnetization might be possible later in time, even though initially the field was normal.

Reversals occurred during the Precambrian and have been observed in all subsequent periods. There is no evidence that periods of either polarity are systematically of longer or shorter duration. Dagley et al. (1967) in their monumental work on Icelandic lavas found 73 lava flows whose directions of magnetization were anomalous and of these at least 55 were unambiguously intermediate, being magnetized neither in the normal nor the opposite direction. This was about 5 per cent of the lava flows examined, and indicates that reversal does not simply involve a change in the sense of the dipole. These intermediate directions could be caused either by a continuous change in the direction of the dipole axis (with possibly also a change in strength) or by a reduction in strength of the dipole without change in orientation, thus allowing the non-dipole part of the field to make a proportionally larger contribution and so give the anomalous directions.

Several other workers have succeeded in obtaining a record of the geomagnetic field during a polarity transition (see, e.g., Watkins (1967, 1969)). It appears that during a reversal the intensity of the field first decreases by a factor of 3 or 4 for several thousand years while maintaining its direction. The magnetic vector then usually executes several swings of about 30°. The vector then moves along an irregular path to the opposite polarity direction, the intensity still being reduced, rising to its normal value later. It is not certain whether the field is dipolar during a transition. There do not seem to be any precursors of a reversal or any indication later that a reversal has occurred. A detailed record of a field reversal has been described recently by Dunn et al. (1971).

During a period of about 60 m yr within the upper Carboniferous and Permian (about 235–290 m yr ago) the polarity of the earth's field appears to have been almost always reversed. Until quite recently no normal intervals at all were known within this Kiaman reversed period. McElhinny (1969) and Burek (1970) have now both reported a normal event at about 280 m yr and Creer et al. (1971) another at about 263 m yr. If the field reversal hypothesis is incorrect, it follows that mineral assemblages necessary for self-reversal were abundant in Carboniferous and Triassic rocks (both these periods have many reversals), but were all but missing in all Permian rocks. Such a conclusion is very difficult to believe; it is far more plausible to assume that the field itself alternated very rarely during the Permian. Another interval of fixed polarity (normal) has been postulated in the Cretaceous by a number of workers. More recently Irving and Couillard (1973) have examined all available continental data and concluded that between about 109 and 82 m yr ago there is, in the land record, no evidence of an authentic reversal of the earth's magnetic field. This agrees well with the oceanic record which shows normal polarity between 111·5 and 84·5 m yr (Larson and Pitman, 1972).

To prove that a reversed rock sample has become magnetized by a reversal of the earth's field, it is necessary to show that it cannot have been reversed by any physicochemical process. This is almost impossible to do since physical changes may have occurred since the initial magnetization or may occur during laboratory tests. More positive results can only come from the correlation of data from rocks of varying types at different sites and by statistical analyses of the relation between the polarity and other chemical and physical properties of the rock sample. A number of workers have reported chemical differences between normally and reversely magnetized lava sequences from various parts of the world (see, e.g. Wilson and Watkins (1967)). Reversely magnetized lavas appear to be more highly oxidized than normal ones—no differences in the distribution of other elements involved in the magnetic minerals have been observed. The significance of this correlation is not yet clear—it is very difficult to imagine what physical connection can exist between the polarity of the earth's magnetic field and the state of oxidation of lavas which became magnetized in that field.

Watkins and Haggerty (1968) later examined the magnetic polarity and oxidation state of over 550 specimens in single lavas and dikes in Eastern Iceland. They found no correlation between polarity and the oxidation state in the dikes, although there was a strong correlation between the percentage of reversed polarity and higher oxidation in the lavas. In a later paper Watkins and Ade-Hall (1970) found no correlation between opaque petrological parameters and polarity in specimens of Canary Island lava flows of Miocene to Pliocene age. The proposed correlation thus does not seem to hold on a worldwide scale and a satisfactory explanation of those positive correlations that have been found has not yet been given. The whole problem has recently been reviewed by Smith (1971).

If the origin of reversals is one of the instantaneous self-reversal mechanisms (such as that of the Haruna dacite), then normally and reversely magnetized rocks should be randomly distributed throughout a group of rocks of different ages. If reversals are due to one of the time-dependent self-reversal mechanisms, reversals should be increasingly abundant in older rocks. If, on the other hand, reversals are due to geomagnetic field reversals, normal and reversely magnetized groups of rocks should be exactly the same age over the entire earth; and, unless it so happened that the earth's field suffered more reversals in the past, the proportion of reversed magnetizations should not be greater among older rocks. Although it has been established that self-reversal does occur in some rocks, the stratigraphic distribution of normally and reversely magnetized rocks overwhelmingly supports field reversal.

Four major normal and reversed sequences have been found during the past 3·6 m yr. These major groupings have been called geomagnetic polarity epochs, and have been named by Cox et al. (1964) after people who have made significant contributions to geomagnetism. Superimposed on these polarity epochs are brief fluctuations in magnetic polarity with a duration that is an order of magnitude shorter. These have been called polarity events and have been named after the localities where they were first recognized (see Fig. 6.6). The duration of polarity events is estimated to vary from 0·70 to 0·16 m yr, and the best estimate of the time required for the earth's field to undergo a complete change in polarity is 4600 yr. However, Cox has later questioned the reality of this division into epochs and events. The geomagnetic reversal time scale has now been extended back to about 75 m yr from studies of the marine magnetic anomaly pattern (see § 8.4).

There has been much speculation on the possibility of short intervals of reversed polarity within the present Brunhes epoch of normal polarity which has lasted for the past 0·70 m yr. The first report of a possible reversal (the Laschamp event) was given by Bonhommet and Babkine (1967) who found reversed mag-

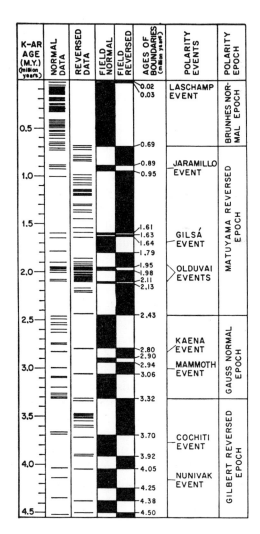

Fig. 6.6 Time scale for geomagnetic reversals. Each short horizontal line shows the age as determined by potassium–argon dating and the magnetic polarity (normal or reversed) of one volcanic cooling unit. Normal polarity intervals are shown by the solid portions of the 'field normal' column, and reversed polarity intervals by the solid portions of the 'field reversal' column. The duration of events is based in part on palaeomagnetic data from sediments and profiles (after Cox (1969a))

netizations in two different formations of the Chaine des Puys (Auvergne, France) volcanic chain of Quaternary rocks. Age determinations, by the C^{14} method, of the lava flows of the Puys de Laschamp place the event between 7000–9000 yr ago while K–Ar ages indicate an upper limit of 20 000 yr for the end of the event. Denham and Cox (1971) have since made palaeomagnetic measurements of late Pleistocene sediments from Mono Lake, California. They found no evidence for a reversal in polarity during the time interval 30 400–13 300 yr ago, the ages being determined by C^{14} dating. If the Laschamp event occurred during this time, its duration can have been no longer than 1700 yr, which is the largest sampling gap in the data. The data do indicate, however, an unusually large excursion of the field (peak-to-peak amplitude 25°) 24 000 yr ago which the authors attribute to secular variation due to local eastward drift of the non-dipole field. This is in contrast to the predominantly westward drift that has characterized the field during the past few centuries. The authors conclude that either the Laschamp event is simply a large excursion of the field lasting less than 1700 yr or else that it occurred during the interval 13 300 to 8700 yr ago lasting less than 4600 yr. Mörner et al. (1971) have reported the discovery of a reversely magnetized sediment section in an oriented core from Gothenburg, south-west Sweden, which is about 12 500 yr old. The date is well documented and the remanent magnetization shown to be stable so that reversely magnetized material of that age does exist. It has not been established, however, that the event does represent a genuine field reversal. The authors tentatively suggest that this reversal is related to the Laschamp event but more work needs to be done to confirm the reversal in other cores from other areas.

Bucha (1970) and Barbetti and McElhinny (1972) have found evidence for a large excursion of the geomagnetic field about 30 000 yr ago. Bucha's analysis was carried out on Pleistocene sediments from Czechoslovakia, and Barbetti and McElhinny's on aboriginal fireplaces at Lake Mungo, Australia. Bucha's studies also confirm the conclusions of Denham and Cox (1971) that no reversals or excursions lasting longer than 1700 yr occurred between about 30 000 and 13 000 yr ago. The Laschamp event and that about 30 000 yr ago thus appear to be unrelated. Barbetti and McElhinny suggest that unusually large excursions of the geomagnetic field ($\sim 120°$ or more away from the axial dipole) are unlikely to be due to secular variation but are a manifestation of some fundamental property of the dynamo.

Finally Smith and Foster (1969) have examined the magnetic record of seven cores of deep-sea sediments and established the existence of a short period of reversed polarity (the Blake event) in the upper part of the Brunhes epoch. The reversed zone in the cores correlates well with palaeontological boundaries and is estimated to have existed between 108 000 and 114 000 yr ago (± 10 per cent).

Hide (1966) has suggested that changes in the radial velocity of the fluid motions in the earth's core might in some cases be impressed from outside. Horizontal temperature variations of only a few degrees and topographical features, 'bumps' only a few kilometres high at the core-mantle interface might affect core motions, perhaps causing reversal (see also § 5.4). Gradual changes in the radius of the core and in the strength of the mechanism that drives core motions might also produce occasional reversals. Hide thus suggested that there are two types of reversals: 'forced' reversals due to changes impressed from outside the core, and 'free' reversals that would arise even in the absence of impressed changes. Each type of reversal would be characterized by its own time-scale. Major geological events are associated with large-scale motions in the mantle. If these motions penetrate to a sufficient depth to produce horizontal variations in the physical conditions that prevail at the core-mantle boundary, then 'forced' reversals should be strongly correlated with other worldwide geological phenomena. 'Free' reversals, however, should show no such correlations, being determined by random processes in the fluid core.

Opdyke *et al.* (1966) found a polarity record in deep-sea sediments going back to the Gilbert epoch (3·6 m yr ago) in which the pattern of reversals was remarkably similar to that found in igneous rocks on land (see Fig. 6.7). Thus polarity studies can provide a method for determining rates of sedimentation and for

Fig. 6.7 Correlation of magnetic stratigraphy in seven cores from the Antarctic. Minus signs indicate normally magnetized specimens; plus signs, reversely magnetized. Greek letters denote faunal zones. Inset: source of cores (after Opdyke *et al.* (1966))

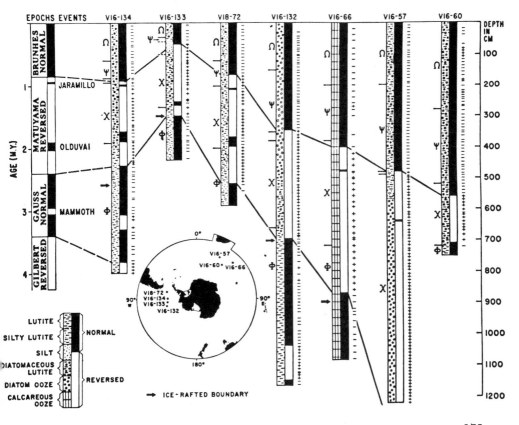

establishing world-wide correlations among different deep-sea sediments. An excellent review of the palaeomagnetic studies of deep-sea cores has recently been given by Opdyke (1972).

No two substances could be more different or have more different histories than the lavas of California and the pelagic sediments of the Pacific. The lavas were poured out, hot and molten, by volcanos and magnetized by cooling in the earth's field (TRM); the ocean sediments accumulated grain by grain by slow sedimentation and by chemical deposition in the cold depths of the ocean (DRM). If these two materials yield the same pattern of reversals then it must be the result of an external influence working on both and not due to a recurrent synchronous change in the two materials. The evidence is compelling that reversals of the earth's field are the cause of the reversals of magnetization.

Cox (1968) has developed a probabilistic model for reversals in which it is assumed that polarity changes occur as the result of an interaction between steady oscillations and random processes. The steady oscillator is the dipole component of the field and the random variations are the components of the non-dipole field. The random variations serve as a triggering mechanism that produces a reversal whenever the ratio of the non-dipole to dipole fields exceeds a critical value (Fig. 6.8).

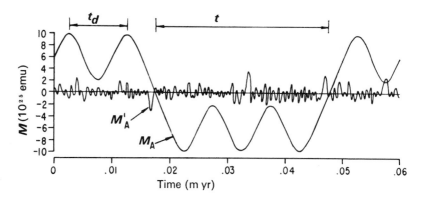

Fig. 6.8   Probabilistic model for reversals. t_d is the period of the dipole field and t the length of a polarity interval. A polarity change occurs whenever the quantity $(M_A + M'_A)$ changes sign, where M_A is the axial moment of the dipole field and M'_A is a measure of the non-dipole field (after Cox (1968))

For the interval $3.5 < t < 10$ m yr, which is not adequately covered by potassium argon dating, information about reversals may be obtained from the magnetic anomaly pattern over the mid-oceanic ridges, as interpreted by the theory of sea floor spreading (see § 8.4). The statistical properties of the polarity intervals from $3.32-10.6$ m yr are remarkably similar to those for the interval $0-3.3$ m yr (see Table 6.1). For both, the total length of time the field was normal is about equal to that when the field was reversed. Moreover, the mean lengths of polarity

Range of ages, m yr	0–3·32	3·32–10·6	0–10·6
Normal polarity intervals			
Time field was normal, m yr	1·66	3·90	5·56
% of total time	50	53	52
Number of normal intervals	8	17	25
Mean length of normal intervals, m yr	0·21	0·23	0·22
Reversed polarity intervals			
Time field was reversed, m yr	1·66	3·46	5·12
% of total time	50	47	48
Number of reversed intervals	7	17	24
Mean length of reversed intervals, m yr	0·24	0·20	0·21
All polarity intervals			
Number of intervals	15	34	49
Mean length, m yr	0·22	0·22	0·22

(After Cox (1968))

intervals are nearly the same. From the statistical homogeneity of these two sets of polarity intervals, Cox concluded that the statistical properties of the field have been constant for the past 10·6 m yr and that the two sets of data can be combined to obtain an estimate of the probability p of a polarity change during one oscillation of the dipole field.

Figure 6.9 shows the cumulative distribution obtained by Cox for all the polarity intervals together with curves of a theoretical cumulative distribution,

$$P(t \leqslant t_c) = \sum_{i=1}^{y} p(1-p)^{i-1} \tag{6.1}$$

where t_d is the period of the dipole field, y an integer and $t_c = yt_d$. Almost all the

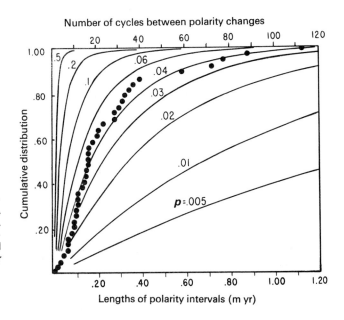

Fig. 6.9 Cumulative distribution of polarity intervals. Circles are observational data. Curves are theoretical cumulative distributions obtained from equation (6.1) (after Cox (1968))

observational data lie between the curves for $p=0.03$ and $p=0.06$. In the above analysis it is assumed that polarity reversals are related to periodic changes in the dipole moment. If these changes are not periodic the basic mechanism of Cox for reversals may still be valid, but the frequency distribution based on the period of the dipole field would not be correct. A distribution that does not depend on the period of the dipole field is the Poisson approximation to the equation for Bernoulli trials:

$$P(t) = \lambda \exp(-\lambda t) \qquad (6.2)$$

where the parameter $\lambda \ (=p/t_d)$ characterizes the observed variations in the length of polarity intervals. The observational data fit the two distributions given by equations (6·1) and (6·2) equally well.

For the interval $0.02 < t < 3.32$ m yr ago, Cox (1969a) obtained a good fit to the observed frequency of polarity intervals from equation (6.1) when p has the value 0·055 (see Fig. 6.10). Reversal time-scales going back 75 m yr have been obtained from marine magnetic profiles on the assumption that the rate of sea floor spreading was constant (Heirtzler et al. (1968)). The best fit of equation (6.1) to the data of Vine (1968) for the past 10·6 m yr is obtained with $p=0.043$.

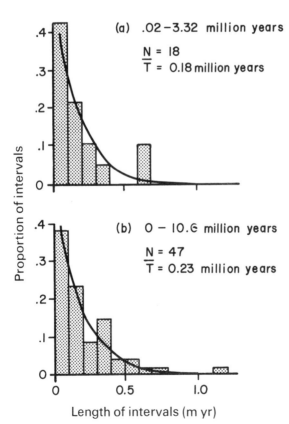

Fig. 6.10 Histograms of lengths of polarity intervals (a) on the reversal time scale of Fig. 6.6, based on ages obtained by potassium–argon dating and (b) from the reversal time scale of Vine (1968) based on the Eltanin 19 magnetic profile. Solid curves, distribution function of equation (6.1) for $p=$ 0·05. N is the number of intervals (after Cox (1969a))

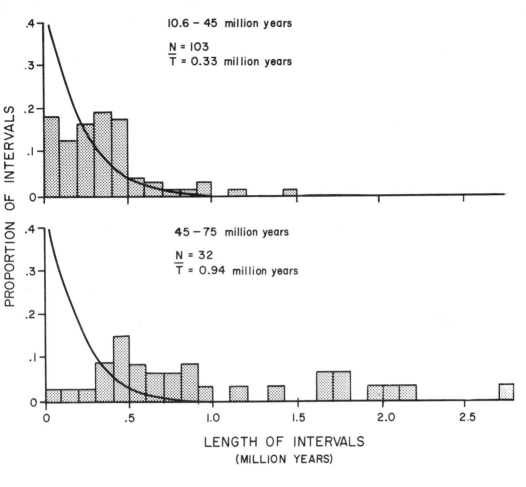

Fig. 6.11 Histograms of lengths of polarity intervals. Both histograms are from the time scale for intervals of Heirtzler *et al.* (1968) based on magnetic profiles. Solid curves, distribution function of equation (6.1) for $p=0.05$. N is the number of intervals (after Cox (1969a))

Cox found that the apparent average duration of polarity intervals was greater during the time $10.6 < t < 45$ m yr ago than during the past 10.6 m yr, while during the time $45 < t < 75$ m yr the average length was longer still (Fig. 6.11). Part of this difference may be due to variations in the spreading rate. However, this would account only for an apparent change in the probability p and not for the observed change in the shape of the distribution. The latter change may, however, be more apparent than real if, as is more than possible, a small number of additional short events occurred which have not yet been detected. As we go farther back in time an average length of 10^7 yr has been reported for polarity events in the early Palaeozoic. It appears that the average length of polarity intervals (and hence the value of p) has changed during the earth's history, reflecting changes in the physical conditions in the earth's core.

Nagata (1969) claims that the two basic assumptions in the theory of Cox, namely that the main field can be represented by a steady dipole oscillator, and that a trigger effect takes place whenever the non-dipole field becomes sufficiently large relative to that of the dipole, have not yet been reasonably substantiated. Nagata offers an alternative interpretation of the distribution function of reversals, based on the hypothesis that the main dipole field is steadily maintained only as long as the convection pattern in the core is asymmetric, as is the case in the Bullard-Gellman-Lilley dynamo, but collapses when the convection pattern becomes symmetric, as in the Bullard-Gellman model. However, Nagata obtains the same mathematical expression as Cox with $\lambda = \frac{1}{2}t_0$ where t_0 is the average lifetime for the polarity of a steady dipole. The average length of polarity intervals is $2t_0$.

Parker (1969) has also developed a statistical model for reversals. He showed that a fluctuation in the distribution of the cyclonic convective cells in the core can produce an abrupt reversal of the geomagnetic field. The simplest fluctuation leading to a reversal is a general absence of cyclones below about latitude $25°$ for a time comparable to the lifetime ($\sim 10^3$ yr) of an individual cell.

In both Cox's and Parker's models, cyclonic convection cells in the core produce reversals by a two-step mechanism. At any instant they are randomly distributed through the core; reversals occur when, through random processes, they arrive at certain critical configurations. However, there is a fundamental difference between the models. In that of Parker (and Nagata) the occurrence of a reversal depends only on the spatial distribution of the cyclones, and not on the intensity of the dipole field. In the model of Cox the occurrence of reversals depends upon both the distribution of the cyclones and on the field strength of the cyclone disturbances (i.e. the non-dipole field) relative to the dipole field. Different conclusions have been drawn from the statistical models of Cox and Parker on the frequency of occurrence of short polarity intervals. However, Cox (1970) has shown that both models are consistent with the same probability function given by equation (6.2).

Kono (1971, 1972) has shown that Cox's model is not compatible with palaeo-intensity data for the last 10^7 yr and discusses various mathematical models which satisfy the statistical properties of field variations during this time interval. In Cox's model the geomagnetic field is represented in the core by a large geocentric dipole and several off-centred dipoles; in Kono's final model, there are ten dipoles in the core with equal moments and directions either parallel or anti-parallel to the rotation axis (see Fig. 6.12 for a schematic representation of Cox's and Kono's models). Inversion of each dipole takes place in a stochastic manner leading to variations in the polarity of the overall field. Kono stresses that his model may not be realistic physically, and may not be unique.

Possible biological effects of reversals of the earth's magnetic field were mentioned in § 1.6. Uffen (1963) pointed out that if the magnetic field of the earth vanishes or is greatly reduced during a reversal, the earth would lose some of its magnetic shielding against cosmic rays, and as a consequence of the increase in radiation, mutation rates should increase. Uffen argued on palaeontological grounds that rates of evolution were exceptionally high at times when the earth's

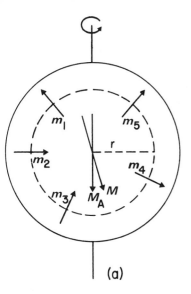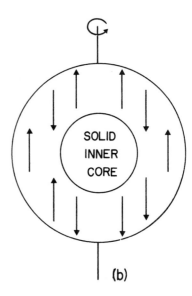

Fig. 6.12 Schematic representation of (a) the model of Cox (1968) and (b) the mathematical model of Kono (1972). The arrows indicate dipoles distributed in the core (after Kono (1972))

magnetic field was undergoing many changes in polarity. The estimated time interval of $\sim 10^4$ yr when the earth's dipole field is greatly reduced during a reversal is almost instantaneous on the geological time-scale but is very many generations for most living organisms. There are, of course, other causes of mutation, but the exposure to increased radiation would be sudden on the geologic time-scale and could contribute to both heritable variation and natural selection. The most striking unexplained fact from palaeontology is that biological evolution has progressed in bursts. Quite adequate explanations have been given for the extinction of whole species by climatic or other environmental changes but the sudden appearance of new forms of life on which the selection pressures of the time would act has not been adequately explained.

A number of investigations (see e.g. Harrison and Funnell (1964); Hays and Opdyke (1967); Watkins and Goodell (1967)) have shown that the extinction of certain species of marine organisms occurred at about the time of the most recent reversal of the earth's magnetic field. However, the reality of any correlation is, in principle, statistical, and there is just not sufficient data. Moreover, as a number of authors (e.g. Waddington (1967); Black (1967)) have pointed out, at the currently observed cosmic ray and solar particle intensities, the additional dosages produced at sea level during a period of complete removal of the geomagnetic field would be negligible, and even complete dumping of the energetic particles in the Van Allen radiation belts would not give rise to the necessary increased dosages. A further possibility is that the solar wind might produce ozone in the high atmosphere which would absorb radiation and produce large changes in climate. Black (1967), however, has shown that any ozone produced in this way is but a small fraction of that produced by ultraviolet light.

The only possibility of a causal connection between magnetic reversals and the extinction of certain species is that there is a climatic change at the time of a reversal. There is some evidence (Harrison (1968)) that the magnetic field of the earth has some control over the temperature of the upper atmosphere; also removal of the earth's magnetic field would cause large increases of ionization at certain levels in the upper atmosphere. What climatic changes, if any, would be brought about by such effects is not known. Wollin *et al.* (1971) found a correlation between variations in magnetic intensity, inclination and climatic changes in deep-sea sediment cores over the last 470 000 yr, and tentatively concluded that magnetism may modulate climate. However, a causal connection between magnetic reversals and climatic fluctuations seems unlikely since the time-scale of climatic variations seems to be a good deal shorter than the interval between reversals.

Glass and Heezen (1967) pointed out that the great field of tektites covering Australia, Indonesia and a large part of the Indian Ocean fell 700 000 yr ago at about the time of the last magnetic reversal (see also § 1.6). They suggested that the fall of the body from which the tektites were formed killed the now extinct radiolaria and gave a jolt to the earth, disturbing the motions in the core and causing the dynamo to reverse. But as Bullard (1968) pointed out, it is 'difficult to believe that the fall of a large meteorite could selectively kill certain species of radiolaria all over the world and yet spare the kangaroos near its point of fall'.

6.4 Statistical analysis of palaeomagnetic data

It is convenient in palaeomagnetic studies to represent the data in terms of that geocentric dipole that would produce the measured field direction. Such poles are called virtual geomagnetic poles (VGP). If (θ, ϕ) are the latitude and longitude of the observatory or sampling site S where the declination and inclination of the magnetic field are D, I, respectively, the latitude θ' and longitude ϕ' of the virtual geomagnetic pole V may be found from the following equations

$$\cot p = \tfrac{1}{2} \tan I \tag{6.3}$$

$$\sin \theta' = \sin \theta \cos p + \cos \theta \sin p \cos D \tag{6.4}$$

$$\sin (\phi - \phi') = \frac{\sin p \sin D}{\cos \theta'} \tag{6.5}$$

p is the magnetic co-latitude, and equation (6.3) is the same equation as (5.7). Equations (6.4) and (6.5), which give θ' and ϕ', follow directly from the spherical triangle PSV (see Fig. 6.13).

If non-dipole components are present, the VGP's calculated from field directions at different sites will not, in general, coincide; their scatter may be taken as a measure of the departure of the observed field from a dipole field.

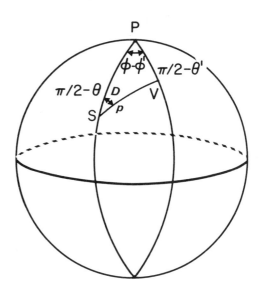

Fig. 6.13

Palaeomagnetic data consisting of a set of directions of magnetization, or of the corresponding set of VGP's, may be expressed as a set of coterminal unit vectors or as a set of points on a unit sphere. A statistical analysis of such a set of data has been developed by Fisher (1953). The set of data is assumed to be randomly drawn from a population with the vectors distributed with azimuthal symmetry about their mean direction. Further, the density of the vectors decreases with increasing angular displacement ψ from the mean direction according to the probability density function.

$$\mathscr{P} = \frac{\kappa}{4\pi \sinh \kappa} \exp\left(\kappa \cos \psi\right) \tag{6.6}$$

This function is analogous to a symmetrical Gaussian distribution on a plane. The quantity κ is called the precision parameter and is a measure of the tightness of the group of vectors about their mean direction. Large values of κ indicate tight grouping while $\kappa = 0$ corresponds to a uniform distribution of points over the unit sphere. The probability density function $\mathscr{P}$ has the following meaning: the proportion of the total points expected in a small area dS on the unit sphere at an angular distance ψ from the mean direction is $\mathscr{P} \, dS$. The quantity $\kappa/4\pi \sinh \kappa$ is a normalizing factor to make the integral of $\mathscr{P}$ over the whole sphere equal to unity.

Fisher also showed that the direction of the vector sum of N unit vectors in a sample is the best estimate of the mean direction of the population, and that the best estimate k of the precision parameter κ is given, for $\kappa > 3$, by

$$k = \frac{N-1}{N-R} \tag{6.7}$$

where R is the length of the vector sum of the N unit vectors. At a probability

183

level of $(1-P)$, the true mean direction of the population lies within a circular cone about the resultant vector R with semi-vertical angle $\alpha_{(1-P)}$ given by

$$\cos \alpha_{(1-P)} = 1 - \frac{N-R}{R}\left\{\left(\frac{1}{P}\right)^{1/(N-1)} - 1\right\} \tag{6.8}$$

In palaeomagnetic analyses, P is usually taken as 0·05, i.e. there is one chance in 20 that the true mean direction of the population lies outside the circle of confidence specified by α_{95} and the direction of R. For small values of α,

$$\alpha_{95} \simeq \frac{140°}{\sqrt{(kN)}} \tag{6.9}$$

A more detailed discussion of Fisher's statistical method with particular reference to its application to palaeomagnetic data has been given by Watson (1956), Watson and Irving (1957) and Cox and Doell (1960). Cox (1969b) has given a method for calculating confidence limits for the precision parameter κ.

Equations (6.3), (6.4) and (6.5) will map the circle of confidence about the field direction into a closed curve around the virtual geomagnetic pole which will be an oval rather than a circle. The ellipticity of the oval of confidence depends only on the inclination of the field. For vertical inclinations a circle of confidence of radius α about the mean field direction maps into another circle about the pole with radius 2α: for zero inclination the semi-axes of the oval are $1/2\alpha$ and α. If α_{95} is estimated, then there is a 95 per cent probability that the VGP lies within the confidence oval.

6.5 Polar wandering and continental drift

If it is assumed that the geomagnetic field at the earth's surface averaged over several thousands of years (to eliminate the superficial secular variation) can be represented by a geocentric dipole with its axis along the axis of rotation, it is possible, by measuring the present direction of magnetization of a suite of rocks, to deduce the position of the earth's rotational axis relative to the location of the rocks at the time when they were laid down. The measured declination will give the azimuth of the land mass at the time the rock became magnetized, and the measured inclination will give the geographical latitude (see equation (5.7)). Opdyke and Henry (1969) have studied the inclinations of cores from all the world's oceans in an attempt to test this hypothesis. Inclination versus latitude plots were constructed and the best fitting geocentric dipole calculated. This dipole was found to be at 89°N, 211°E, which is not significantly different from

184

the pole of rotation. They concluded that for the past 2 m yr at any rate the earth's field has closely approximated a geocentric axial dipole.

If these ancient pole positions are plotted in chronological order on the surface of the earth for different areas we obtain the results shown schematically in Fig. 6.14. What is the meaning of these 'polar wandering paths'? If it is agreed that the magnetic field has always been a dipole field, then it is clear that the magnetic pole could not have followed several paths at the same time. The simplest explanation now generally held is that the continents must all have moved

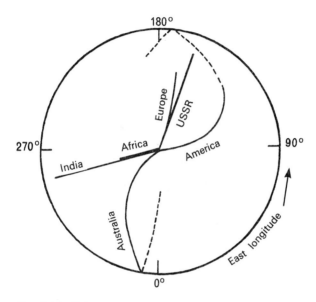

Fig. 6.14 Schematic polar movement paths for several continental areas (after Wilson (1965))

somewhat independently of each other, except for Europe and the USSR which have similar polar wandering paths. Figure 6.15 shows in more detail the palaeo-magnetic results for Africa, South America and Australia during the Palaeozoic. Figure 6.15(a) shows a comparison of the results from South America and Africa —the lower Palaeozoic portions of the polar wandering curves coincide quite remarkably when Africa and South America are brought together as shown. Figure 6.15(b) shows a comparison between the African and Australian curves— Africa and Australia have been brought into the relative positions shown which are required to match the upper Palaeozoic curves in each case. Figure 6.15(c) is a combination of the other two. When the polar wandering curves for different continents fail to match, it is an indication that relative movement has taken place between the continents and the time when this relative movement took place can be estimated. These questions will be discussed in more detail in Chapter 8 in the framework of the newer concepts of sea floor spreading and plate tectonics.

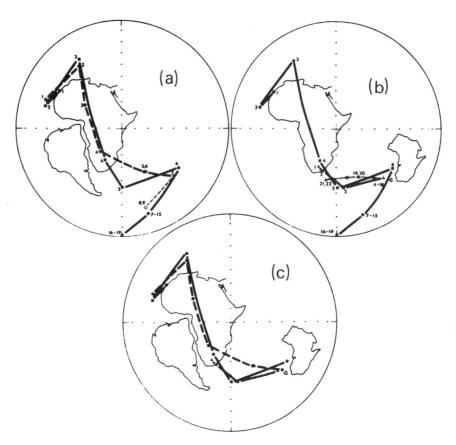

Fig. 6.15 Comparison of Phanerozoic data for Africa (squares), South America (circles) and Australia (triangles). Africa is in its present coordinates and open symbols represent the present South Pole. (a) South America is moved adjacent to Africa to make the Lower Paleozoic portions of their curves coincide; (b) Australia is moved so that its Upper Paleozoic curve coincides with the African one. (c) A combination of (a) and (b) showing the fit of the Paleozoic curves for the three continents (after McElhinny (1967))

References

Barbetti, M. and M. McElhinny (1972) Evidence of a geomagnetic excursion 30,000 yr B.P., *Nature*, **239**, 327.
Black, D. I. (1967) Cosmic ray effects and faunal extinctions at geomagnetic field reversals, *Earth Planet. Sci. Letters*, **3**, 225.

Bonhommet, N. and J. Babkine (1967) Sur la présence d'aimantations inversées dans la Châine des Puys, *Comp. Rend. Acad. Sci.*, **264**, 92.

Brunhes, B. (1906) Recherches sur le direction d'aimantation des roches volcaniques, *J. Phys.*, **5**, 705.

Bucha, V. (1970) Geomagnetic reversals in Quaternary revealed from a palaeomagnetic investigation of sedimentary rocks, *J. Geomag. Geoelect.*, **22**, 253.

Bullard, E. C. (1968) Reversals of the earth's magnetic field, *Phil. Trans. Roy. Soc.*, **A 263**, 481.

Burek, P. J. (1970) Magnetic reversals; their application to stratigraphical problems, *Bull. Amer. Assoc. petrol. Geol.*, **54**, 1120.

Chevallier (1925) L'aimantation des laves de l'Etna et l'orientation du champ terrestre en Sicile du XIIe au XVIIe siècle, *Ann. Phys.*, **4**, 5.

Cox, A. (1968) Lengths of geomagnetic polarity intervals, *J. geophys. Res.*, **73**, 3247.

Cox, A. (1969a) Geomagnetic reversals, *Science*, **163**, 237.

Cox, A. (1969b) Confidence limits for the precision parameter κ, *Geophys. J.*, **17**, 545.

Cox, A. (1970) Reconciliation of statistical models for reversals, *J. geophys. Res.*, **75**, 7501.

Cox, A., R. R. Doell and G. B. Dalrymple (1964) Reversals of the earth's magnetic field, *Science*, **144**, 1537.

Cox, A. and R. R. Doell (1960) Review of palaeomagnetism, *Bull. geol. Soc. Amer.*, **71**, 645.

Creer, K. M., J. G. Mitchell and D. A. Valencio (1971) Evidence for a normal geomagnetic field polarity event at 263 ± 5 m yr B.P. within the late Palaeozoic reversed interval, *Nature Phys. Sci.*, **233**, 87.

Dagley, P., R. L. Wilson, J. M. Ade-Hall, G. P. L. Walker, S. E. Haggerty, T. Sigurgeirsson, N. D. Watkins, P. J. Smith, J. Edwards and R. L. Grasty (1967) Geomagnetic polarity zones for Icelandic lavas, *Nature*, **216**, 25.

Denham, C. R. and A. Cox (1971) Evidence that the Laschamp polarity event did not occur 13,300–30,400 years ago, *Earth Planet. Sci. Letters*, **13**, 181.

Dunlop, D. J. (1968) Monodomain theory, experimental verification, *Science*, **162**, 256.

Dunlop, D. J. (1972) Magnetite: behaviour near the single domain threshold, *Science*, **176**, 41.

Dunn, J. R., M. Fuller, H. Ito and V. A. Schmidt (1971) Palaeomagnetic study of a reversal of the earth's magnetic field, *Science*, **172**, 840.

Fisher, R. A. (1953) Dispersion on a sphere, *Proc. Roy. Soc.*, **A 217**, 295.

Glass, B. P. and B. C. Heezen (1967) Tektites and geomagnetic reversals, *Nature*, **214**, 372.

Hargraves, R. B. and W. M. Young (1969) Source of stable remanent magnetism in Lambertville diabase, *Amer. J. Sci.*, **267**, 1161.

Harrison, C. G. A. (1968) Evolutionary processes and reversals of the earth's magnetic field, *Nature*, **217**, 46.

Harrison, C. G. A. and B. M. Funnell (1964) Relationship of palaeomagnetic reversals and micropalaeontology in two late Cenozoic cores from the Pacific Ocean, *Nature*, **204**, 566.

Hays, J. D. and N. D. Opdyke (1967) Antarctic radiolaria, magnetic reversals and climatic change, *Science*, **158**, 1001.

Heirtzler, J. R., G. O. Dickson, E. M. Herron, W. C. Pitman III, and X. Le Pichon (1968) Marine magnetic anomalies, geomagnetic field reversals and motions of the ocean floor and continents, *J. geophys. Res.*, **73**, 2119.

Hide, R. (1966) Free hydromagnetic oscillations of the earth's core and the theory of the geomagnetic secular variation, *Phil. Trans. Roy. Soc.*, **A 259**, 615.

Hide, R. (1967) Motions of the earth's core and mantle and variations of the main geomagnetic field, *Science*, **157**, 55.

Irving, E. and R. W. Couillard (1973) Cretaceous normal polarity interval, *Nature Phys. Sci.*, **244**, 10.

Johnson, E. A., T. Murphy and O. W. Torreson (1948) Pre-history of the earth's magnetic field, *Terr. Mag. Atmos. Elec.*, **53**, 349.

Kono, M. (1971) Intensity of the earth's magnetic field during the Pliocene and Pleistocene in relation to the amplitude of mid-ocean ridge magnetic anomalies, *Earth Planet. Sci. Letters*, **11**, 10.

Kono, M. (1972) Mathematical models of the earth's magnetic field, *Phys. Earth Planet. Int.*, **5**, 140.

Larson, E. E., M. Ozima, M. Ozima, T. Nagata and D. Strangway (1969) Stability of remanent magnetization of igneous rocks, *Geophys. J.*, **17**, 263.

Larson, R. L. and W. C. Pitman (1972) Worldwide correlation of Mesozoic anomalies and its implications, *Bull. geol. Soc. Amer.*, **83**, 3645.

McElhinny, M. W. (1967) The palaeomagnetism of the southern continents—a survey and analysis, *Intern. Union Geol. Sci. U.N.E.S.C.O. Symp. Continental Drift, Montevideo, Oct. 1967.*

McElhinny, M. W. (1969) The palaeomagnetism of the Permian of south-east Australia and its significance regarding the problem of intercontinental correlation, *Spec. Publ. Geol. Soc. Austral.*, **2**, 61.

Mörner, N.-A., J. P. Lanser and J. Hospers (1971) Late Weichselian palaeomagnetic reversal, *Nature Phys. Sci.*, **234**, 173.

Nagata, T. (1969) Length of geomagnetic polarity intervals, *J. Geomag. Geoelec.*, **21**, 701.

Néel, L. (1951) L'inversion de l'aimantation permanente des roches, *Ann. Geophys.*, **7**, 90.

Néel, L. (1955) Some theoretical aspects of rock magnetism, *Phil. Mag. Supp. adv. Phys.*, **4**, 191.

Opdyke, N. D. (1972). Palaeomagnetism of deep-sea cores, *Rev. Geophys. Space Phys.*, **10**, 213.

Opdyke, N. D., B. Glass, J. D. Hays and J. Foster (1966) Palaeomagnetic study of Antarctic deep-sea cores, *Science*, **154**, 349.

Opdyke, N. D. and K. W. Henry (1969) A test of the dipole hypothesis, *Earth Planet, Sci. Letters*, **6**, 139.

Parker, E. N. (1969) The occasional reversal of the geomagnetic field, *Astrophys. J.*, **158**, 815.

Smith, P. J. (1967) The intensity of the ancient geomagnetic field: a review and analysis, *Geophys. J.*, **12**, 321.

Smith, P. J. (1971) Field reversal or self-reversal?, *Nature*, **229**, 378.

Smith, J. D. and J. H. Foster (1969) Geomagnetic reversal in Brunhes normal polarity epoch, *Science*, **163**, 565.

Strangway, D. W. (1970) *History of the Earth's Magnetic Field* (McGraw-Hill).

Strangway, D. W., E. E. Larson and M. Goldstein (1968) A possible cause of high magnetic stability in volcanic rocks, *J. geophys. Res.*, **73**, 3787.

Uffen, R. J. (1963) Influence of the earth's core on the origin and evolution of life, *Nature*, **198**, 143.

Vine, F. J. (1968) Magnetic anomalies associated with mid-ocean ridges, in: *The History of the Earth's Crust* (ed. R. A. Phinney) (Princeton Univ. Press).

Waddington C. J. (1967) Palaeomagnetic field reversals and cosmic radiation, *Science*, **158**, 913.

Watkins, N. D. (1967) Unstable components and palaeomagnetic evidence for a geomagnetic polarity transition, *J. Geomag. Geoelec.*, **19**, 63.

Watkins, N. D. (1969) Non-dipole behaviour during an upper Miocene geomagnetic polarity transition in Oregon, *Geophys. J.*, **17**, 121.

Watkins, N. D. and H. G. Goodell (1967) Geomagnetic polarity change and faunal extinction in the Southern Ocean, *Science*, **156**, 1083.

Watkins, N. D. and S. E. Haggerty (1968) Oxidation and polarity variation in Icelandic lavas and dikes, *Geophys. J.*, **15**, 305.

Watkins, N. D. and J. Ade-Hall (1970) Absence of correlation between opaque petrology and natural remanence polarity in Canary Island lavas, *Geophys. J.*, **19**, 351.

Watson, G. S. (1956). Analysis of dispersion on a sphere, *Mon. Not. Roy. Astr. Soc. geophys. Suppl.*, **7**, 153.

Watson, G. S. and E. Irving (1957) Statistical methods in rock magnetism, *Mon. Not. Roy. Astr. Soc. geophys. Suppl.*, **7**, 289.

Wilson, R. L. (1965) Palaeomagnetism, *Die. Natur.*, **11**, 286.

Wilson, R. L. and N. D. Watkins (1967) Correlation of petrology and natural magnetic polarity in Columbia Plateau basalts, *Geophys. J.*, **12**, 405.

Wollin, G., D. B. Ericson and W. B. F. Ryan (1971) Magnetism of the earth and climatic changes, *Earth Planet. Sci. Letters*, **12**, 175.

7

The Thermal History of the Earth

7.1   Introduction

The temperature at a given point on the earth's surface depends mainly on the radiation from the sun which reaches it and on the angle with the surface at which the radiation arrives. The average solar heat flux reaching the ground on the continents is of the order of 10^{-2} cal/cm² s. The heat flow from the interior of the earth is of the order 10^{-6} cal/cm² s,* and is thus negligible in comparison and has no influence on the atmospheric temperature and climate.

The surface of the earth undergoes daily and annual changes of temperature. It is easy to show that such periodic changes decrease exponentially with depth, and that there is a time lag in the phase of the temperature wave. Measuring z vertically downwards from the earth's surface considered as the plane $z=0$, the one dimensional heat conduction equation is

$$\frac{\partial T}{\partial t} = \kappa \frac{\partial^2 T}{\partial z^2} \tag{7.1}$$

where T is temperature, and κ is the thermal diffusivity. Let us look for a solution of the form

$$T = f(z)\, e^{i\omega t} \tag{7.2}$$

which has a period $2\pi/\omega$. It follows by direct substitution in equation (7.1) that f satisfies the ordinary differential equation

$$\frac{d^2 f}{dz^2} = \frac{i\omega}{\kappa} f \tag{7.3}$$

---

* Heat flow has conventionally been measured in μcal/cm² s. Thus the heat flow through the surface of the earth is of the order 1 μcal/cm² s which has been defined as 1 heat-flow unit (HFU). In the SI system of units, heat flow should be measured in milliwatts/metre² (mW/m²). The conversion factor is 1 μcal/cm² s $\simeq 41\cdot87$ mW/m². Since most of the existing literature uses the older system of units it has been retained in this chapter for convenience.

The solution of this equation which is finite as $z \to \infty$ is

$$f = A \exp\left\{-z\sqrt{\left(\frac{i\omega}{\kappa}\right)}\right\} = A \exp\left\{-z(1+i)\sqrt{\left(\frac{\omega}{2\kappa}\right)}\right\} \qquad (7.4)$$

and the complete solution of equation (7.1) which has the value $T_0 \sin \omega t$ when $z = 0$ is thus

$$T = T_0 \exp\left\{-z\sqrt{\left(\frac{\omega}{2\kappa}\right)}\right\} \sin\left\{\omega t - z\sqrt{\left(\frac{\omega}{2\kappa}\right)}\right\} \qquad (7.5)$$

If $\kappa = 0.0049$ cm²/s, a daily range of 20°C at the surface is reduced to 1.4° at a depth of 30 cm, at which depth the time lag is about 10 h. Diurnal changes become negligible at a depth of about 1 m while seasonal changes do not penetrate more than a few tens of metres. If, however, there have been climatic changes extending over thousands of years, temperature perturbations may extend to a depth of several hundred metres, and thus affect measurements of the temperature gradient of the heat flow from the earth's interior. In this respect the effect of the cold of the last ice-age, which ended about 11 000 yr ago, is still appreciable and corrections must be made for it.

The two heat flow probes left on the lunar surface by Apollo 15 have recorded a difference in surface temperature of $\sim 280°C$ between lunar night and noon. At a depth of 30 cm the temperature varies by about 7°C, and at a depth of 1 m by about 1.5°C. Langseth has estimated that the heat flow from the moon's interior is about one-half the average heat flux through the earth's crust. This is considerably higher than was expected (the volume of the moon is only about one-fiftieth that of the earth), and higher than that predicted by thermal calculations based on geochemical models of the lunar interior—perhaps the moon has a significantly higher proportion of radioactive materials (whose decay is thought to be the main heat source). It is very disappointing that the heat flow experiments planned for Apollo 16 could not be carried out.

7.2 Heat flow measurements

The heat flow through the earth's surface is given by the product of the temperature gradient and the thermal conductivity. No account of the techniques of obtaining heat flow measurements will be given. Details of measurements on land can be found in Misener and Beck (1960) and Beck (1965), and at sea by Langseth (1965). The annual variation of temperature in the bottom waters of the deep oceans is negligible. In shallow seas, however, it is several degrees, and this has, until recently, prevented meaningful heat flow measurements from being made. However, Steinhart et al. (1969) have now obtained 145 reliable measurements in Lake Superior by correcting the temperature gradient for climatic variations knowing the annual cycle of water temperature. Measurements have also been made by Von Herzen and Vacquier (1967) in Lake Malawi, central Africa, and by Sclater *et al.* (1970) in Lake Titicaca, Peru.

Compared with other geophysical measurements the number of heat flow determinations is still quite small. Lee and Uyeda (1965) reviewed and analysed all available heat flow data up to the end of 1964. The data were revised by Lee and Clark (1966) and additional data have been compiled by Simmons and Horai (1968) who hope to keep the list up to date by publishing additional summaries.

Fig. 7.1   (a) Heat flow in 5°×5° grid. In each grid element, the upper value is the number of observations, and the lower value is the arithmetic mean (with quality of data weighting) in units of 10^{-6} cal/cm² s (after Lee (1970))

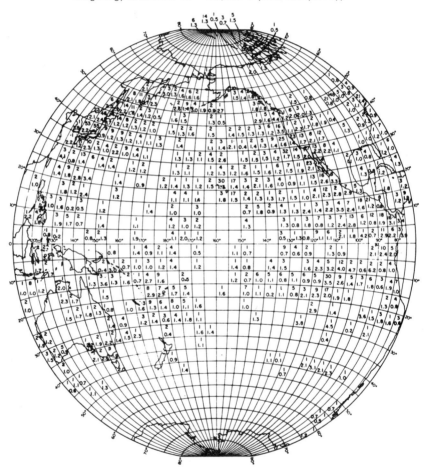

The geographical distribution of heat flow measurements is very uneven, about 80 per cent being obtained in the oceans. Large gaps exist in the data from Asia, Africa, South America and Antarctica; see Fig. 7.1(a), (b). Care must be taken therefore in the interpretation of spherical harmonic analyses that have been carried out on the existing data and of correlations that have been made with other geophysical data (see § 4.4). Of more interest is the recognition now of heat flow provinces which are correlated with major geological provinces. Higher and more scattered values of heat flow are found in active tectonic regions, and lower, more uniform values in stable areas. The data also appear to be consistent with the newer concepts of sea floor spreading and plate tectonics (see § 8.5).

The average heat flux through continental and oceanic regions is not significantly different, the mean value being about $1 \cdot 5 \times 10^{-6}$ cal/cm^2 s.* Table 7.1

Fig. 7.1 (b) Heat flow in 5°×5° grid. In each grid element, the upper value is the number of observations, and the lower value is the arithmetic mean (with quality of data weighting) in units of 10^{-6} cal/cm^2 s (after Lee (1970))

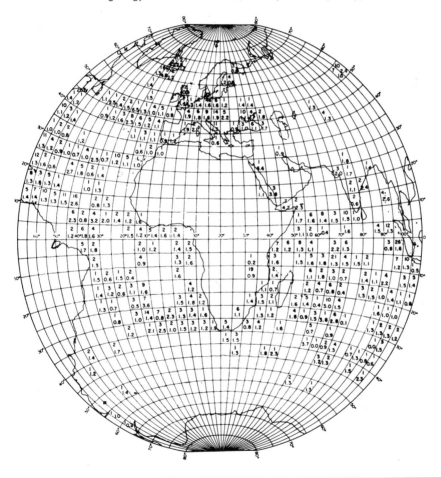

---

* In the rest of this section, all heat flow measurements will be given in units of 10^{-6} cal/cm^2 s, the units being omitted to avoid repetition.

Table 7.1   Statistics of heat-flow values for various tectonics regions

Tectonic region	N	$\bar{q}$	s.d.
Precambrian shields	214	0·98	0·24
Post-Precambrian non-orogenic areas	96	1·49	0·41
Palaeozoic orogenic areas	88	1·43	0·40
Mesozoic-Cenozoic orogenic areas	159	1·76	0·58
Ocean basins	683	1·27	0·53
Mid-oceanic ridges	1065	1·90	1·48
Ocean trenches	78	1·16	0·70
Continental margins	642	1·80	0·93

N is the number of data.
$\bar{q}$ is the arithmetic mean with quality weighting in μcal/cm^2 s.
s.d. is the standard deviation from the mean with quality weighting in μcal/cm^2 s.
(After Lee (1970))

which has been compiled by Lee (1970) gives the heat flow values for various tectonic regions—these provinces are somewhat differently defined from those of Langseth and Von Herzen (1970).

A review of oceanic heat flow measurements up to 1964 has been given by Von Herzen and Langseth (1966); the same authors discuss the most recent data and its tectonic significance (Langseth and Von Herzen (1970)). No detailed account will be given here. The deepest part of active trenches and the landward wall are often characterized by low heat flows confined to a narrow zone, whereas the areas landward of the trenches have high heat flows averaging greater than 2·2 (see Fig. 7.2). In contrast there is no significant discontinuity of

Fig. 7.2   Heat flow versus distance from the trench axis for the western Pacific Ocean. Distance is positive towards the oceans and negative towards the continents. The dashed curve is based on the arithmetic mean with quality weighting of values in each 100 km distance interval (after Lee (1970))

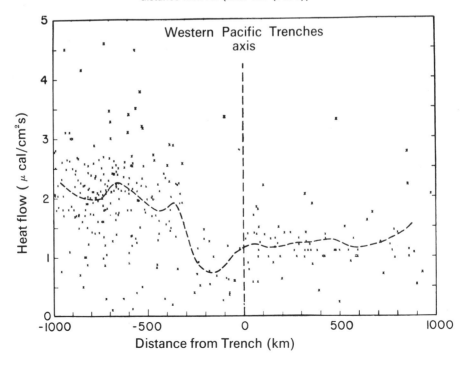

the heat flow associated with inactive or stable coastlines. The average heat flow through the broad deep ocean basins is 1·3. Some ocean basins show significant variations from this mean, especially the northern basins of the Atlantic, where heat flow averages 1·1, and the central basin of the Indian Ocean, where heat flow averages 1·5–1·7. The axis of mid-ocean ridges is characterized by very high heat flows (see Fig. 7.3). The width of the high heat flow zone is much smaller in

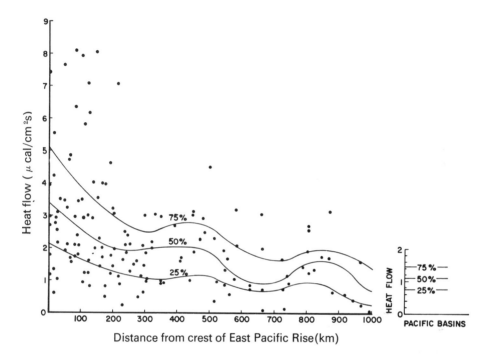

Fig. 7.3 Heat flow values versus distance from the crest of the East Pacific Rise (50°S to 20°N). 75-, 50-, and 25-percentile lines are given for values from the East Pacific Rise and the Pacific basins. The 25-percentile line, for example, separates the data points into 75 per cent above and 25 per cent below it (after Lee and Uyeda (1965))

the Atlantic and the Indian oceans than in the Pacific. The flanks adjacent to the high heat flow zones are often characterized by low heat flows, approximately one half the basin average. Lee (1970) has plotted heat flow against the age of the sea floor; in the light of the recent developments in plate tectonics, the age of the sea floor is a more fundamental parameter than distance from the ridge axis because of different spreading rates along the ridge.

Polyak and Smirnov (1968) showed that the average heat flow through continental orogenic belts decreases with the age of the orogeny. The average values for the provinces of Precambrian folding, Caledonian, Hercynian and Mesozoic are 0·93, 1·11, 1·24 and 1·42, respectively. The average heat flow for provinces in the North Pacific also decreases with the age of the province (see Fig. 7.4); the mean heat flow through the provinces younger than 10 m yr is 2·82, whereas the mean heat flow through those older than middle Cretaceous is 1·15 (Sclater and Francheteau (1970)). However, the time-scales for the thermal decay of heat flow through continents and oceans are an order of magnitude different. The

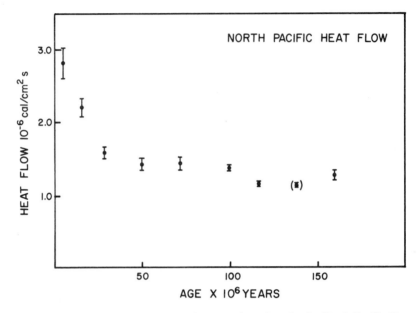

Fig. 7.4 Plot of mean heat flow against age of province for the North Pacific. The length of the bar gives the magnitude of the respective standard error. The mean value for the youngest province has been plotted for a mean age of 5 m yr in order to account for the paucity of observations in the crestal regions (after Sclater and Francheteau (1970))

continents take about 1000 m yr to reach a constant value of about 1·0 while the oceans reach a value of 1·1 after only 100 m yr. This, coupled with the fact that the chemical composition of continental and oceanic crustal rocks is radically different, indicates that the processes leading to the observed heat flow through the continents and oceans are different. The near equality of oceanic and continental heat flows may thus be fortuitous and need not reflect any deep seated mantle process (see also § 8.5).

Birch *et al.* (1968) and Roy *et al.* (1968) divided North America into three regions and showed that in each of them there is a linear relationship between the heat flow Q and the concentration A of radioactive elements in the surface rocks, i.e.

$$Q = a + bA \qquad (7.6)$$

The radioactivity measured at the surface is assumed to be constant to a depth b (although different in each region); a is the heat flow from the lower crust and upper mantle. Lachenbruch (1968) later showed that to allow for differential erosion, equation (7.6) is true only if the crustal heat sources decrease exponentially with depth. Swanberg (1972) measured the radioactive heat generation for a wide variety of rocks from the Idaho batholith using γ-ray spectrometry and found that in fact the heat generation did decrease exponentially with depth. However, as Lachenbruch (1970) pointed out, equation (7.6) does not determine the heat source distribution uniquely; the exponential model is only one model that is not inconsistent with the linear relationship.

The three heat flow provinces in the United States (with the relevant values of a and b) are

(i)   the Eastern United States ($b = 7\cdot5$, $a = 0\cdot8$)

(ii)   the Sierra Nevada ($b = 10\cdot0$, $a = 0\cdot4$)

(iii)   the Basin and Range province ($b = 9\cdot4$, $a = 1\cdot4$)

The units for b are km and those for a 10^{-6} cal/cm^2 s, respectively. The variable values of the surface heat flow in each province reflect the variable radioactivity of the upper crust. The large variations in the value of the intercept a reflect variations in the heat flow from the mantle, and give important additional information to any attempt to explain the mechanism of plate tectonics.

Sclater and Francheteau (1970) found that the heat flow data from North America showed the same correlation with age as that found by Polyak and Smirnov (1968). Radioactive age determinations and heat flow measurements from Africa, Australia and India by Hamza and Verma (1969) indicate that this decrease of heat flow with age is probably true for all continents. Figure 7.5

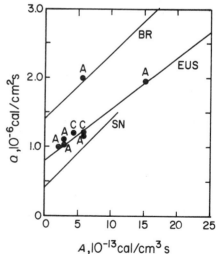

Fig. 7.5 Heat flow and heat production relations for the Basin and Range (BR), Eastern United States (EUS), and Sierra Nevada (SN). Data points labelled A are from Australia and C are from Canada (after Roy *et al.* (1972))

shows the linear relationship (7.6) between heat flow and the radioactive heat generation of plutonic rocks for North America with additional data from Australia and Canada. The similarity of the slopes gives a value for b between about 7 and 10 km. A comprehensive review of continental heat flow data has recently been given by Roy et al. (1972). Using data on heat flow versus heat production in granitic rocks and all published measurements up to September 1970, they produced the first contour map of heat flow in the Western United States (Fig. 7.6). There are three bands of high heat flow and a single discontinuous band of subnormal values coastward of the largest band of high heat flow in the Basin

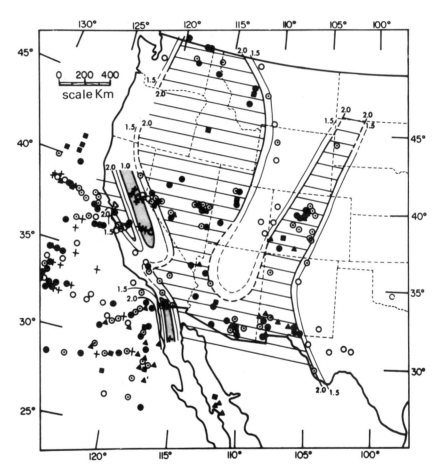

Fig. 7.6 Contour map of the heat flow in the Western United States. The contours delineate regions of high and low heat flow with average ratios of flux that would be measured in rocks with heat production within the range of granodiorite. Pluses represent heat flow values in the range 0–0.99; open circles 1.0–1.49, dotted circles 1.5–1.99; solid circles 2.0–2.49; solid triangles 2.5–2.99; solid rectangles > 3.0. Units are 10^{-6} cal/cm^2 s (after Roy et al. (1972))

and Range, Columbia plateau, and Northern Rocky Mountains. Particularly noteworthy is the value of a for the Sierra Nevada Mountains which is 0·4, as compared to 1·4 in the high heat flow regions and 0·8 in the eastern United States.

In 1971, Sass *et al.* published 150 new heat flow determinations in the western United States, which more than doubled the existing amount of heat flow data from that region. These new results confirm that, although the heat flow in the western United States is in general higher than normal, a number of regions (of the order of several hundred km² in area) of lower heat flow can now be identified.

Combs and Simmons (1973) have recently obtained 26 new heat flow measurements in the north central United States. Although all the stations were located in the stable continental interior they were able to distinguish two distinct provinces—the Interior Lowlands characterized by a regional mean of 1·4 and the northern Great Plains by a value of about 2·0. Both regions consist of a thick sedimentary sequence overlying the Precambrian basement and both have been subjected to only mild tectonic disturbance since Precambrian time. The lower heat flow values are associated with negative seismic travel-time anomalies and the high values with positive station residuals. Lower Pn velocities are also associated with high heat flows in the northern Great Plains. All evidence indicates that the difference in regional heat flows reflects lateral differences in the temperature of the upper mantle.

Further linear relationships have been proposed between other geophysical thermal parameters. Horai and Nur (1970) investigated possible correlations between heat flow Q, thermal conductivity K and geothermal gradient G. They found a positive correlation between Q and K in many major continental tectonic provinces. However, as Jacobs (1970) has pointed out, there are pitfalls in such 'arithmetic' which must be examined with care.

7.3 The earth's internal sources of heat

All thermal histories of the earth are based on either a hot or on a cold origin; should a cold origin lead at some later stage in the evolution of the earth to a molten state, the subsequent thermal histories, whatever the initial origin, would be the same. A cold origin is usually preferred and quite low initial temperatures (well below the melting point of the silicate rocks that form the mantle) are generally assumed. In this case, sources of heat within the earth are needed to explain both the present day heat flow and the relatively high internal temperatures. Possible heat sources are the radioactive decay of both long-lived and short-lived isotopes and sources due to the evolution of the earth (such as the formation of the core).

The radioactive isotopes which contribute significantly to the present heat-production within the earth are U^{238}, U^{235}, Th^{232} and K^{40}, all of which have half-lives comparable to the age of the earth (Table 1.1). Because of the comparatively short time of accretion of the earth ($\sim 10^8$ yr), the temperature increase due to the radioactive decay of these long-lived isotopes is small, of the order of 150°C (MacDonald (1959)). During the first 1000 m yr, the temperature increase would only be about 700°C (assuming no heat escape), while the total heat produced over the life-time of the earth (~ 4500 m yr), if trapped within the earth, would raise the temperature by ~ 1800°C. Thus long-lived radioactive isotopes can account for part of the initial heating up of the earth, but other sources are needed as well.

Short-lived radioactive isotopes could have contributed to the initial heat of the earth if the time between the formation of the elements and the aggregation of the earth was short compared with the half-lives of the isotopes. The important short-lived isotopes are U^{236}, Sm^{146}, Pu^{244} and Cm^{247}, all of which have half-lives sufficiently long to have heated up the earth during the 10^7–10^8 yr after the initial formation. These four isotopes would have contributed about 20 times the heat produced by K^{40} during this period. MacDonald (1959) estimated that if all this heat was retained by the earth, a temperature increase of the order of 2000–3000°C would be possible. The decay of three shorter-lived radionuclides Al^{26}, Cl^{36} and Fe^{60} would have significantly heated up accreting planetary bodies for a period of about 5–15 m yr after the termination of nucleosynthesis in the primitive solar system (Fish et al. (1960)). Al^{26} is the most important. It decays to Mg^{26} with a half-life of 0·74 m yr, and would remain as a significant source of heat for about 10 m yr. If the earth accreted within 20 m yr of the termination of nucleosynthesis, the heat released through the decay of Al^{26} could be the dominant cause of its high internal temperature. If, as is more probable, the time of accretion was of the order of 10^8 yr, the decay of Al^{26} would have had a negligible effect on the earth's thermal history. Moreover, no trace of any anomaly in the Mg^{26}/Mg^{24} ratio has been found in meteorites or in lunar and terrestrial samples (Schramm et al. (1970)).

The temperature of the material within the aggregating earth would also increase because of adiabatic compression. Although data (particularly on the variation with pressure of the coefficient of thermal expansion) are rather uncertain, a rise in temperature of several hundred degrees from this source is possible. Another source of available energy is the potential energy due to the mutual gravitational attraction of the particles of the dust cloud. This energy, upon aggregation, is either converted into internal energy or radiated away. It is extremely difficult to estimate the contribution from this source because of our lack of knowledge of the physical processes of accretion. The result depends quite critically on the temperature attained at the surface of the aggregating earth and on the transparency of the surrounding atmosphere to radiation.

Comparatively low surface temperatures (of the order of a few hundred degrees) have been suggested, mainly because the atmosphere of the primitive earth was assumed transparent so that the thermal energy of the impinging particles was immediately re-radiated into space. Ringwood (1960) has argued, however, that during these early years, the primitive earth would have a large reducing atmosphere. In the presence of these reducing agents (chiefly carbon and methane), the accreting material would be reduced to metallic alloys, principally of iron, nickel and silicon. The outer regions of the earth would thus be metal rich and dense (referred to zero pressure) compared with the interior. Such a state is gravitationally unstable, and convective overturn would follow, leading to a sinking of the metal rich outer regions to the centre. This would release further heat due to the energy of gravitational rearrangement. Ringwood believes the whole process is likely to be catastrophic, since the overturn would be accelerated as the initial temperature rose. Safronov (1965) believes that the earth accumulated mainly by the impact of asteroid-sized bodies, not dust particles. A few per cent of the energy from the impact of large bodies would be released in the form of seismic waves and heat the whole of the interior, where heat cannot easily escape into space. Thus a relatively high initial temperature of the earth seems probable even if the accumulation time was as long as 100 m yr (Safronov (1969)). Again the impact of large bodies would produce large craters beneath which a zone of increased temperature would be formed; such zones could extend to a depth of between 10 and 100 km and remain hot or at least warm for millions of years. Initial inhomogeneities in temperature thus seem probable—they would be places where partial melting and magmatic differentiation would have first begun.

Another possible source of internal heat is the dissipation of the earth's rotational energy as it slows down through tidal interaction with the moon (and to a lesser extent with the sun). Part of this energy is dissipated by tides in the oceans and part in the interior of the earth by earth tides. It is difficult to estimate how the energy loss is divided between ocean and earth tides—probably most of it is dissipated in the shallow seas. If the heat generated by tides were evenly distributed over the whole volume of the earth, the increase of temperature would be unimportant (probably less than 100°C). However, if there is a region of low viscosity in the upper mantle, most of the energy of tidal deformation should be dissipated there, resulting in an increase of temperature by perhaps as much as 1000°C. This would have occurred when the earth and moon were relatively close together, and the earth's rotation was much faster than it is now. The present-day loss of rotational energy could account for but a fraction of the total heat flow.

It is generally assumed that the earth formed by accretion from approximately homogeneous material and that it later differentiated into crust, mantle and core. The formation of the core would release a large amount of gravitational energy as a result of the concentration of the high density nickel-iron in the centre of the earth. Tozer (1965) has estimated that the release of heat arising from core formation from an originally undifferentiated earth would be about 500 cal/g. About 6 per cent of this would melt the Ni–Fe phase, while the rest would raise the mean temperature of the earth about 1500°C. Birch (1965) has made a similar estimate and obtained essentially the same result. It would appear that core formation plays a major role in the thermal history of the earth. This question is discussed further in § 10.5.

It would seem from the above discussion that the temperature in an originally cold earth would, at some time in its history, rise above the melting temperature. However, Urey (1962) has put forward convincing evidence that the earth, accumulating at low temperatures, has at no time become completely molten. The earth has lost most of its hydrogen, helium and other gaseous materials, and this must have taken place at low temperatures since otherwise many fairly volatile elements such as Hg, As, Cd and Zn would have been lost as well, which is not the case. Urey argues that any heat must have been lost before this separation and thus was not available for producing high temperatures in an accumulating earth. Radioactive heating would raise the temperature, reducing the viscosity, so that convection would then occur. This would dissipate the heat, and no general and complete melting of the earth is likely to have taken place. It must be confessed, however, that it is not possible at the moment to decide whether or not the earth passed through a completely molten state some time in its history.

7.4 The earth's inner core

All evidence indicates that at least part of the earth's core is now liquid, while the mantle is solid, and any theory of the thermal history of the earth must satisfy these two conditions. It has also been generally believed that the core contains a solid inner core beginning at a depth of approximately 5000 km (see § 3.5) although no definite proof of this was obtained until Julian *et al.* (1972) identified the phase PKJKP on seismograms. We are thus faced with the problem of giving a physical explanation of how the earth, if ever completely molten, could have cooled to leave the mantle and inner core solid while the outer core remained liquid. Assuming the core to consist of iron and the mantle of silicates, Jacobs (1953a) offered the following explanation of this point.

At the boundary between the silicate mantle and the iron core there will be a discontinuity in the melting-point–depth curve, although the actual temperature must be continuous across the boundary. The form of this discontinuity could, mathematically, take any of the three cases shown in Fig. 7.7. Case (i),

Fig. 7.7   Possible forms of the melting-point–depth curve in the neighbourhood of the mantle-core boundary

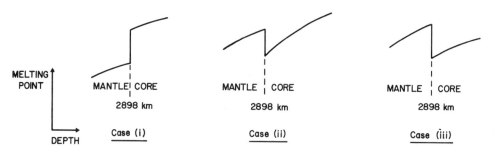

in which the melting-point curve in the core is always above that in the mantle, is impossible; the actual temperature curve must lie below the melting-point curve in the mantle, above the melting-point curve in the core, and yet be continuous across the boundary. Cases (ii) and (iii) are both possible. In Case (iii), the melting-point curve in the core never rises above the value of the melting point in the mantle at the mantle-core boundary, while in Case (ii) it exceeds this value for part of the core. Considering first Case (ii), the melting-point curve will be of the general shape shown in Fig. 7.8.

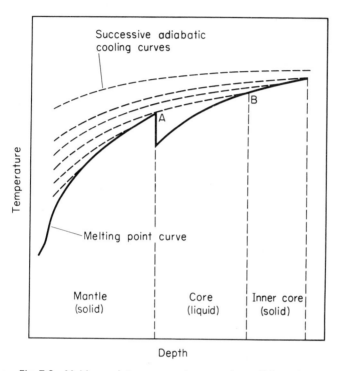

Fig. 7.8   Melting-point curve and successive adiabats in the earth's interior (after Jacobs (1953a))

As the earth cooled from a molten state, the temperature gradient would be essentially adiabatic, there being strong convection currents and rapid cooling at the surface. Solidification would commence at that depth at which the curve representing the adiabatic temperature first intersected the curve representing the melting-point temperature. Solidification would thus begin at the centre of the earth, and not at the boundary of the core and mantle as has been supposed. A solid inner core would continue to grow until a curve representing the adiabatic temperature intersected the melting-point curve twice, once at A, the boundary of the core and mantle, and again at B, as shown in Fig. 7.8. As the earth cooled still further, the mantle would begin to solidify from the bottom upward. The

203

liquid layer between A and B would thus be trapped. The mantle would cool at a relatively rapid rate, leaving this liquid layer essentially at its original temperature, insulated above by a rapidly thickening shell of silicates and below by the already solid (iron) inner core.

In the above discussion no specific values of the temperatures are postulated, and the behaviour of the adiabatic and melting-point curves need not be known exactly. If they vary qualitatively as shown, the above argument gives a physical explanation for the existence of a solid inner core. It follows by similar reasoning that if the melting-point–depth curve in the neighbourhood of the core-mantle boundary is as shown in Case (iii) of Fig. 7.7, then as the earth cooled from a molten state, the entire core would be left liquid. Finally, if the earth had a cold origin and never became completely molten, then as the temperature increased with time, either Case (ii) or Case (iii) could lead to a liquid outer core with a solid inner core. However, if the earth were never completely molten, it is difficult to explain the differentiation into core and mantle and hence a melting-point–depth curve of the form given in Fig. 7.8.

The above arguments break down if the recent melting-point and adiabatic temperature curves of Higgins and Kennedy (1971) are correct, for in their curves the adiabatic gradient is steeper than the melting-point gradient in the outer core. The shape of these two curves will thus be discussed in more detail.

Simon (1937) formulated a semi-empirical equation for the dependence of the melting-point on pressure which has often been applied to geophysical problems. Simon's equation is

$$p = A\left\{\left(\frac{T_m}{T_{m0}}\right)^c - 1\right\} \tag{7.7}$$

where A is a constant, related to the internal pressure, T_m the melting-point temperature at pressure p, T_{m0} the melting point at atmospheric pressure, and c a numerical constant. Gilvarry (1956), under certain assumptions, has given a theoretical justification for this equation. The Simon equation has been applied to the earth's core on the assumption that it is pure iron (Simon (1953); Jacobs (1953a); Gilvarry (1957)). Bullard (1954) has also used it to estimate melting temperatures in both the mantle and core.

Another empirical expression for the melting-point at high pressures has been proposed by Kraut and Kennedy (1966). They showed that the Simon equation for metals almost invariably predicts higher melting temperatures at higher pressures than are measured in the laboratory, and that for most metals there is a linear relationship between T_m and the fractional change in volume $\Delta V/V_0$ resulting from the applied pressure. The Simon equation appears to hold for Van der Waals solids, but some other formulation seems necessary to describe the melting behaviour of silicates and other ionic compounds.

If the linear relationship between T_m and $\Delta V/V_0$ holds for iron, a melting-point–depth curve may be obtained for the core of the earth provided the initial slope of the melting curve of iron and also the relationship between pressure and the specific volume of iron along its melting curve are known. Such data have

204

been provided by the experimental work on the melting of iron by Sterrett *et al.* (1965) and by the results of shock-wave experiments on the density of iron at high pressures (Van Thiel (1966)). Higgins and Kennedy (1971) have thus re-estimated the melting-point gradient in the core of the earth. Figure 7.9 shows their linear relationship between T_m and $\Delta V/V_0$ with the various pressure inter-

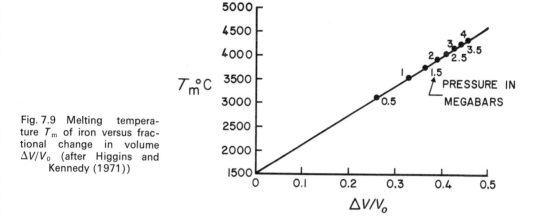

Fig. 7.9 Melting temperature T_m of iron versus fractional change in volume $\Delta V/V_0$ (after Higgins and Kennedy (1971))

vals marked on the curve, and Fig. 7.10 gives a plot of the melting curve of iron versus depth in the earth. Their melting-point gradient in the earth's core is much less steep than earlier estimates, there being an increase of only 500°C in the melting-point across the outer core.

Fig. 7.10 Melting-point curve of iron and adiabatic temperature curves of Jacobs (1971b) and Higgins and Kennedy (1971) versus depth in the earth's interior

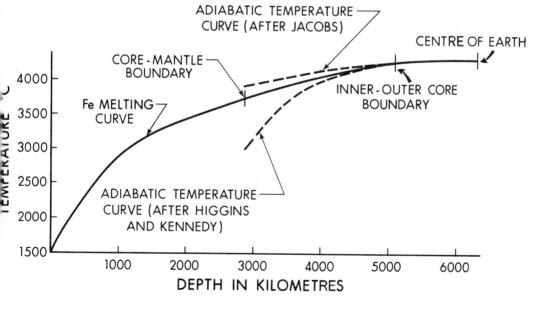

Let us consider now the adiabatic temperature gradient. The increase in temperature dT for a reversible adiabatic increase of pressure dp is given by

$$dT = \frac{T\alpha}{\rho c_p} dp \tag{7.8}$$

where α is the volume coefficient of thermal expansion, ρ the density, and c_p the specific heat at constant pressure. Assuming hydrostatic equilibrium, the variation of pressure with depth z is

$$\frac{dp}{dz} = g\rho \tag{7.9}$$

Hence the adiabatic temperature gradient is given by

$$\frac{dT}{dz} = \frac{g\alpha T}{c_p} \tag{7.10}$$

Introducing Grüneisen's parameter γ defined by the equation

$$\gamma = \frac{\alpha k_s}{\rho c_p} \tag{7.11}$$

where k_s is the adiabatic incompressibility, the adiabatic gradient becomes

$$\frac{dT}{dz} = \frac{g T \rho \gamma}{k_s} = \frac{g T \gamma}{\phi} \tag{7.12}$$

where $\phi = k_s/\rho = V_P^2 - \frac{4}{3}V_S^2$ (see equation 3.13) and is known from seismic data. There have been many estimates of the adiabatic temperature gradient using equations (7.10) or (7.12). The most recent estimate by Higgins and Kennedy (1971) uses equation (7.12) and a value of 1·75 for γ (Knopoff and Shapiro (1969)). They obtained an alternative estimate based on a relationship derived by Valle (1952).

Both the melting-point and adiabatic gradients were found to be extremely flat in the inner core. Assuming that the inner core is solid and the outer core fluid, so that the adiabatic and melting temperatures are the same at the inner-outer core boundary, Higgins and Kennedy found that these temperatures differ by only 15°C at the centre of the earth. On the other hand, they found a very sharp curvature in the adiabatic gradient in the outer core—much steeper than the melting-point gradient—the adiabatic gradient being about 1250°C across the outer core, compared with 500°C for the melting-point gradient (see Fig. 7.10). This is just the opposite to what has usually been supposed, and the mechanism which Jacobs (1953a) put forward for the formation of a solid inner core and liquid outer core would no longer be valid. If actual temperatures were distributed along the adiabatic curve of Higgins and Kennedy throughout the outer core, it too would be solid and there would be no liquid outer core. Higgins and Kennedy thus concluded that the actual temperature gradient in the outer core is much less than the adiabatic gradient. If this is the case, it would pose a substantial inhibition to radial components of convection in the outer core and the

206

question of the generation of the motions which drive the geomagnetic dynamo would have to be reconsidered (see § 5.6).

The adiabatic gradient is quite sensitive to the assumed value of γ for iron at high pressure—in fact for a liquid, γ may be no more than a dimensionless combination of thermodynamic parameters and have no real connection with Grüneisen's parameter γ for a solid (Knopoff and Shapiro (1969)). Jacobs (1971a) has re-estimated the adiabatic gradient in the core by a different method (Jacobs (1953b)), assuming with Higgins and Kennedy a temperature of 4250°C at the inner-outer core boundary, and still finds it slightly less than the melting-point gradient (see Fig. 7.10). Since there are difficulties in accepting the adiabatic gradient of Higgins and Kennedy in the outer core, and since their adiabatic gradient is greater than the melting-point gradient, and that of Jacobs is less, it would seem perhaps that in the outer core there is little difference between these two gradients, and that in the inner core they are essentially the same. It is not possible at the moment to refine the estimates of these two gradients because of the inherent uncertainty in the variation of the parameters involved.

As already mentioned, acceptance of the melting-point and adiabatic gradients of Higgins and Kennedy in the core implies that the liquid in the outer core is thermally stably stratified, thereby inhibiting convection. A possible explanation of what Kennedy has called the core paradox has been given by Busse (1972) who proposed that the liquid in the outer core is a slurry made up of extremely fine iron particles suspended in an iron-rich liquid. Elsasser and Malkus independently made the same suggestion at a recent international conference on the core-mantle interface (see Cox and Cain (1972)). At the same conference, Kennedy gave as his opinion that the size of the iron particles would need to be of the order of a few microns or less for such a slurry to be stable. His experience in the heat treatment of steels suggested that rapid grain growth in ferrous alloys takes place at temperatures as low as 1200°C, and that such a slurry would not be stable at temperatures around 4500°C. Malkus (1972) has shown quantitatively that microscopic convection can exist only if the particulate solid is of sufficiently small size; a convection driven dynamo would not be possible for particles greater than one micron in diameter, and a precession driven dynamo would probably fail for particles greater than ten microns in diameter.

In the previous discussion it has been assumed that the inner core is the solid phase corresponding to the liquid outer core. Analysis of the periods of free oscillations of the earth, especially of the radial overtones, strongly supports this assumption. Birch (1972) has pointed out however that most discussions of the melting of iron neglect the fact that iron exists in several crystalline forms, and that the melting curve defines a condition of equilibrium between the liquid phase and one of the solid phases. Four crystalline phases of iron are known (α, γ, δ and ε). Only at low pressures is the α phase in equilibrium with the melt. Our knowledge of the thermodynamic properties of the γ and ε-phases (of most geophysical interest) is very rudimentary. Birch estimates that the γ melting temperatures are about 700° higher than those of Higgins and Kennedy (1971), with ε melting temperatures still higher; he does not believe that present evidence is sufficient to predict melting temperatures of iron at core pressures to within 500°.

Birch (1972) also pointed out some of the problems involved in calculating adiabats, e.g. the usual (unverified) assumption that the Grüneisen parameter γ is independent of temperature and the deflection of an adiabat on passing through a phase boundary. Birch conjectured that the γ-phase is suppressed or confined to low pressures and that the ε-phase will be the stable solid phase at core pressures. In this case, he estimated that the adiabat originating at the melting temperature of the inner core–outer core boundary lies entirely in the liquid phase, though never far from the freezing curve, supporting Jacobs' earlier (1971b) conclusions.

It must not be forgotten that application of all these results to the earth is further complicated by the fact that pure iron is too dense to satisfy the seismic data for the core which must contain some lighter alloying elements, probably silicon or sulphur (see § 10.5). This would reduce melting temperatures below those of pure iron; in the vicinity of the core, the eutectic temperature could well be some 1500°C lower than the melting temperature of pure iron.

If the melting point and adiabatic gradients are virtually the same throughout the entire core (Jacobs (1971b); Birch (1972)), then perhaps parts of the mantle may from time to time become soluble in the outer core, or material from the outer core diffuse into the lower mantle, thereby giving rise to 'bumps' on the core-mantle boundary as has been suggested by Hide (1969) in another connection (§ 5.4). The position and shape of the boundary may thus change over the course of time and be instrumental in initiating and dictating motions in the outer core. If the topography of either boundary of the outer core has a direct influence on core motions, then changes in the frequency of reversals of the earth's magnetic field may well be random as the shape of these boundaries changes randomly. It is not easy to estimate the time-scale for producing such bumps on the core boundary, but it seems likely that it is greater than the (geologically) short time-scale of reversals found for the last 20 m yr (see § 6.3). If this is the case, then there could not be a one-to-one correspondence between the production of bumps and individual reversals. However, changes in the topography of the core boundary could indirectly affect the frequency of reversals of the earth's magnetic field, perhaps accounting for such intervals in the earth's history as the Kiaman (a span of some 50–60 m yr, some 300 m yr ago) when reversals have only very rarely been observed.

7.5   The thermal history of the earth

The classical approach to the earth's thermal history is to formulate it as an initial boundary value problem with calculations based on the theory of heat conduction in a solid. The equation of heat conduction for a sphere with no lateral differences in its physical properties is

$$\rho c \frac{\partial T}{\partial t} = \frac{1}{r^2} \frac{\partial}{\partial r} \left(K r^2 \frac{\partial T}{\partial r} \right) + H(r, t) \qquad (7.13)$$

where $H(r, t)$ is the rate of production of heat by radioactivity per unit time and volume, c the specific heat and K the thermal conductivity. $H(r, t) = \sum_i h_i(r) \, e^{-\lambda_i t}$ where $h_i(r)$ is the initial heat production per unit volume and λ_i the decay constant of the ith radioactive substance. For the solution of equation (7.13) to be acceptable it must give a present temperature distribution that is within the limits deduced from other geophysical evidence. It must also give a present surface flow comparable to that which is observed.

Holmes (1915) was the first to calculate the radial distribution of temperature. He assumed constant density, specific heat, thermal conductivity and rate of radioactive heat production. He showed that about three-quarters of the surface heat flow is due to radioactivity, the remainder coming from the original heat content. Slichter (1941) carried out extensive studies of steady temperature distributions for several earth models, and Jacobs and Allan (1954, 1956), using digital computers, carried out detailed calculations of an analytical solution of equation (7.13), again assuming constant specific heat and conductivity, but taking into account the previously neglected time dependence of radioactive heat production. Jacobs and Allan examined a number of different earth models in which K and H were constant within discrete shells and for which different initial temperature distributions were considered. A feature of their results was that, for any reasonable distribution of radioactivity, temperatures at depths below about 500 km would still be increasing. Apart from the assumption that conduction is the dominant mode of heat transfer, the main uncertainty in a solution of equation (7.13) for the temperature distribution in the earth is the unknown variation of thermal conductivity with depth.

Two main processes account for heat conduction in solids. Below about 1000°K, energy-transfer is mainly by thermo-elastic waves, i.e. phonon or lattice conduction. Lattice conduction may be considered as the propagation of anharmonic lattice waves through a continuum, or as the interaction between quanta of vibrational thermal energy. Above about 1000°K, radiative heat transfer or photon conduction begins to dominate. Photon conduction is transmission or absorption and re-radiation of electromagnetic energy. Above the Debye temperature (which is about 500°K for the mantle) the lattice conductivity of most materials decreases with increasing temperature as $1/T^m$, where m lies between 1 and 2. The effect of increasing pressure is to increase the lattice conductivity. In the earth, pressure and temperature effects on lattice conductivity seem to counterbalance each other, although at moderate depths the decrease with increasing temperature is the dominant effect. Lubimova (1958) has estimated that the conductivity reaches a minimum at a depth of about 200 km. The effect of this minimum is for the temperature to be increased just below it, and this may well be a contributing factor in the formation of the low velocity layer.

Photon conduction depends on T^3 and could play a dominant role at greater depths in the earth's mantle where the temperatures are higher. The radiative contribution to thermal conductivity is given by

$$K_r = \frac{16}{3} \frac{n^2 \sigma T^3}{\varepsilon} \qquad (7.14)$$

where n is the refractive index, σ the Stefan-Boltzmann constant and ε the opacity or coefficient of absorption of the material. ε (and hence K_r) are strongly dependent on the wavelength of the radiation. It has been known for some time that the thermal conductivity of glasses and ceramic materials increases at high temperatures because of radiation. More recently Kanamori et al. (1968) found that the thermal diffusivity of some common minerals, including quartz and olivine, starts to increase about 450°C. In equation (7.14), both n and ε are weighted means over the appropriate band of wavelengths, and depend on both temperature and pressure. The effect of pressure and temperature on the various absorption bands must be evaluated before reliable estimates of radiative heat conduction under mantle conditions can be made.

The importance of radiative transfer in the mantle has now been questioned. Fukao (1969) found that the sum of the lattice conductivity and photon conduction of olivine is approximately constant ($\simeq 0.012$ cal/cm s deg) in the temperature range from 300–1300°K. Pitt and Tozer (1970a) have made optical absorption measurements on natural and synthetic ferromagnesium minerals at high pressures (up to 5×10^4 bars). They believe that, apart from the near surface regions of the earth, pressure rather than temperature will still play the more important role in changing those parts of the absorption spectra relevant to radiative transfer, since it is much the more effective parameter in changing the density of the materials throughout the mantle. In a companion paper, Pitt and Tozer (1970b) combined their results with those of Fukao et al. (1968) to derive radiative thermal conductivities for olivine under typical conditions in the upper mantle. They concluded that radiative conductivity is probably much smaller in the upper mantle than phonon thermal conductivity. Schatz and Simmons (1972) have more recently measured the total thermal conductivity (lattice plus radiative) of a number of earth materials in the temperature range 500–1900°K. Their results for single crystal and polycrystalline forsterite-rich olivines and an enstatite indicate that there is no rapid increase in radiative conductivity with temperature. They predict that the maximum value of the total thermal conductivity at a depth of 400 km in an olivine mantle is less than twice the surface value.

Radiative heat transfer takes place by electromagnetic waves principally in the infra-red region. It has usually been assumed that the earth, particularly the upper mantle, acted as a grey body, i.e. it was a material in which the absorption was independent of wavelength. This is now known to be incorrect since electronic absorption bands in the infra-red exist in many of the materials that make up the mantle and core of the earth. The width and position of these absorption bands in the electromagnetic spectrum depend upon pressure and temperature, so that even in a homogeneous earth it is possible that certain regions undergo sudden changes in radiative heat transfer properties. Furthermore, wherever a first order phase transformation exists the transmission of radiation is greatly reduced, and even the lattice conductivity is somewhat reduced because of an increasing probability of phonon scatter due to more lattice imperfections forming as precursors of the transition. Beck (personal communication) has pointed out that since a number of transition zones exist within the earth, it is quite possible that these shield the outer regions from radiation from regions below these zones, thus leading to a build up of heat and higher temperatures in and below the transition zones, while above these zones it may be relatively cool. Although phonon conductivity will allow some transfer of heat it appears that there is a strong possibility of the temperature gradient undergoing sudden changes in the transition zones, and that there might be 'hot' and 'cold' zones in both the upper and lower mantle.

At still higher temperatures another mechanism of heat transfer, exciton conductivity, may become important. Neutral atoms may be excited by radiation which has not enough energy to cause free electrons. The energy of an excited atom can be transferred to a neighbouring atom and hence a flow of heat set up. The resulting contribution to the thermal conductivity is given by

$$K_{\text{ex}} = K_0 \, e^{-E/kT} \tag{7.15}$$

where K_0 is a constant and E the excitation energy. A more detailed discussion of exciton conductivity has been given by Lubimova (1967).

MacDonald was the first to apply finite differences to solve numerically the heat-conduction equation, and he constructed more realistic models of the earth's thermal history (MacDonald (1959, 1963, 1964)). Temperatures differ widely between his different earth models. However, all models indicate that at a given depth in the upper mantle, the temperature beneath the oceans is greater than that beneath the continents. This is a consequence of the difference in the crustal concentration of radioactive materials and the near equality of surface heat flows. Also if the mantle is as radioactive as chondritic meteorites, the radioactivity must be concentrated in the upper few hundred km; a deeper distribution would lead to temperatures exceeding the melting point of the crust some time during the earth's history. Reynolds et al. (1966) have modified MacDonald's calculations to include the effects of melting.

Although calculations based on the theory of heat conduction are relatively straightforward, the results are not appliable to the real earth. The data required are poorly known, and conduction does not describe all the processes of heat and mass transfer within the earth. Large-scale convection is extremely efficient in transporting heat, and such convective heat transfer will dominate thermal lattice and radiative heat conduction even for a small velocity of the order of 10^{-2} cm/yr. Both MacDonald (1963) and Knopoff (1964) have presented evidence against large-scale convection in the mantle, but since their arguments depend quite critically on the assumed rheological behaviour of the earth, they cannot be regarded as conclusive.

Most theories of the origin of the earth assume initially a uniform distribution of the radioactive elements. However, geochemical data indicate that these elements are now concentrated towards the earth's surface. Only if such a fractionation were completed during the earth's early history would thermal calculations based on an estimate of the present distribution of radioactivity be justified.

An investigation of the thermal history of the earth taking into account convection and fractionation of radioactive elements is extremely difficult. The difficulties are two-fold: mathematical and physical. A mathematical treatment of the problem entails the formulation and solution of the complete field equations of a multi-component, multi-phase and radioactive continuum of varying properties. The physical difficulties arise mainly from our lack of understanding of the earth's rheological behaviour and fractionation processes. The most detailed discussion of this problem is due to Lee (1967, 1968) who developed mathematical techniques to treat the earth's thermal history beyond simple heat-conduction theory, taking into account latent heat, convection, and fractionation of radioactive elements. Lee showed that large-scale convection within the earth is unlikely, and that heat transfer by small-scale penetrative convection is also unimportant. Such convection is of great importance, however, as a means of moving the radiogenic heat sources upward.

Lee constructed a number of models of the mantle for a range of thermal data and initial conditions. All models with stationary heat sources were unsatisfactory; on the other hand, several models with non-stationary heat sources gave acceptable results. A chondritic earth model is unacceptable since it produces too much heat in addition to other geochemical objections. Many difficulties can be resolved if we accept Ringwood's (1966a, b) theory that the primordial materials forming the earth had a composition similar to that of Type I carbonaceous chondrites. Lee thus used an Orgeuil and a modified Orgueil model but found that they still produced present surface heat flows that were too high (by about 30 per cent). This implies that either the earth has less radioactivity than these Orgueil models or that the initial temperature was lower. Wasserburg *et al.* (1964) proposed the adoption of the terrestrial ratio $K/U = 1 \times 10^4$ rather than the value 8×10^4 found in chondrites. Since the earth has probably lost a considerable amount of volatiles (including K, and perhaps also Th and U) during the final stages of accretion, Lee favours the Wasserburg model of radioactivity (which in addition gives the required value of the present surface heat flow). The heat production of various radioactive earth models throughout geologic time is shown in Fig. 7.11. Figure 7.12 shows the present temperature distribution in the mantle for a non-fractionated model (1A) and a fractionated model (1B). Calculations for different times in the earth's history show that for the non-fractionated model, temperatures rose rapidly and approached the melting point of dunite after about 1500 m yr. After 4500 m yr the temperature of most of the mantle would be at the melting point of dunite. This model also fails in another respect: it gives a surface heat flow only about one-half of that observed. When fractionation of radioactive elements is allowed (model 1B), the rapid rise of temperature is halted because the upward migration of the radioactive

212

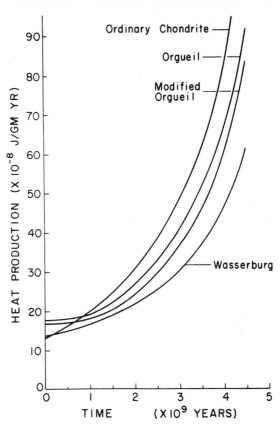

Fig. 7.11 Heat production by various radioactive earth models throughout geologic time (after Lee (1967))

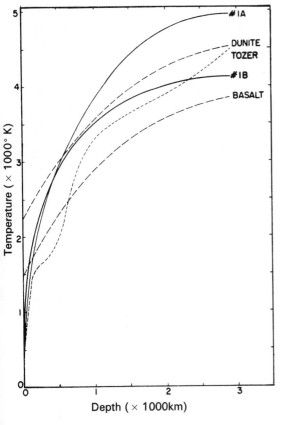

Fig. 7.12 Present temperature distribution within the mantle for a non-fractionated model (#1A), and for a fractionated model (#1B). Dashed curves are melting temperatures for dunite and basalt. The short-dashed curve is the present temperature estimated by Tozer (1959) (after Lee (1967))

elements depletes the heat sources in the lower mantle and increases the surface heat loss. In this model the temperature is below the melting point of dunite at all times.

Hanks and Anderson (1969) have investigated the early thermal history of the earth using a different approach. Most theories of the earth's main magnetic field ascribe its origin to motions in the fluid core of the earth (see § 5.5). They thus examined the thermal history of the earth with the additional constraint that core formation must have occurred before the emplacement of the oldest known rocks possessing remanent magnetism. To fulfill this constraint they found that the earth must have accreted in a period less than 0·5 m yr (4500 m yr ago)—it is probable that the accretion period was less than 200 000 yr and that large-scale differentiation of the earth upon accretion took place. This would imply a hot origin. Another consequence is an early enrichment of radioactive elements in the outer regions of the earth, in contrast to Lee's conclusion of a continuous differentiation of the mantle throughout geologic time. The temperatures in the earth's deep interior thus reflect residual accretional energy (including that of core formation), and are not primarily a result of radioactive heating. Levin (1972) disagrees with such a short accumulation time, believing that the terrestrial planets could not have accreted in less than 100 m yr.

Wang (1972) has suggested another method for estimating temperatures in the lower mantle based on the distribution of seismic velocities and densities within the earth and high pressure shock-wave data. An estimate is first made of the composition of the mantle which does not depend on a knowledge of the temperature. The method makes use of the empirical relation between the bulk sound velocity and density proposed earlier by Wang (1970), who showed that the lower mantle has a mean atomic weight of 21·3. The next step is the construction of an equation of state for material of this composition using empirical high pressure-density data for rocks. This equation is then used to derive the isentropic relationship between pressure and temperature in the lower mantle. The final step is the determination of temperature by comparing the density-pressure in the lower mantle with that predicted by the equation of state. Wang estimates a temperature of 2800°K at a depth of 1300 km increasing almost linearly with depth to 3300°K at 2800 km, with an average gradient of about 0·33°/km. These estimates are appreciably lower than most others that have been proposed. Wang places the uncertainty in his estimates as $\pm 800°C$, largely because of uncertainties in the density in the lower mantle and in the formulation of the high pressure equation of state.

Very little work has been done on convection as a mode of heat transfer in the mantle although it has been realized that convection may play a dominant role in the thermal regime in the earth. Tozer (1967, 1970a, b) has made some significant contributions in this regard, although the value of his work has unfortunately not been fully recognized or appreciated.

As a result of both experimental work and theoretical considerations, Tozer proposed that convective motions in the earth are such as to minimize the mean temperature across any spherical shell in which the conduction solution is unstable, and that it is possible to use data from laboratory model experiments to find that minimum. Tozer also showed that it is possible to estimate mean temperatures on level surfaces without knowing the exact details of the velocity distribution; such temperatures below a depth of about 800 km are controlled by the viscosity dependence on temperature. For all plausible viscosity-temperature relationships, the temperature is always that which gives a viscosity $\simeq 10^{20}$–10^{21} poise. Thus the prevalent idea that any convection theory for the mantle must be very imprecise because of uncertainty in the viscosity-temperature relationship is not true; rather the viscosity itself is constrained to lie within very narrow limits.

In a later paper, Tozer (1972) showed that for a very large range of physically plausible values of the material parameters, the temperature distribution in bodies larger than about 800 km in radius is very different from that predicted by conduction theory. In any body the average temperature rises with depth according to the conduction or state of rest solution until either the centre of the body is reacted or the kinematic viscosity has fallen to a value $\simeq 10^{20}$ stoke, whichever is reached first. Once a body is large enough for its central viscosity to be incapable of stabilizing a state of rest solution, the steady central temperature becomes comparatively independent of the radius and surprisingly low (see Fig. 7.13). Thus of all bodies in the solar system, small objects (radius $\lesssim 800$ km) best

Fig. 7.13   The central temperature T_c as a function of the external radius R for two values of the heat source density H. Curve 1, $H = 1 \cdot 6 \times 10^{-14}$ cal/cm^3 s, Curve 2, $H = 5 \times 10^{-15}$ cal/cm^3 s. On the left the steeply rising curves are steady state of rest solutions; unstable where dotted. Note the relative independence of T_c on R when $R > 800$ km and the suprisingly low values of T_c (after Tozer (1972))

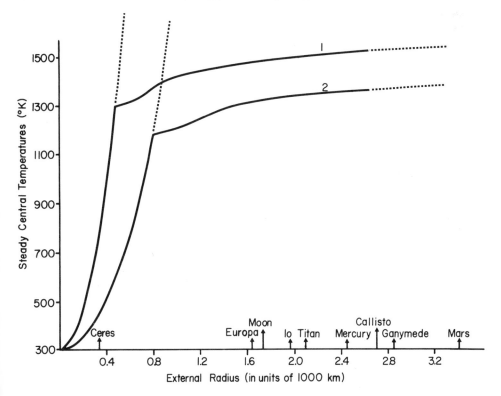

preserve conditions existing at their birth. The magnitude of the velocities, the non-hydrostatic stresses and the viscosity in the convective zones are all relatively insensitive to the choice of values of the material parameters.

A difficult feature of any calculations of the thermal regime of the earth based on conduction theory alone is that the thermal time constant is much greater than the age of the earth, with the result that temperatures in the deep interior depend on the thermal conditions existing at the time of formation of the earth. This difficulty is largely overcome by the convection theory. A stabilization temperature of less than about 2500°K and the great increase in heat transport above it, result in the present thermal conditions being virtually independent of the initial conditions, and make the temperature distribution far less sensitive to details of the heat source distribution.

7.6   The electrical conductivity of the mantle

The electrical conductivity of the earth depends very strongly on temperature, and estimates of the present temperature distribution in the earth have been inferred from the electrical conductivity. Electrical conductivities of solids from metal conductors to good insulators cover an enormous range, more than $10^{24}:1$ at ordinary temperatures, and estimates of conductivity within a factor of about 10 are satisfactory. At the lower temperatures and pressures of the crust the electrical properties are usually dominated by the amount and salinity of the interstitial water; large contrasts exist between the conductivities of sedimentary, igneous and metamorphic rocks. In the crust the effect of pressure is relatively much greater on account of its effect on the porosity; the conductivity of porous rocks is determined by electrolytic conduction in the pores. Thus direct measurements of the conductivity of rocks at atmospheric pressure and varying temperatures are of little use in determining the conductivity of the deep crust and mantle. Pressures corresponding to only a few tens of km burial reduce the conductivity by factors up to a thousand, especially if the rocks are saturated with water.

Estimates of the electrical conductivity of the earth's mantle have been deduced from geomagnetic variations of external origin, from the secular variation and from core-mantle coupling. Considerable work has been done on the problem of electric currents induced in a spherical conductor when the conductivity σ is a function of r only, where r is the distance from the centre of the earth. In particular Lahiri and Price (1939) developed methods for both periodic and aperiodic fields for the case when the conductivity varies as any power, positive or negative, of r. The assumption that σ is a function of r only is not true near the surface of the earth, where, for example, the average conductivity of the oceans is very much higher than that of continental areas. Also the conductivity of the uppermost part of the mantle may not be laterally uniform. However, for variations for which the depth of penetration of the induced currents is greater than about 100 km, calculations based on a model of the earth in which σ is a function of r only should give useful results for interpreting the observations.

Preliminary calculations show that the currents induced by the quiet day solar daily variation S_q and the storm time variation D_{st} (see § 5.7) flow mainly in the earth's mantle and, to a lesser extent, in the crust. The conductivity of dry crustal rocks is usually within the range $10^{-16}–10^{-14}$ emu, while that of moist earth and sediments is in the range $10^{-13}–10^{-12}$ emu, and that of sea water about 4×10^{-11} emu (1 emu $= 10^{-11}$ mho/m). These conductivities are not sufficiently high for the earth's crust to effectively screen the mantle from the S_q and D_{st} variations. On the other hand earth currents induced by more rapid variations such as micropulsations with periods of the order of 100 s or less, will flow mainly in the crustal layers, particularly in the sea. In the models of Lahiri and Price, the currents induced by D_{st} and S_q do not penetrate appreciably beyond a depth of about 1200 km, so that the knowledge of σ obtained from these variations is restricted to an outer shell of this thickness. The range of admissible values obtained by them is represented by the extreme distributions d and e in Fig. 7.14. These curves intersect at a depth of about 600 km, and indicate a conductivity of the order of $1 \cdot 5 \times 10^{-12}$ emu at that depth. A knowledge of σ

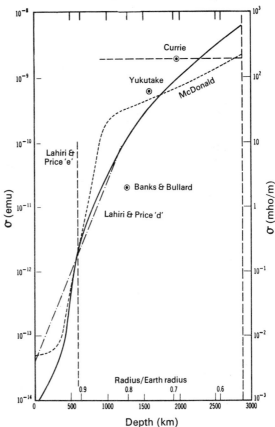

Fig. 7.14   Electrical conductivity of the mantle. The solid line is the curve pre-ferred by Stacey (1969). Also shown are distributions by Lahiri and Price (1939), McDonald (1957) and data by Yukutake (1965), Banks and Bullard (1966) and Currie (1968) (after Stacey (1969))

217

at greater depths could theoretically be obtained from slower variations of external origin such as the 27-day and 6-month variations, but it is difficult to make satisfactory analyses of these variations (see Eckhart *et al.* (1963)). Banks and Bullard (1966) used the small annual variation in the geomagnetic field to estimate a conductivity of about 2×10^{-11} emu at a depth of 1275 km, and Yukutake (1965) obtained a value of about 6×10^{-10} emu at a depth of 1600 km, using magnetic variation data associated with the 11-year sunspot cycle. These two points are also plotted in Fig. 7.14. In a later paper, Banks (1969) estimated the electrical conductivity of the upper mantle by comparing the measured response of the earth to magnetic variations in the frequency range 0·003–0·25 c/day with the theoretical response of particular conductivity distributions. In this frequency range, line spectra at frequencies of 1 and 2 c/yr and 1, 2 and 3 c/27 days were used. Banks found that the conductivity increases sharply by two orders of magnitude at a depth of around 400 km—the width of the region in which this jump occurs is not more than 200 km. This sudden increase may be caused by a phase transition from the olivine to the spinel form of peridotite (Akimoto and Fujisawa (1965)).

Parker (1970) has applied the general method of Backus and Gilbert (see § 3.9) to the inverse problem of the electrical conductivity of the mantle. He reworked the data of Banks (1969)—their two models are compared in Fig. 7.15—and assessed the uncertainties in the determination. Great improvement in the accuracy of the model would be possible if more precise data were available.

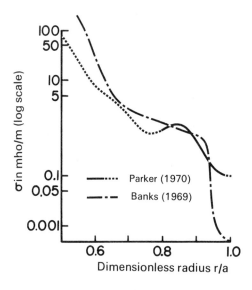

Fig. 7.15 Electrical conductivity models of the earth according to Banks (1969) and Parker (1970). Sections of the latter distribution are shown dotted to emphasize their uncertainty (after Parker (1970)

In this respect the work of Bailey (1970) is interesting. Instead of the traditional inversion methods, Bailey found a direct method whereby the conductivity structure is constructed from the data by solving a non-linear integro-differential equation. The procedure requires precise data with frequencies from zero to infinity. The perfection required of the data makes the method rather impractical, although uniqueness of the solution has been demonstrated for models with analytic behaviour; the existence of a solution has not been proved. A recent review of electrical conductivity determinations down to a depth of about 1500 km has been given by Banks (1972).

Estimates of σ at greater depths have been derived from investigations of the rate of change of the secular variation. In such studies the source field and the conductivity distribution are both unknown; all we have is the observed secular variation. McDonald (1957) assumed a spatially random distribution of secular variation sources at the core surface and allowed for geometrical spreading as well as electromagnetic screening. He obtained a conductivity of the order of 10^{-9} emu near the base of the mantle. Combining his results with an upper mantle curve intermediate between the Lahiri and Price d and e distributions, he derived a conductivity profile for the entire mantle, which is also shown in Fig. 7.14. Currie (1968) reconsidered the secular variation spectrum and obtained a slightly higher mean value for the conductivity of the lower mantle. Some evidence on the conductivity of the mantle may also be obtained from the small changes in the length of the day if these are assumed to result from electromagnetic core-mantle coupling (see § 5.4).

Silicate minerals such as olivine behave essentially as insulators at room temperature but become semi-conductors at higher temperatures. The basic physics of semi-conductors is given in texts on solid state physics (see e.g. Kittel (1966)) and will not be discussed here. Semi-conductors are characterized by the existence of an energy gap between the highest filled (valence) electron states and the next available state (the conduction band). An electron in a completely full band cannot give electrical conduction. However, in a semi-conductor the energy gap between the filled valence states and empty conduction states is sufficiently small, relative to the available thermal energy kT at temperature T, that a few electrons are excited into the conduction band. The conductivity is proportional to the number of electrons excited and is a strong function of temperature. A material in which this occurs is called an intrinsic semi-conductor. At sufficiently high temperatures conduction by ions (ionic conductivity) becomes important. Finally there may be conduction at all temperatures by impurities (impurity conductivity). In this case the presence of foreign atoms with a misfitting valency in a crystal lattice produces either excess electrons or 'holes' (missing electrons) which migrate through the lattice when an emf is applied. The general form of the conductivity σ as a function of temperature T is the sum of three terms of the form

$$\sigma = \sigma_0 \, e^{-E/kT} \qquad (7.16)$$

where σ_0 and E are constant (different for each process and probably pressure dependent) and k is Boltzmann's constant.

Tozer (1959) in a detailed review article, suggests that at normal pressure the conduction mechanism is impurity conductivity at temperatures below about 900–1000°C, intrinsic electronic conductivity from 1000–1400°C, and mainly ionic at higher temperatures. The effect of pressure could modify these statements considerably, however (see e.g. Lubimova (1967)). Impurity conductivity is dominant in the first hundred kilometres where the temperature and electrical conductivity are comparatively low. At depths greater than about 1000 km the pressure inhibits ionic conductivity, the ions being held more firmly in place in the crystal lattices at high pressure. In the lower mantle intrinsic electronic conductivity is thus probably predominant: ionic conductivity is probably important in the depth range 100–1000 km. Tozer (1959) has used conductivity estimates to obtain the present temperature distribution in the mantle. His results were later (1967) modified by McKenzie using more recent experimental data of Akimoto and Fujisawa (1965). These temperature distributions based on conductivity estimates are shown in Fig. 7.16.

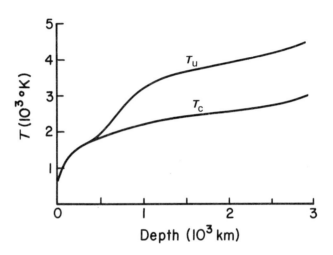

Fig. 7.16 Temperature distribution within the mantle. T_u according to Tozer (1959); T_c corrected for phase change (after McKenzie (1967))

References

Akimoto, S. and H. Fujisawa (1965) Demonstration of the electrical conductivity jump produced by the olivine–spinel transition, *J. geophys. Res.*, **70**, 443.

Bailey, R. C. (1970) Inversion of the geomagnetic induction problem, *Proc. Roy. Soc.*, **A315**, 185.

Banks, R. J. (1969) Geomagnetic variations and the electrical conductivity of the upper mantle, *Geophys. J.*, **17**, 457.

Banks, R. J. (1972) The overall conductivity distribution of the earth, *J. Geomag. Geoelec.*, **24**, 337.

Banks, R. J. and E. C. Bullard (1966) The annual and 27 day magnetic variations, *Earth Planet. Sci. Letters*, **1**, 118.

Beck, A. E. (1965) Techniques of measuring heat flow on land, in: *Geophys. Monograph Series No. 8* (Amer. Geophys. Union).

Birch, F. (1965) Energetics of core formation, *J. geophys. Res.*, **70**, 6217.

Birch, F. (1972) The melting relations of iron, and temperature in the earth's core. *Geophys. J.* **29**, 373.

Birch, F., R. F. Roy and E. R. Decker (1968) Heat flow and thermal history in New England and New York, in: *Studies of Appalachian Geology: Northern and Maritime* (Interscience).

Bullard, E. C. (1954) The interior of the earth, in: *The Earth as a Planet* (Univ. Chicago Press).

Busse, F. H. (1972) Comment on the paper 'The adiabatic gradient and the melting point gradient in the core of the earth', by G. Higgins and G. C. Kennedy, *J. geophys. Res.*, **77**, 1589.

Combs, J. and G. Simmons (1973) Terrestrial heat-flow determinations in the North Central United States, *J. geophys. Res.*, **78**, 441.

Cox, A. and J. C. Cain (1972) International conference on the core-mantle interface, *Trans. Amer. Geophys. Union, EθS*, **53**, 591.

Currie, R. G. (1968) Geomagnetic spectrum of internal origin and lower mantle conductivity, *J. geophys. Res.*, **73**, 2779.

Eckhart, D., K. Larner and T. Madden (1963) Long-period magnetic fluctuations and mantle electrical conductivity estimates, *J. geophys. Res.*, **68**, 6279.

Fish, R. A., G. G. Goles and E. Anders (1960) The record in the meteorites, III. On the development of meteorites in asteroidal bodies, *Astrophys. J.*, **132**, 243.

Fukao, Y. (1969) On the radiative heat transfer and the thermal conductivity in the upper mantle, *Bull. Earthquake Res. Inst.*, **47**, 549.

Fukao, Y., H. Mizutani and S. Uyeda (1968) Optical absorption spectra at high temperatures and radiative thermal conductivity of olivines, *Phys. Earth Planet. Int.*, **1**, 57.

Gilvarry, J. J. (1956) Equation of the fusion curve, *Phys. Rev.*, **102**, 325.

Gilvarry, J. J. (1957) Temperatures in the earth's interior, *J. Atmos. Terr. Phys.*, **10**, 84.

Hamza, V. M. and R. K. Verma (1969) The relationship of heat flow with age of basement rock, *Bull. Volcanol.*, **33**, 123.

Hanks, T. C. and D. L. Anderson (1969) The early thermal history of the earth, *Phys. Earth Planet. Int.*, **2**, 19.

Hide, R. (1969) Interaction between the earth's liquid core and solid mantle, *Nature*, **222**, 1055.

Higgins, G. and G. C. Kennedy (1971) The adiabatic gradient and the melting point gradient in the core of the earth, *J. geophys. Res.*, **76**, 1870.

Holmes, A. (1915) Radioactivity in the earth and the earth's thermal history, *Geol. Mag.*, **2**, 60 and 102.

Horai, K. and A. Nur (1970) Relationship among terrestrial heat-flow, thermal conductivity and geothermal gradient, *J. geophys. Res.*, **75**, 1985.

Jacobs, J. A. (1953a) The earth's inner core, *Nature*, **172**, 297.

Jacobs, J. A. (1953b) Temperature–pressure hypothesis and the earth's interior, *Can. J. Phys.*, **31**, 370.

Jacobs, J. A. (1970) Geophysical numerology, *Nature*, **227**, 161.

Jacobs, J. A. (1971a) Boundaries of the earth's core, *Nature*, **231**, 170.

Jacobs, J. A. (1971b) The thermal regime in the earth's core, *Comments on Earth Sciences: Geophysics*, **2**, 61.

Jacobs, J. A. and D. W. Allan (1954) Temperature distribution within the earth's core, *Nature*, **173**, 258.

Jacobs, J. A. and D. W. Allan (1956). The thermal history of the earth, *Nature*, **177**, 155.

Julian, B. R., D. Davies and R. M. Sheppard (1972) PKJKP, *Nature*, **235**, 317.

Kanamori, H., N. Fujii and H. Mizutani (1968) Thermal diffusivity measurement of rock-forming minerals from 300°–1100°K, *J. geophys. Res.*, **73**, 595.

Kittel, C. (1966) *Introduction to Solid-state Physics* (3rd ed.) (Wiley).

Knopoff, L. (1964) The convection current hypothesis, *Rev. Geophys.*, **2**, 89.

Knopoff, L. and J. N. Shapiro (1969) Comments on the inter-relationship between Grüneisen's parameter and shock and isothermal equations of state, *J. geophys. Res.*, **74**, 1439.

Kraut, E. A. and G. C. Kennedy (1966) New melting law at high pressures, *Phys. Rev.*, **151**, 668.

Lachenbruch, A. H. (1968) Preliminary geothermal model of the Sierra Nevada, *J. geophys. Res.*, **73**, 6977.

Lachenbruch, A. H. (1970) Crustal temperature and heat production: implications of the linear heat-flow relation, *J. geophys. Res.*, **75**, 3291.

Lahiri, B. N. and A. T. Price (1939) Electromagnetic induction in non-uniform conductors, and the determination of the conductivity of the earth from terrestrial magnetic variations, *Phil. Trans. Roy. Soc.*, **A 237**, 509.

Langseth, M. G. (1965) Techniques of measuring heat flow through the ocean floor, in: *Geophys. Monograph Series No. 8* (Amer. Geophys. Union).

Langseth, M. G. and R. P. Von Herzen (1970) Heat flow through the floors of the world oceans, in: *The Sea*, Vol. 4, Part I, 299 (ed. A. E. Maxwell) (Wiley-Interscience).

Lee, W. H. K. (1967) *The Thermal History of the Earth* (Ph.D. thesis, Univ. Cal. Los Angeles).

Lee, W. H. K. (1968) Effects of selective fusion in the thermal history of the earth's mantle, *Earth Planet. Sci. Letters*, **4**, 270.

Lee, W. H. K. (1970) On the global variations of terrestrial heat flow, *Phys. Earth Planet. Int.*, **2**, 332.

Lee, W. H. K. and S. Uyeda (1965) Review of heat flow data, in: *Geophys. Monograph Series No. 8* (Amer. Geophys. Union).

Lee, W. H. K. and S. P. Clark Jr. (1966) Heat and volcanic temperatures, *Handbook of Physical Constants* (Geol. Soc. Amer. Mem. 97).

Levin, B. J. (1972) Origin of the earth, *Tectonophysics*, **13**, 7.

Lubimova, E. A. (1958) Thermal history of the earth with consideration of the variable thermal conductivity of its mantle, *Geophys. J.*, **1**, 115.

Lubimova, E. A. (1967) Theory of thermal state of the earth's mantle, in: *The Earth's Mantle* (ed. T. F. Gaskell) (Acad. Press).

MacDonald, G. J. F. (1959) Calculations on the thermal history of the earth, *J. geophys. Res.*, **64**, 1967.

MacDonald, G. J. F. (1963) The deep structure of continents, *Rev. Geophys.*, **1**, 587.

MacDonald, G. J. F. (1964) Dependence of the surface heat flow on the radioactivity of the earth, *J. geophys. Res.*, **69**, 2933.

Malkus, W. V. R., Convection at the melting point; a thermal history of the earth's core, *J. Geophys. Fluid Dyn.* (in press).

McDonald, K. L. (1957). Penetration of the geomagnetic secular field through a mantle with variable conductivity, *J. geophys. Res.*, **62**, 117.

McKenzie, D. P. (1967) The viscosity of the mantle, *Geophys. J.* **14**, 297.

Misener, A. D. and A. E. Beck (1960) The measurement of heat flow over land, in: *Methods and Techniques in Geophys.*, **1** (Interscience Publ).

Parker, R. L. (1970) The inverse problem of electrical conductivity in the mantle, *Geophys. J.*, **22**, 121.

Pitt, G. D. and D. C. Tozer (1970a) Optical absorption measurements on natural and synthetic ferromagnesian minerals subjected to high pressure, *Phys. Earth Planet. Int.*, **2**, 179.

Pitt, G. D. and D. C. Tozer (1970b) Radiative heat transfer in dense media and its magnitude in olivines and some other ferromagnesian minerals under typical upper mantle conditions, *Phys. Earth Planet. Int.*, **2**, 189.

Polyak, B. G. and Ya. B. Smirnov (1968). Relationship between terrestrial heat flow and the tectonics of continents, *Geotectonics*, **4**, 205.

Reynolds, R. T., P. E. Fricker and A. L. Summers (1966) Effect of melting upon thermal models of the earth, *J. geophys. Res.*, **71**, 573.

Ringwood, A. E. (1960) Some aspects of the thermal evolution of the earth, *Geochim. cosmochim. Acta*, **20**, 241.

Ringwood, A. E. (1966a) Chemical evolution of the terrestrial planets, *Geochim. cosmochim. Acta*, **30**, 41.

Ringwood, A. E. (1966b) Composition and origin of the earth, in: *Advances in Earth Sciences* (ed. P. M. Hurley) (M.I.T. Press).

Roy, R. F., D. D. Blackwell and F. Birch (1968) Heat generation of plutonic rocks and continental heat flow provinces, *Earth Planet. Sci. Letters*, **5**, 1.

Roy, R. F., D. D. Blackwell and E. R. Decker (1972) Continental heat flow, in: *The Nature of the Solid Earth* (ed. E. C. Robertson) (McGraw-Hill).

Safronov, V. S. (1965) Sizes of the largest bodies fallen on planets in process of their formation, *Astron. Zh.*, **42**, 1270.

Safronov, V. S. (1969) *Evolution of the Preplanetary Cloud and the Formation of the Earth and Planets* (Nauka, Moscow).

Sass, J. H., A. H. Lachenbruch, R. J. Monroe, G. W. Greene and T. H. Moses (1971) Heat-flow in the western United States, *J. geophys. Res.*, **76**, 6376.

Schatz, J. F. and G. Simmons (1972) Thermal conductivity of earth materials at high temperatures, *J. geophys. Res.*, **77**, 6966.

Schramm, D. N., F. Tera and G. J. Wasserburg (1970) The isotopic abundances of ^{26}Mg and limits on ^{26}Al in the early solar system, *Earth Planet Sci. Letters*, **10**, 44.

Sclater, J. G. and J. Francheteau (1970) The implications of terrestrial heat flow observations in current tectonic and geochemical models of the crust and upper mantle of the earth, *Geophys. J.*, **20**, 509.

Sclater, J. G., V. Vacquier and J. H. Rohrhirsch (1970) Terrestrial heat flow measurements on Lake Titicaca, *Earth Planet. Sci. Letters*, **8**, 45.

Simmons, G. and K. Horai (1968) Heat flow data 2, *J. geophys. Res.*, **73**, 6608.

Simon, F. E. (1937) On the range of stability of the fluid state, *Trans. Far. Soc.*, **33**, 65.

Simon, F. E. (1953). The melting of iron at high pressures, *Nature*, **172**, 746.

Slichter, L. B. (1941) Cooling of the earth, *Bull. geol. Soc. Amer.*, **52**, 561.

Stacey, F. D. (1969) *Physics of the Earth* (John Wiley and Sons Inc.)

Steinhart, J. S., S. R. Hart and T. J. Smith (1969) Heat flow, *Carn. Inst. Ann. Rept. 1967–68*, 360.

Sterrett, K. F., W. Klement Jr. and G. C. Kennedy (1965) The effect of pressure on the melting of iron, *J. geophys. Res.*, **70**, 1979.

Swanberg, C. A. (1972) Vertical distribution of heat generation in the Idaho batholith, *J. geophys. Res.*, **77**, 2508.

Tozer, D. C. (1959) The electrical properties of the earth's interior, in: *Physics and Chemistry of the Earth*, Vol. 3 (Pergamon Press).

Tozer, D. C. (1965) Thermal history of the earth I. The formation of the core, *Geophys. J.*, **9**, 95.

Tozer, D. C. (1967) Towards a theory of thermal convection in the mantle, in: *The Earth's Mantle* (ed. T. F. Gaskell) (Acad. Press).

Tozer, D. C. (1970a) Factors determining the temperature evolution of thermally convecting earth models, *Phys. Earth Planet. Int.*, **2**, 393.

Tozer, D. C. (1970b) Temperature, conductivity, composition, and heat flow, *J. Geomag. Geoelec.*, **22**, 35.

Tozer, D. C. (1972) The present thermal state of the terrestrial planets. *Phys. Earth Planet. Int.*, **6**, 182.

Urey, H. C. (1962) Evidence regarding the origin of the earth, *Geochim. cosmochim. Acta*, **26**, 1.

Valle, P. E. (1952) Adiabatic temperature gradients in the earth's interior, *Ann. di Geofisica*, **5**, 41.

Van Thiel, M. (ed.) (1966) *Compendium of Shock Wave Data* (Univ. California, Livermore 50108).

Von Herzen, R. P. and M. G. Langseth (1966) Present studies of oceanic heat flow measurements, in: *Physics and Chemistry of the Earth*, Vol VI (Pergamon Press).

Von Herzen, R. P. and V. Vacquier (1967) Terrestrial heat-flow in Lake Malawi, Africa, *J. geophys. Res.*, **72**, 4221.

Wang, C.-Y. (1970) Density and constitution of the mantle, *J. geophys. Res.*, **75**, 3264.

Wang, C.-Y. (1972) Temperature in the lower mantle, *Geophys. J.*, **27**, 29.

Wasserburg, G. J., G. J. F. MacDonald, F. Hoyle and W. A. Fowler (1964) Relative contribution of uranium, thorium and potassium to heat production in the earth, *Science*, **143**, 465.

Yukutake, T. (1965). The solar cycle contribution to the secular change in the geomagnetic field, *J. Geomag. Geoelec.*, **17**, 287.

8

Plate Tectonics

8.1 Introduction

This chapter is concerned with large-scale horizontal movements of the earth's surface and gives an account of what has been called *plate tectonics*. This is a fairly recent development and grew out of Hess' (1962) concept of *sea floor spreading* and older ideas on *continental drift*. It also represents a triumph for the synthesis of new advances in many different areas of geophysics and has unified the whole field of the earth sciences. The relevant developments in the different disciplines are discussed in the following sections. Supporting evidence for large-scale horizontal movements comes from studies of palaeoclimatology, palaeobotany and the life sciences—it is impossible in this book to give an account of all these additional pertinent data. The suggestion of continental drift is old but attracted little attention until the time of Wegener who wrote a long series of papers on the topic, and in 1915 published the first edition of his book '*Die Entstehung der Kontinente und Ozeane.*' This book ran to four editions, each being a complete revision involving new material. The third edition was translated into English (1924) and also the fourth (1966). Wegener proposed that until the Mesozoic there was only one supercontinent, Pangaea, which later broke up. By and large there was strong opposition to Wegener's ideas although he had a few supporters —in particular du Toit (1937) who advocated not one, but two primitive super-continents, Gondwanaland in the south and Laurasia in the north. At the time there was little geophysical evidence to support continental drift and ideas about the possible causes of motion were vague and often untenable. One strong argument in favour of continental drift is the similar shape of the land masses on the two sides of the South Atlantic. Bullard *et al.* (1965) have shown by numerical methods that the best fit is not the coast lines but the 500 fathom contour. Actually as we shall see later the term continental drift is a misnomer; differential motion takes place between different blocks of surface material (plates) and a plate may contain both continents and oceans.

It is important to distinguish between the terms *crust* and *mantle* on the one hand and *lithosphere* and *asthenosphere* on the other. The crust is the outermost layer of the earth of variable thickness separated from the underlying mantle by a discontinuous jump in the velocities of P and S waves (the Mohorovičić discontinuity, see § 3.5). The lithosphere includes the crust and part of the upper mantle. It is distinguished, not by composition, but by strength. It is relatively cool, rigid and brittle whereas the underlying weaker asthenosphere is hot (with temperatures near the melting point) and able to flow. Below the asthenosphere the mantle is much stronger and more resistant to flow; this region is sometimes referred to as the *mesosphere*.

Essentially Hess' (1962) concept of sea floor spreading is that, as a result of convection in the earth's mantle, the deep ocean basins are recent and ephemeral features of the earth's surface, constantly being regenerated at mid-ocean ridge crests which are situated over mantle upwellings, and destroyed beneath the marginal trenches and island arcs which are the 'sinks' in the system. Supporting the idea of sea floor spreading is the discovery that the ocean floors are young. All oceanic islands (Wilson (1963a)) and all cores from the sea floor range in maximum age from recent to Pleistocene along the mid-ocean ridges to upper Jurassic near some continental margins. The thickness of sediments also increases from zero along the crests of the mid-ocean ridges to several kilometres near some coasts. This suggests a greater age and more time for sediments to accumulate farther from the spreading ridges (Ewing and Ewing, 1967).

8.2 Faults

The stresses acting on a body can be represented by three normal stresses (the *principal stresses*) acting along three directions (the *principal axes*) at right angles to one another. The maximum stress acting on a body lies along one of these directions and the minimum stress along another. The three planes which each contain two of the principal stresses, and hence are normal to the third, are *planes of no shear*. In any body subject to non-hydrostatic stress (i.e. one in which the principal stresses are not all equal) two planes exist parallel to which shear is a maximum. Each of these planes contains the direction of the intermediate principal stress and bisects one of the angles between the other two principal axes. It might be expected that faulting would occur parallel to these *planes of maximum shear stress*. However, because of internal friction, planes of actual faulting make smaller angles of some 20°–30° with the direction of maximum stress.

The earth's surface is a plane of no shear, so that of the principal stresses one is always nearly vertical and two are horizontal. There are three possibilities depending upon whether the direction of greatest, intermediate or smallest principal stress is vertical (E. M. Anderson (1951)).

(*i*) *Normal faults* occur when the direction of greatest principal stress is vertical and there is relief of pressure in all horizontal directions. The fault planes dip downward under the downthrust blocks at angles steeper than 45°. At any depth the vertical pressure is the hydrostatic pressure which is greater than that in any other direction. The pressure across the fault planes are therefore less than hydrostatic and dykes may be intruded along them. To raise up folded mountains horizontal pressures must exceed the vertical force of gravity, and thus normal faults are not likely to be primary features in orogeny.

(*ii*) *Reverse or thrust faults* occur when the direction of least stress is normal to the surface and the pressure in all horizontal directions is greater than the hydrostatic pressure. The dip of these fault planes is shallower than 45° and one faulted slab rides over another so that the surface is shortened by thickening. Because the vertical or hydrostatic pressure is less than the pressure across the fault planes, intrusions do not tend to follow these planes.

(*iii*) *Transcurrent faults* occur when the direction of intermediate stress is normal to the surface and the directions of both the largest and smallest principal stresses are parallel to the surface. In this case the fault planes will be approximately vertical and the two sides of faults will be horizontally displaced relative to one another. The terms *wrench faults* and *strike-slip faults* have been used as synonymous names for transcurrent faults, but now that another class of faults (*transform faults*), also with horizontal displacements, has been recognized (Wilson (1965a)), it is convenient to use the term strike-slip fault to refer to both transcurrent and transform faults collectively.

It is now realized that some parts of the earth's surface have moved large distances relative to others (e.g. the Great Glen Fault across Scotland, the Alpine fault in New Zealand, the San Andreas fault in California and the fracture zones in the Northwest Pacific Ocean). Such large offsets are impossible on Anderson's theory of faulting and their existence has often been disputed. The situation is completely changed by Hess' (1962) concept of sea floor spreading which violates an assumption implicit in Anderson's theory, namely that the crust of the earth is conserved. If sea floor spreading takes place, then other types of faults (transform faults) must exist which may have large displacements. Wilson (1965a) has stressed that there are in fact two types of faulting, that in which crust is conserved and faults terminate by dying out, and that in which faults end abruptly by changing into *spreading zones* where crust is created and *subduction zones* where it is re-absorbed back into the mantle.

The differences between transcurrent and transform faults are listed in Table 8.1. Possible types of transform faults are shown in Fig. 8.1 and their appearance after a period of growth in Fig. 8.2. Sykes (1967) has shown that the direction of motion and distribution of earthquakes (see § 8.3) both indicate that fracture zones have the properties of transform, not transcurrent faults. Transform faults or fracture zones terminate abruptly at the coasts of oceans with which they are connected. Their terminal points once were together, and have been called conjugate points by Wilson (1963b, 1965b). Figure 8.3 is a sketch map of the South Atlantic Ocean and shows that this concept agrees with the fit of coastlines.

Table 8.1   Distinctions between transcurrent and transform faults

Property of faults	*Transcurrent faults*	*Transform faults*
1. Change in length of offset during faulting	Increases	Remains constant
2. Direction of fault motion	Tends to increase the offset	Opposite to the direction for a transcurrent fault
3. Distribution of motion and of earthquakes along fault trace	All along the trace	Confined to the offset part of trace
4. Behaviour at ends of a fault (i.e. on intersecting a bounding coast)	Continues on to the continent until it dies out	Stops abruptly at the continental slope
5. Change in area of surrounding crust	None	Must be an increase or decrease at both ends
6. Near vertical fault with horizontal motion	Yes	Yes

(After Wilson (1970))

Wilson has given many examples of other types of transform faults, e.g. the sinistral and dextral fault systems that join the West Indies and Southern Antilles to the mountains on adjacent continents are concave arc to concave arc faults. Wilson's new concept of transform faults plays a key role in the theory of plate tectonics, and was necessary before it could be properly formulated.

Besides junctions of pairs of transform faults, trenches and mid-ocean ridges, triple points exist where three plates and three boundaries meet. Herron and

Fig. 8.1   Diagram illustrating the six possible types of dextral transform faults: (a) ridge to ridge; (b) ridge to concave arc; (c) ridge to convex arc; (d) concave arc to concave arc; (e) concave arc to convex arc; (f) convex arc to convex arc. Note that the direction of motion in (a) is the reverse of that required to offset the ridge (after Wilson (1965a))

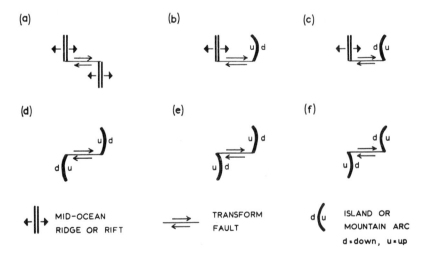

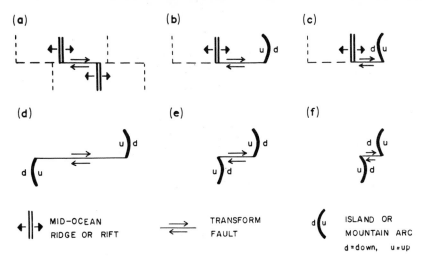

Fig. 8.2   Diagram illustrating the appearance of the six types of dextral transform faults shown in Fig. 8.1 after a period of growth. Expansion of the ridges has pushed their initial positions out to the positions now shown by dashed lines. The dashed positions are seismically inactive, although visible in the topography (after Wilson (1965a))

Fig. 8.3   Sketch map of the South Atlantic Ocean illustrating three ways of determining conjugate points. A and A′ are determined to be conjugate because the coasts fit together. Bb and bB′ are the Rio Grande and Walvis Ridges and Ccc′C′ is a fracture zone. B and B′ and C and C′ are also conjugate pairs (after Wilson (1965b))

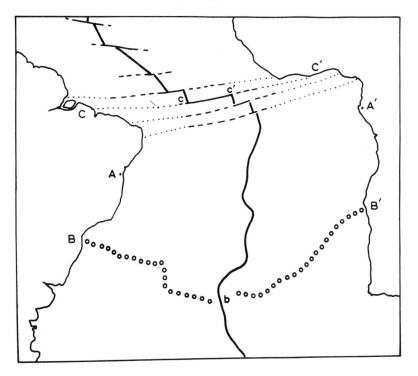

Heirtzler (1967) and Raff (1968) proposed that the Galapagos rift, an east Pacific rise, forms a triple point of active ridge axes. This has since been confirmed during a cruise of the USNS *De Steiguer* during April–May 1970, and Hey *et al.* (1972) place the triple point of the Pacific, Cocos and Nazca plates at 2°11′N, 102°10′W. The concept of a triple point of three ridge axes has also been invoked to explain the origin of the Great Magnetic Bight of the North-east Pacific (Raff (1968); Pitman and Hayes (1968); Vine and Hess (1970)). Falconer (1972) has produced evidence that the Indian, Antarctic, and Pacific plates intersect in a ridge-fault-fault triple junction at 61°30′S, 161°E. The problem of triple points of various combinations of plate boundaries has been discussed by McKenzie and Morgan (1969). They analysed all possible forms of triple junctions with velocity vector triangles (see Fig. 8.4) and showed that they can be stable or unstable, depending on whether they are able or unable to maintain

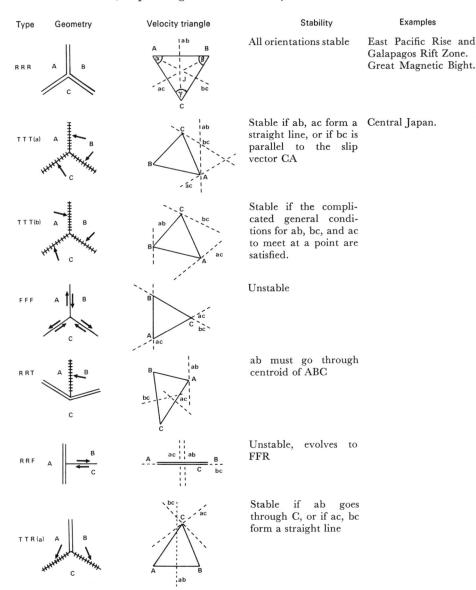

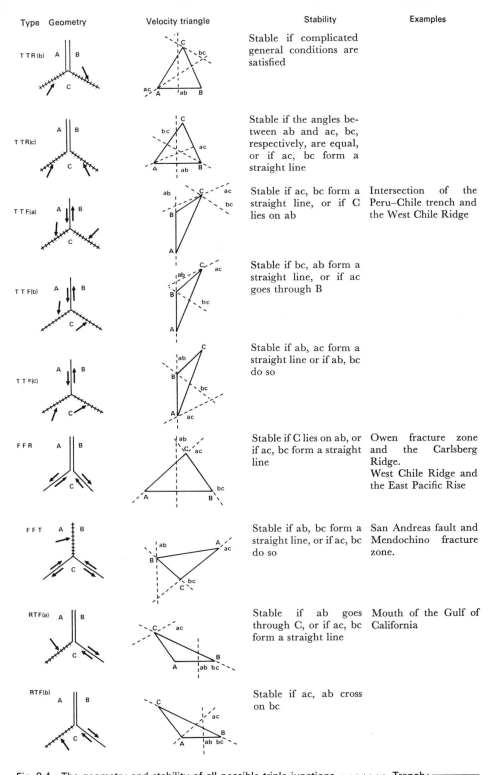

Type	Geometry	Velocity triangle	Stability	Examples
T T R (b)			Stable if complicated general conditions are satisfied	
T T R(c)			Stable if the angles between ab and ac, bc, respectively, are equal, or if ac, bc form a straight line	
T T F(a)			Stable if ac, bc form a straight line, or if C lies on ab	Intersection of the Peru–Chile trench and the West Chile Ridge
T T F(b)			Stable if bc, ab form a straight line, or if ac goes through B	
T T F(c)			Stable if ab, ac form a straight line or if ab, bc do so	
F F R			Stable if C lies on ab, or if ac, bc form a straight line	Owen fracture zone and the Carlsberg Ridge. West Chile Ridge and the East Pacific Rise
F F T			Stable if ab, bc form a straight line, or if ac, bc do so	San Andreas fault and Mendochino fracture zone.
R T F(a)			Stable if ab goes through C, or if ac, bc form a straight line	Mouth of the Gulf of California
R T F(b)			Stable if ac, ab cross on bc	

Fig. 8.4 The geometry and stability of all possible triple junctions. ++++++ Trench; ═══════ Ridge; ──── Transform fault. Dashed lines ab, bc and ac in the velocity triangles join points the vector velocities of which leave the geometry AB, BC and AC, respectively, unchanged. The relevant junctions are stable only if ab, bc and ac meet at a point. This condition is always satisfied by RRR; in other cases the general velocity triangles are drawn to demonstrate instability. Several of the examples are speculative (after McKenzie and Morgan (1969))

their geometry as they evolve. They applied their analysis to a number of problems such as the evolution of the San Andreas Fault.

McKenzie (1969b) has pointed out that the existence of plate margins and seismic belts shows that the relative motion between plates is taken up along weak zones in the lithosphere whose positions change slowly (if at all) with time. Under these conditions there is little relationship between the type of faulting and the orientation of the principal stresses. The nature of faulting is intimately related to the relative motion between the plates and the strike of the plate boundary, i.e. to the displacement rather than to the stress field. Since the type of faulting is not closely related to the stress field, very little is known about the orientation of the principal axes within the plates.

8.3 The seismic evidence

Figures 3.1 and 3.2 show maps of the seismicity of the world for the period 1961–1967, from which a number of important conclusions can be drawn:

(a) To a first approximation it is possible to divide the earth's surface into a number of essentially aseismic plates or blocks that are bounded by active seismicity associated with ridge crests, faults, trenches and mountain systems.

(b) The seismic belts are in general continuous. Even for this very short time interval there are few, if any, substantial gaps in the major belts which nearly always terminate on intersection with another belt. There are some minor exceptions; a few shocks occur in normally stable blocks and scattered activity is found occasionally (e.g. in southern Africa).

(c) The seismic belts are generally narrow. Those shown in Figs. 3.1 and 3.2 are, in general, much narrower than those reported earlier by Gutenberg and Richter (1954) using poorer data and less refined techniques.

Le Pichon (1968) has divided the earth's surface into six blocks bounded by belts of tectonic activity (see Fig. 8.5). Although the number of blocks in the real earth must be greater, six is adequate for a first approximation. From the orientation of fracture zones and rates of sea floor spreading based on the marine geomagnetic anomaly pattern (see § 8.4), Le Pichon computed the movement of each block relative to an adjoining block.

The concept of rigid plates moving around on the earth's surface and interacting at their boundaries has been remarkably successful in explaining tectonics and the evolution of ocean basins. Tension results in a rise, compression in a trench, and lateral motion in a transform fault and fracture zone. The boundary between two oceanic plates can be a deep oceanic trench, an oceanic rise or a strike-slip fault, depending on whether the plates are approaching, receding from, or moving past one another; the forces involved are respectively compressional,

232

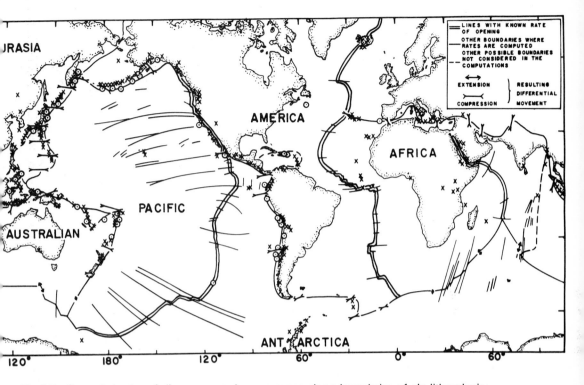

Fig. 8.5 Computed rates of divergence and convergence along boundaries of six lithospheric blocks (after Le Pichon (1968)). Computed movements were derived from rates of spreading determined from magnetic data and from orientations of fracture zones along features indicated by double lines. The extensional and compressional symbols in the legend represent rates of 10 cm/yr; other similar symbols are scaled proportionally. Symbols appearing as diamonds represent small computed rates of extension for which the arrowheads coalesced. Historically active volcanoes (Gutenberg and Richter (1954)) are denoted by crosses. Open circles represent earthquakes that generated tsunamis (seismic sea waves) detected at distances of 1000 km or more from the source (after Isacks, Oliver and Sykes (1968))

tensional and shearing. The plates are perhaps as little as 0–10 km thick at ocean ridges and as much as 200 km or more thick far from the ridges. Seismic results indicate that the South American lithospheric plate is very thick (about 200–300 km)—no low velocity, high seismic attenuation channel for shear waves has been found to depths of about 300 km (James (1971)). Similar observations indicate that the underthrust Pacific plate is only 50–60 km thick. When a thick continental plate is involved, compression can also result in high upthrust and folded mountain ranges: the Himalayas probably resulted from the collision of the subcontinent of India with Asia. Tension can result in a wide zone of crustal thinning, normal faulting and volcanism; it can also create a narrow rift of the kind found in the Gulf of California and the Red Sea.

The characteristics of plate tectonics on a sphere were determined simultaneously and independently by McKenzie and Parker (1967) and Morgan (1968). Any relative motion of two rigid plates on the surface of a sphere is equivalent to a rotation about some axis. This is merely Euler's theorem. If one of the two plates is taken as fixed the movement of the other corresponds to a rotation about some pole, and all relative velocity vectors between the two plates must lie along small circles that are lines of latitude with respect to that pole. On the assumption that in this way crustal blocks can be regarded as rigid plates, Morgan (1968) and Le Pichon (1968) deduced instantaneous centres of relative movement between blocks separated by active ridge crests in three ocean basins. Chase (1972) has since developed a least squares method for finding all the relative motions of any number N plates interacting simultaneously, thus ensuring internal consistency of the results.

Transform faults conserve crust and are lines of pure slip. Therefore, they are always parallel to the relative velocity vector between two plates. There is no geometrical reason why ridges or trenches should lie along lines of longitude with respect to the rotation pole and in general they do not. The pole position itself has no tectonic significance—it is merely a construction point. It must be stressed that plate motion is relative; there is no coordinate system within which absolute plate motion can be defined. We can only specify motion in relation to a particular plate or plate boundary that is arbitrarily chosen as being fixed. Any frame can be used as a reference frame and no special frame is favoured by the observations. Moreover, in dealing with phenomena which have deep roots, we must be careful to distinguish the motion of two plates relative to one another from the motion of each plate relative to the underlying mantle.

The general conclusion that most of the world's earthquakes are predominantly strike-slip was based on unreliable data. At the zones of divergence there are two kinds of focal mechanisms (Sykes (1967)). For those shocks along the central rift of the ridges there is predominantly normal faulting, in agreement with the concept of freshly formed thin lithosphere being pulled apart. Along the fracture zones where seismicity is high between intersections with ridges, there is predominantly strike-slip faulting along a near vertical surface (Fig. 8.6). More important is the finding that the sense of motion is that expected from a transform fault (Wilson (1965a)) and opposite to that expected on the assumption that the ridges had been offset by transcurrent faulting. Some of the largest active strike-slip faults on land, such as the San Andreas fault, may be interpreted as transform faults. It appears that part of Northern and Central California west of the San Andreas Fault has moved northwest more than 1100 km while the southern San Andreas Fault has moved only about 500 km. D. L. Anderson (1971) has shown that this discrepancy can be reconciled if we drop the concept of a single San Andreas Fault and assume that the two segments of the fault originated at different times and in different ways and may now be moving at different rates.

At the zones of convergence focal mechanisms are more complex. In the most active seismic zone that lies primarily beneath the landward slope of a trench, the focal mechanisms consistently indicate large-scale underthrusting of slabs of lithosphere beneath the arc. Beneath the seaward slope of a trench earthquakes are shallow and relatively few. The focal mechanisms of these shocks consistently

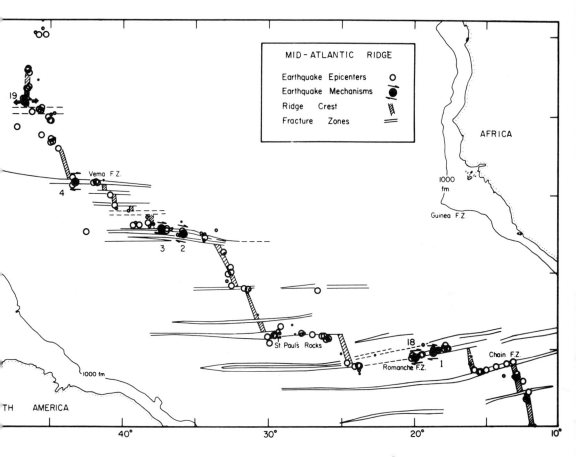

Fig. 8.6 Relocated epicentres of earthquakes (1955–1965) and mechanism solutions for six earthquakes along the equatorial portion of the Mid-Atlantic Ridge. The ridge crest and fracture zones are from Heezen *et al.* (1964a, b). The sense of shear displacement and the strike of the inferred fault plane are indicated by the orientation of the set of arrows beside mechanisms 1, 2, 3, 4 and 18. These solutions are characterized by a large component of strike-slip motion. The horizontal projection of the axis of maximum tension is shown by heavy arrows for event 19, a solution with a large component of normal faulting. Large circles denote the more precise epicentral computations; smaller circles, the poorer determinations (after Sykes (1968))

indicate tension, approximately normal to the axis of the trench (Stauder (1968a,b)). Although contortions and even disruption of the descending plate undoubtedly occur in some regions, stresses due to the bending of lithospheric plates do not seem to play an important role in the generation of earthquakes in the mantle.

Earthquakes at greater depth have different mechanisms from those of shallow shocks. Isacks and Molnar (1969) used focal mechanism data to support the idea that at intermediate depths the lithosphere sinks into the asthenosphere under its own weight, but below about 300 km encounters resistance with stronger or denser material. The results of their more detailed and comprehensive analysis (1971)

235

of 204 focal-mechanism solutions strongly support the idea that those portions of the lithosphere that descend into the mantle behave as slab-like stress guides that align the earthquake generating stresses parallel to the inclined seismic zones. At intermediate depths, extensional stresses parallel to the dip of the zone are predominant in those zones that are characterized either by gaps in the seismicity as a function of depth or by an absence of deep focus earthquakes. On the other hand, stresses parallel to the dip of the zone tend to be compressional where the zone extends below about 300 km. The double-couple or shear dislocation model of the source mechanism (see § 3.1) is adequate to explain the data. The parallelism between the axes of compression or tension of the double-couple solutions and the seismic zones is the primary evidence that the earthquakes in the mantle occur inside descending lithospheric plates in response to stresses within the plates. The non-parallelism between the nodal planes (one of which is the fault-plane) and the seismic zones shows that the solutions cannot be interpreted as simple shearing parallel to the zones.

Consider now the down-dip compression between depths of about 300–700 km. The compressional stress inside the slab is the resultant of the downward directed forces applied to the upper portions of the slab and the upward directed resisting forces applied to the lower parts of the slab. The cause of this resistance could be an increase in strength in the surrounding mantle or a buoyancy effect if the density of the slab were lower than that of the surrounding mantle material. The range of depths in which down-dip compression is predominant is approximately between 350–400 and 650–700 km, at which depths major phase changes are believed to occur (Anderson (1967)). Figure 8.7 (after Isacks and Molnar (1969)) illustrates possible stress distributions in a descending slab of lithosphere. In (a)

Fig. 8.7 A model showing plausible distributions of stresses within slabs where gravitational forces act on excess mass within the slabs. A filled circle represents down-dip extension, an unfilled circle represents down-dip compression, and the size of the circle qualitatively indicates the relative amount of seismic activity. In (a) the slab sinks into the asthenosphere, and the load of excess mass is mainly supported by forces applied to the slab above the sinking portion; in (b) the slab penetrates stronger material, and part of the load is supported from below, part from above; the stress changes from extension to compression as a function of depth. In (c) the entire load is supported from below, and the slab is under compression throughout. In (d) a piece has broken off. A gap in seismicity as a function of depth would be expected for (b) and (d), whereas no deep earthquakes occur beneath (a). The horizontal dashed lines in the figures indicate possible phase changes in the upper mantle near 350–400 km and 650–700 km (Anderson (1967)) (after Isacks and Molnar (1969))

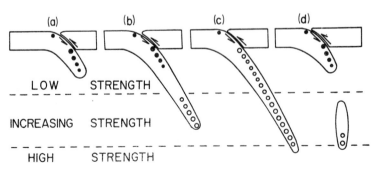

the slab sinks into the asthenosphere and the load of excess mass is mainly supported by forces applied to the slab above the sinking portion. The stresses are tensile throughout the slab and no deep earthquakes occur; in (b) the slab penetrates stronger material, and part of the load is supported from below (where the stresses are compressional), and part from above (where the stresses are tensile); in (c) the entire load is supported from below and the slab is under compression throughout; in (d) a piece has broken off. The horizontal lines indicate possible phase changes in the upper mantle. A break in the seismicity as a function of depth would be expected in cases (b) and (d). In (b) the lithosphere is continuous but the deviatoric stress vanishes in changing from compression to extension. In (d) the lithosphere is discontinuous and the break in seismicity corresponds to a break in the lithosphere; a piece of lithosphere has become detached and is sinking independently. There is some evidence that at least in some regions a piece of lithosphere may have become detached in this manner (e.g. under Peru, Western Brazil and the New Hebrides). In a more recent paper, Barazangi *et al.* (1973) confirm that this is indeed the case beneath the New Hebrides island arc. The interpretation of observations of deep earthquakes in New Zealand is ambiguous, but other evidence indicates the detachment of pieces of the lithosphere beneath New Zealand as well.

The above results place severe constraints on any theory of mantle convection. The predominance of compressive stresses at great depths, and the correspondence of the depth of the sharp cut-off in world seismicity as a function of depth with a major discontinuity or transition region in the mantle (Engdahl and Flinn (1969)) suggest, but do not prove, that lithospheric material does not penetrate any deeper than about 700 km. Again since the descending slabs would be impenetrable, the configuration of the inclined seismic zones places strong restrictions on the movement of material in the asthenosphere: any flow of material must turn down near the zones and cannot pass through them.

In 1968 Brune developed a quantitative method for estimating the rate of slip along major fault zones represented by the occurrence of earthquakes. The method gave excellent results in California along the San Andreas fault where there is good geodetic and geologic control. It also indicated rates of underthrusting under island arcs comparable to the rates suggested for sea floor spreading based on magnetic anomalies (see § 8.4). However, for deep earthquakes the computations indicated that the seismicity was not sufficient to account for such large slip rates and therefore a good deal of motion must be occurring as creep. In a further study of seismic slip rates, Davies and Brune (1971) obtained additional quantitative agreement with rates of slip predicted by magnetic anomalies and sea floor spreading. Their study suggests that worldwide interactions must in part be controlling the occurrence of major earthquakes. Their data also illustrate one of the difficulties, common to most areas of geophysics—that of trying to make deductions without having an adequately long time sample. For the world taken as a whole, and for many major fault zones, much of the seismic slip occurred during the episode of high seismicity around 1900. If seismic data had been available only for the last few decades anomalously low rates of slip would have been deduced.

8.4 The magnetic evidence

Magnetic anomalies observed in the ocean basins (see Fig. 8.8) are quite unlike those observed over land or on the continental shelves. They have long wavelengths that appear as linear or greatly elongated features on contoured anomaly

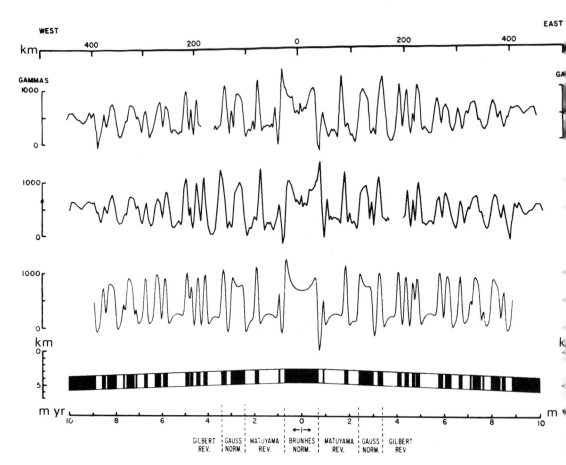

Fig. 8.8 Magnetic profile over the Pacific–Antarctic ridge. The middle curve is the Eltanin 19 magnetic profile—east is to the right. The upper profile is that of Eltanin 19 reversed—west is to the right. The lower profile is the theoretical profile for the model shown below. Normally magnetized bodies are black and reversely magnetized are white, and all are 2 km thick. With a spreading rate of 4·5 cm/yr the magnetized blocks correlate with the known history of reversals to the Gilbert epoch (after Pitman III and Heirtzler (1966))

maps. There are two particularly outstanding features of oceanic magnetic anomalies: a central anomaly associated with the axis of a ridge, and a remarkable striped pattern of anomalies parallel to the ridge axis and symmetric about it. These anomalies retain their characteristic shape and spacing for thousands of km along their length and are very different from those observed over continents. They are very difficult to simulate if one assumes any reasonable lithologic contrasts within the oceanic crust or plausible geologic structures. An explanation for their origin was given in 1963 by Vine and Matthews who suggested that variations in the intensity and polarity of the earth's magnetic field may become 'fossilized' in the oceanic crust and that, as a result of sea floor spreading, would be preserved in the resulting short wavelength anomalies in the earth's magnetic field. Thus the 'conveyor belt' concept of Hess, in which upwelling mantle material beneath the oceanic ridges is carried away in both directions, can also be thought of as a tape recorder. As new oceanic crust forms and cools through the Curie temperature at the centre of an oceanic ridge, the permanent component of its magnetization will assume the ambient direction of the earth's magnetic field. A spreading rate of a few cm/yr and a duration of 700 000 yr for the present polarity would imply a central block of crust a few tens of km in width in which the magnetization is uniform and normal. The adjacent blocks would be of essentially reversed polarity and the width and polarity of successive blocks more distant from the central block would depend on the reversal time scale of the earth's magnetic field in the past. The aeromagnetic survey over the Reykjanes Ridge (Fig. 8.9) in 1963 was one of the first detailed surveys over the known axis of a mid-ocean ridge. It clearly shows that magnetic anomalies are parallel to the ridge axis and symmetric about it, and supports the theory of sea floor spreading, and the hypothesis of Vine and Matthews for the origin of the anomalies.

By 1966 a geomagnetic reversal time-scale had been deduced empirically for the past 3·5 m yr from detailed palaeomagnetic and geochronological studies (see § 6.3). It was then possible, from a knowledge of the distance from the ridge axis at which particular reversals occurred, to show that the magnetic anomalies associated with ridge crests during this period could be explained in terms of symmetrical spreading at essentially constant rates. The same sequence of magnetic anomalies away from ridge crests is observed for thousands of km in all major ocean basins, implying that spreading and reversals of the earth's magnetic field have occurred for a significant part of geologic time. To assign provisional ages to the older magnetic anomalies and thus to the underlying oceanic crust and reversals of the earth's magnetic field, Heirtzler et al. (1968) assumed a constant spreading rate of 2 cm/yr in the South Atlantic. One of the chief objectives of Leg III of the JOIDES (*Joint Oceanographic Institutions for Deep Earth Sampling*) project was to test the spreading hypothesis and proposed geomagnetic reversal time-scale. Eight sites were drilled across the mid-Atlantic ridge in the South Atlantic at a latitude of approximately 30°S. At all but one site the complete sedimentary column was penetrated and basalt basement recovered. The palaeotological age of the oldest sediment recovered at each site, i.e. immediately overlying or incorporated within the basalt, was found to be directly proportional to the distance of the site from the ridge axis (see Fig. 8.10, and Maxwell et al. (1970)). Thus spreading about the South Atlantic ridge seems to have been continuous throughout the Cenozoic and at an essentially constant rate of 2 cm/year, as was assumed by Heirtzler et al. in assigning ages to the magnetic

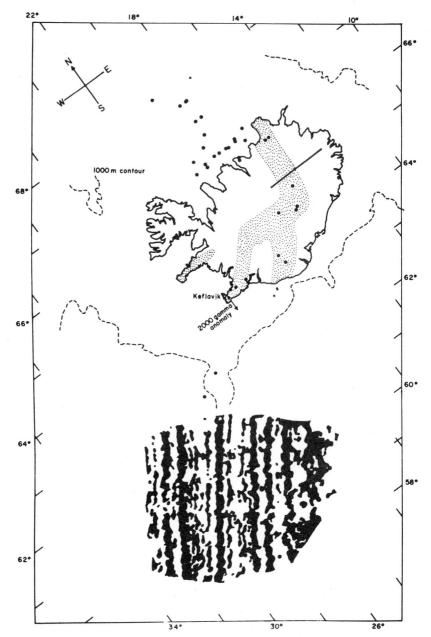

Fig. 8.9   Magnetic survey over the Reykjanes Ridge southeast of Iceland. Positive magnetic anomalies are shown in solid black. The belts of Quaternary volcanoes in Iceland are shaded. Epicentres north of the survey area are represented by solid dots (after Heirtzler *et al.* (1966)

anomalies. This leads to the surprising conclusion that perhaps 50 per cent of the present deep sea floor, i.e. one-third of the surface area of the earth, has been created during the last 1·5 per cent of geologic time. It seems likely that all other

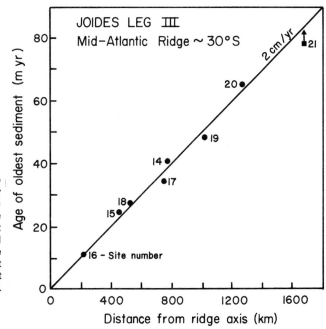

Fig. 8.10 The age of the oldest sediments recovered by JOIDES Leg III at each site across the mid-Atlantic ridge in the South Atlantic at approximately 30°S plotted against the distance of the site from the ridge axis. Basalt basement was not reached at site 21 (after Maxwell *et al.* (1970))

oceanic areas are no greater than Mesozoic in age (i.e. that they were formed within the past 225 m yr). The oldest sediments recovered so far by the JOIDES drilling ship are middle Jurassic in age (i.e. about 160 m yr old).

In some regions, notably along the continental margins, magnetic anomalies are all but non-existent. In the north Atlantic a continuous smooth zone extends from Newfoundland to the Bahamas in the western basin, and from Ireland to the Cape Verdes in the eastern basin (see e.g. Hertizler and Hayes (1967); Vogt *et al.* (1970); Rona *et al.* (1970)). The origin of such zones is uncertain. Possible explanations are that they were formed

(*i*) when the geomagnetic field did not reverse for a long period of time
(*ii*) when the proto-Atlantic basin was at a low geomagnetic latitude
(*iii*) when some process, such as sediment burial of the early spreading axis prevented formation of the highly magnetized pillow basalt layer or caused its magnetic erasure near the axis.

Few magnetic studies in the South Atlantic have been published until recently. Mascle and Phillips (1972) have now examined aeromagnetic profiles from Project Magnet and found two types of magnetic smooth zones symmetrically

241

placed with respect to the mid-Atlantic ridge. The first borders the continental margins off Argentina and the Union of South Africa, in a position similar to that found in the North Atlantic and along the northwestern African margin. The second type, characterized by a less smooth pattern, lies in the deeper part of the South Atlantic, principally over the abyssal plains (parts of the Argentine–Brazil and Angola–Cape basins). From the limited amount of data available, Mascle and Phillips favour a diagenetic origin for the marginal smooth zones resulting from burial beneath a thick sedimentary cover. For the origin of the abyssal smooth zones they suggest sea floor spreading during middle-late Cretaceous time when the polarity of the earth's magnetic field was relatively constant.

Magnetically smooth areas have also been found in the Pacific (Hayes and Pitman (1970)) and other oceans. Geophysical studies have also been carried out over inland or marginal seas. No magnetic lineations indicative of sea floor spreading and geomagnetic reversals have been identified in any of them. Some areas are clearly continental. However, under other areas seismic refraction results have revealed a crust not greatly different from that under the major ocean basins.

There is one curious feature of the marine geomagnetic anomaly pattern that is as yet unresolved: the amplitudes are not constant but decrease with distance from the ridge. The central anomaly is generally almost twice as large as that on either side. Recent studies (Haggerty and Irving (1970); Irving (1970); Schaeffer and Schwarz (1970)) on rocks dredged from the mid-Atlantic ridge have shown that the intensity of the natural remanent magnetization (NRM) of rocks from the axial part of the ridge is more than an order of magnitude larger than that of rocks from the flanks (see Fig. 8.11). The intensity falls to about one-twentieth within 20 km of the axis. Taking a value of 1 cm/yr for the spreading rate, more

Fig. 8.11   Natural remanent magnetization (c.g.s./cm³) of rocks dredged from the mid-Atlantic ridge versus distance from the ridge axis (after Irving (1970))

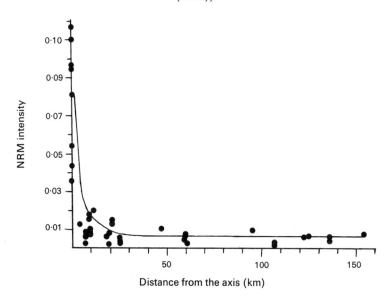

than an order of magnitude change in the NRM occurs in 2 m yr. Several explanations have been put forward to account for these facts.

One of the earliest suggestions was that the rocks producing the older anomalies are deeper and have a thicker sediment cover and thus are partially shielded magnetically. However, there does not seem to be any correlation of amplitude with sediment thickness. Matthews and Bath (1967) suggested that the dikes causing the central anomaly might be injected slightly asymmetrically. Normally magnetized dikes 200 m wide with a standard deviation of 5 km were shown to 'contaminate' the blocks of opposite polarity on either side. Harrison (1968) pointed out, however, that such large deviations would obliterate the short magnetic events, such as the Olduvai event. He offered as an alternative explanation a change in the intensity of the earth's magnetic field. Carmichael (1970) made essentially the same suggestion, but to account for more than an order of magnitude difference between the NRMs on the axis and the flanks, the change in the intensity of the main field would have had to have been far greater than seems possible. Both Harrison and Carmichael also suggested the possibility of a secular decay of the frozen NRM of the dikes due to the demagnetizing action of past reversals. Such an effect would be similar to that of demagnetization by a slowly alternating weak magnetic field, i.e. similar to that of the viscous magnetization acquired at room temperature in the earth's field, and is not likely to be important. Banerjee (1971) proposed the decay of magnetization by diffusion of ferrous ions but with no chemical alteration. Chemical alteration has often been suggested as the cause of many problems in palaeomagnetism but definite proof is difficult to obtain. Banerjee's intrinsic decay mechanism for the NRM due to ionic diffusion of Fe^{2+} ions in quenched titanomagnetites is based on theoretical arguments and depends on uncertain estimates of several parameters.

Haggerty and Irving (1970) had suggested that the high axial remanent magnetization is due to unoxidized titanomagnetite, while the weak flank remanent magnetization is due to destruction of the original TRM during subsequent maghemitization (i.e. oxidation of titanomagnetite to titanomaghemite). Schaeffer and Schwarz (1970) proposed that the CRM produced during the above maghemitization process could explain the observed large differences in NRMs, as CRM is known to be generally weaker than TRM. In a later paper, Ozima (1971) followed up these suggestions and showed that concurrent magnetic processes accompanying maghemitization can account for the above characteristic features of the marine geomagnetic anomaly patterns. Evans and Wayman (1972) have carried out an electron microscope investigation of the magnetic minerals in basalt samples and succeeded in identifying a finely dispersed phase believed to be single domain titanomagnetite grains. In the last few years there has been an increasing amount of evidence favouring single domain magnetite as a carrier of TRM in continental rocks (see § 6.2). Further work needs to be done, however, before the problem of the decay of magnetic anomalies with distance from the axis of the ridge can be considered as solved.

8.5 The thermal evidence

Heat flow measurements have already been discussed in some detail in § 7.2. To account for the equality of the heat flow through old ocean basins and continental shield areas, Sclater (1972a) suggested that the lithosphere is twice as thick under the shields as under the old ocean basins. The contribution to the surface heat flow from the upper mantle is then about 0·5 for the shields and 1·0 for the deep ocean basins. Independent support for a thicker lithosphere under the continents than under the oceans comes from the interpretation of seismic data (Kanamori and Abe (1968); Brune (1969)), electrical conductivity data (Cox *et al.* (1970); Cox (1971)) and from calculations of the elastic deformation of the lithosphere due to surface loading (Walcott (1970)).

Sclater (1972b) has also examined the question of the high heat flow which is associated with many marginal basins of the western Pacific (see Fig. 8.12). High heat flow and shallow water are normally associated with active intrusion and tensional forces; the discovery of high heat flow in a region presumed to be dominated by compressive forces thus presents a real problem. There is a difference, however, between the marginal basins in the north-western and south-western Pacific. All marginal basins in the north-western Pacific, except the west Philippine and Celebes basins, have high flow, whereas in the south-western Pacific it is only the Fiji plateau and the Lau basin that show high heat flows. Sclater (1972b) has shown that the pattern in the south-western basins is compatible with our ideas on sea floor spreading and plate tectonics, but this cannot explain the observations in the north-west Pacific. Karig (1971a) has given convincing evidence that the marginal basins in the north-west Pacific owe their origin and development to large-scale crustal extension landward of the island arc system. Sclater has shown that the intrusion of oceanic crust associated with this extension cannot alone account for the heat flow and elevation data; an extra source of heat or some method of decreasing the elevation of the marginal basins is necessary as well. He makes a number of suggestions and favours a combination of the intrusion model and a thinner lithosphere which thickens with distance from the downgoing slab.

There is no longer any reason to believe that the mean continental and oceanic heat flows are the same—this result followed because the excess heat flow resulting from plate creation was included in the average heat flow. This should not be done since there is no corresponding process on the continents. The boundaries of plates are ridge crests, trenches or transform faults. Since each plate moves more or less as a single unit, active tectonic processes occur mainly along the edges, as indicated by the distribution of earthquake foci (Isacks *et al.* (1968)). As tectonic processes involve changes of energy, it might be expected that the heat flow will fluctuate more near the edge of a plate than in its interior. That this is largely so has been verified by Lee (1970).

Although the areas of anomalous heat flow generally correspond with boundaries of plates defined by seismic and other data, the general distribution of heat flow over a lithospheric plate is not what one would expect. The net heat

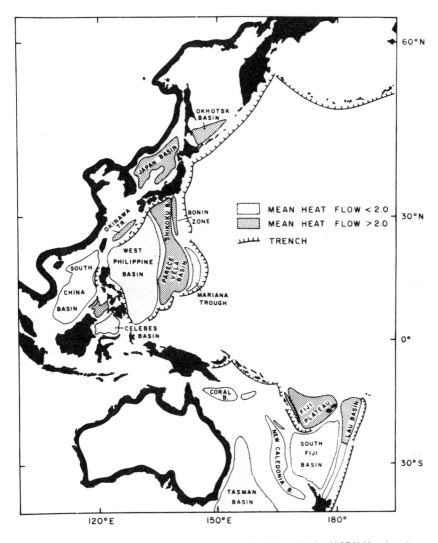

Fig. 8.12 Marginal basins of the western Pacific (from Karig (1971b)), showing regions of high heat flow—above 2·0 μcal/cm² s (after Sclater (1972b))

flow over the region in which lithosphere is being created is lower than that over the region in which it is thrust beneath the surface. McKenzie and Sclater (1968) have considered possible causes of the high heat flow below the active trenches. They suggested that stress heating of material along the plane of shear might be an important source of heat. This coupled with subsequent rapid transport of such heat to the surface by magmatic activity might explain the high heat flow. If magmatic activity also concentrates radioactive elements into the upper part of the lithosphere it might be possible to produce the observed excess. The enrichment of radioactive elements in the upper lithosphere over island arcs or continental margins bounded by active trenches would in turn imply that the heat flow anomaly produced there is long lived, the near surface heat production greatly inhibiting escape of heat from depth.

An Upper Mantle Symposium on the Mechanical Properties of the Upper Mantle was held at Flagstaff, Arizona in 1970; many of the papers have been collected and published in *J. geophys. Res.*, **76**, 1101–1466, 1971. The concept of plate tectonics has led to a somewhat different picture of mantle convection, in which the dominant motions are horizontal near the surface and the cycle is closed in an unknown manner below. Such convection may be acyclic, episodic and not necessarily stationary with respect to the mantle beneath. Schubert and Turcotte (1972) have more recently shown that models for mantle convection which postulate a counter flow in the asthenosphere to balance the motion of lithospheric plates cannot be correct. The return flow must lie deeper in the mantle, possibly below 700 km or even throughout the entire mantle. The upper mantle immediately below the lithosphere must then be moving in the same, rather than in the opposite, direction as the lithosphere.

There has been much controversy over the effect phase changes would have on mantle convection. Knopoff (1964) maintains that the existence of inhomogeneities (whether chemical or involving a phase transition) in the depth range 400–1000 km would inhibit mantle wide convection, and would also decouple any convective motion in the upper mantle from that in the lower mantle. On the other hand Schubert *et al.* (1970) and Oxburgh and Turcotte (1971) have suggested that the olivine-spinel phase change at a depth of about 400 km, far from preventing large-scale convection, may actually intensify deep mantle convection. They also showed that the olivine-spinel phase boundary would be elevated as much as 100 km in the descending slab of lithosphere beneath oceanic trenches relative to the rest of the mantle. The denser spinel phase within the slab would exert an additional body force (comparable in magnitude to the force on the slab due to thermal contraction) and tend to drive the slab down still further into the mantle.

Ringwood (1972) has shown that phase transitions would definitely inhibit large-scale convection currents throughout the mantle. Such currents have often been proposed as the mechanism for transporting passive lithospheric plates by viscous coupling and dragging them down into the mantle beneath oceanic trenches. The situation is very different, however, if the lithosphere plays an active rather than a passive role: in this case the lithosphere sinks into the mantle because it is denser than the surrounding mantle. In this regard, Ringwood supports Schubert and Turcotte's (1971) contention that the olivine-spinel phase change would, by causing a marked increase in the average density of the sinking slab, increase the gravitational forces on the downgoing slab. Phase transformations may thus play a major role in driving the lithospheric slabs into the mantle at depths greater than about 400 km.

A discussion of the thermal regime within a downgoing slab, and the flow and

246

stress fields caused by its motion, requires a knowledge of the mechanical and thermal properties of mantle materials. The sinking slab must be brittle enough to produce earthquakes by fracture and yet sufficiently undeformable to maintain its shape to a depth of 600 km after passing through perhaps 1000 km of mantle (Sykes (1966). Therefore for all purposes, except that of generating earthquakes, it behaves as a rigid plane slab. Since the thermal conductivity of rocks is small the slab will maintain its original temperature distribution even after it has sunk to considerable depths. Estimates of temperature at depth vary greatly depending on the assumptions made regarding the significance for heat transfer of conduction, radiation and convection—at a depth of 500 km there is a difference of about 1000°C betwᴇen estimated geotherms for a conduction model (Lubimova (1967)) and for a convection model (Tozer 1967)).

Oxburgh and Turcotte (1968) have obtained a steady state distribution of isotherms beneath an oceanic ridge computed on the basis of boundary layer theory and assuming constant viscosity independent of temperature. An alternative to this model of broad upwelling is the injection of dikes from a low velocity zone in a narrow axial zone (McKenzie (1967); Sleep (1969)). The distribution of geotherms beneath island arcs and continental margins has also been calculated by McKenzie (1969a), and Oxburgh and Turcotte (1970). McKenzie showed that there is a definite relationship between the temperature at the centre of the sinking slab and the occurrence of intermediate and deep focus earthquakes, and that the temperature structure beneath ridges and trenches is controlled by plate motions.

Details of the thermal regime in a sinking slab depend very much on what assumptions are made, and there has been much argument about the validity of some solutions (see, for example, the exchange of comments in *J. geophys. Res.*, **76**, 605–626, 1971, following a paper by Minear and Toksöz (1970)). It seems, however, that in all probability the sinking slab descends at a greater rate than that at which it can be heated to the temperature of the adjacent mantle (see Fig. 8.13). At a depth of 400 km the temperature in the slab could well be as much as

Fig. 8.13 Temperature distribution in a downgoing lithospheric slab for a 1 cm/yr spreading rate at time 103·7 m yr. Heat transfer is by lattice conduction only and no internal sources are included (after Toksöz *et al.* (1971))

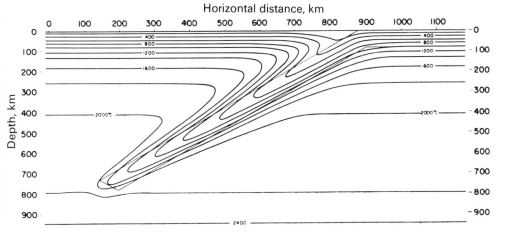

Horizontal distance, km

1000°C lower than that of the surrounding mantle. This temperature difference will decrease with depth but may be significant even at 600–700 km. Thus the slab may overcome the difficulty of the effectively increased adiabatic gradient in the vicinity of a phase transformation and penetrate the transition. Having done so the density contrast between the slab and adjacent mantle will have increased and, as already mentioned, this will help drive the slab to even greater depths.

It should not be forgotten that plate tectonics is a kinematic theory depending on fairly simple ideas. In contrast the equations that govern convection in a medium with variable viscosity have no known analytic solution and have been studied but little. The situation is analogous to the difference between the kinematic and dynamic dynamo problem (see § 5.6). A recent review of mantle convection and the new global tectonics has been given by Turcotte and Oxburgh (1972).

8.6 The mechanics of plate motions

In spite of the impressive evidence in favour of the hypothesis of sea floor spreading and the 'new global tectonics', we are little nearer to understanding the process or processes at depth by which they are initiated and maintained. Thermal convection in some form is probably the only mechanism capable of supplying sufficient energy. Convection in this context should be interpreted in its broadest terms, i.e. as the generation of horizontal temperature gradients which result in gravitational instabilities and the vertical transport of hot and cold materials (McKenzie (1969a)). The cold descending slab of oceanic lithosphere landward of the trench systems is also a source of gravitational energy and convection as defined above.

It cannot be over emphasized that the current seismicity and outline of crustal plates is only a transient one, and there is no *a priori* reason to suppose that it is valid to extrapolate back into geologic time. The concept of, and evidence for, plate tectonics only applies to motions that have taken place during less than 2 per cent of geologic time. It tells us nothing about the slow evolution of plate boundaries or changes in their relative motion. Thus the breakup of Gondwanaland was presumably caused by stresses within the original plate and cannot be understood from geometry alone.

There is no known difference between the motion and behaviour of plates consisting mostly of continental lithosphere (such as the Eurasian plate), and those that consist almost entirely of oceanic material (such as the Pacific plate). There is, however, a remarkable difference in the distribution of seismicity where plate boundaries cross continental regions (see Figs. 3.1 and 3.2). The narrow, sharply defined trenches and the regularly inclined earthquake zones sloping away from trenches show that oceanic lithosphere is easily consumed by subduction. Intracontinental seismic zones associated with mountain ranges are spread over a wide belt, implying that continental lithosphere is hard to consume. It is surprising that plate boundaries have so little relationship with continent-ocean boundaries and that most plates consist of both oceans and continents.

There is abundant evidence that spreading has occurred at ridge crests producing an extension of the crust and an increase in the surface area of the earth. Evidence for crustal destruction in the sinks of the system (the trenches) is less convincing except for the focal mechanism solutions for some earthquakes (see § 8.3). It has thus been argued that spreading could equally well be accounted for by an expansion of the earth (see also § 1.5). A difficulty with all expansion hypotheses is the rate required—sea floor spreading velocities are more than an an order of magnitude greater than had been expected and catastrophic expansion would be required, starting in the Jurassic. Again, spreading in the Atlantic and Pacific Oceans is predominantly in an east–west direction at the equator, leading to increases of up to 16 cm/yr in the equatorial circumference. On the other hand the rate of increase along meridians rarely exceeds 6 cm/yr. Differential rates of increase of this magnitude would produce differences in circumference of more than 1000 km within 10 m yr and the earth would not maintain its approximately spherical shape (Le Pichon (1968)). It thus follows that oceanic crust and upper mantle must be destroyed somewhere.

For many years Vening Meinesz (1962) had maintained that trenches were the site of such destruction but until recently the only firm evidence has come from seismology. Detailed mapping of surface displacements in Alaska after the 1964 earthquake can only be explained by underthrusting of the ocean beneath the continent on an enormous scale (Plafker 1965)). Focal mechanism studies of the main shock and many aftershocks have also confirmed this (Stauder and Bollinger (1966)). A number of other phenomena appear to be related to plate destruction: the occurrence of a trench 3 to 4 km deeper than the surrounding abyssal ocean floor, of a large negative gravity anomaly and smaller positive one over the island arc, of intermediate and deep focus earthquakes, and of a large heat flow anomaly above the descending slab. Thus there is now a considerable body of evidence that the cold lithosphere moves down to great depths beneath the trenches. Such a model also accounts for the propagation of high frequency P and S waves along the plane in the Tonga area (Oliver and Isacks (1967)) and in Japan (Utsu (1967)).

A number of causes of plate motions have been suggested. The older theories postulate large-scale convection throughout at least the upper mantle—viscous forces are then required to couple the plates to the moving mantle below. McKenzie (1968) has shown that a modified version of such a theory is not incompatible with all relevant observations. A closely related theory is that of Elsasser (1969), who suggested that the lithosphere may be considered as a stress guide, and that surface motions are maintained by the cold sinking slabs. The energy source of Elsasser's mechanism is still thermal convection, since it is driven by the temperature induced density contrast between the cold slab and hot mantle through which it sinks, but the force causing the motion is transmitted through the rigid lithosphere and not through viscous coupling between the mantle and the plates. Another mechanism which could maintain motions is the force exerted on the plate boundary by neighbouring plates—the motion of small plates may well be sustained by such forces. This mechanism cannot produce worldwide movements of large plates, however, since it has no source of energy. McKenzie (1969a) has shown that Elsasser's model cannot be correct and that the forces driving plates can only be fully understood if solutions to the full convection equations are obtained; solutions to simplified models give only limited insight into the general problem and may not be internally consistent.

McBirney (1971) has raised a number of questions about the subduction process. If the intrusion of basic magma at ridge crests has been at the same rate in the past, the crustal material of the oceans must have been produced many times over in Phanerozoic time alone. A small fraction of the oceanic crust may become incorporated into the continents, but it seems that most of it is returned to the mantle, and basalts are thus endlessly recycled. McBirney comments that 'geologically and conceptually this is a lot to swallow' and that 'the efficiency of this process is truly remarkable'. Basalts must continue to be erupted with no apparent systematic differences to distinguish them from oceanic basalts of Precambrian age. Moreover, sediments derived from the sialic continental crust must be carried down without perceptibly altering the regenerated basalts that reappear at the ridge crests. McBirney also found no correlation between volcanism and subduction rates or distances from spreading centres. One possible answer to this dilemma is given by McBirney himself who pointed out that igneous and tectonic events occur spasmodically. There is some evidence that sea floor spreading may not be a steady state process and thus may perhaps be reconciled with the episodic events it is indirectly assumed to produce. When considered in detail the magnetic anomalies show considerable variations in spreading rate (Vine (1966); Heirtzler et al. (1968)) and even discontinuities in the record in the north-west Indian Ocean (Le Pichon and Heirtzler (1968)) and the north Atlantic (see also § 8.4).

Protracted worldwide discontinuities in spreading rates have been inferred from a consideration of the distribution of oceanic sediments (Ewing and Ewing (1967); Le Pichon (1968)), and on the basis of palaeomagnetic results from the continents (Briden (1967); McElhinny (1970)). In particular it has been suggested that the present phase of spreading may only have been operative for the past 10 m yr. Recent detailed palaeomagnetic studies indicate that the pole, relative to particular continental areas, moves rapidly within comparatively short periods of time (typically 10–20 m yr) and then remains stationary with respect to that continent for much longer periods of time, e.g. 50–100 m yr (Irving (1966); Briden (1967); DeBoer (1968); McElhinny et al. (1968); McElhinny (1970)). This has been interpreted as an indication of episodes of rapid drift separated by periods of quiescence. It is important to consider the possibility of discontinuous spreading within the framework of plate theory, although it cannot answer all the questions raised by McBirney.

Dickinson and Luth (1971) have suggested an alternative solution which eliminates the need to invoke successive phases of differentiation and re-mixing of the same materials. In essence they proposed that sea floor spreading processes lead to an irreversible evolution of the mantle. The refractory portion of the descending lithosphere falls to the base of the asthenosphere and is thus added progressively to the outer surface of the mesosphere, i.e. as sea floor spreading proceeds, the mesosphere grows and the asthenosphere diminishes. Presumably this has been going on since sea floor spreading began; Dickinson and Luth show, from a semi-quantitative analysis based largely on rates of mass transfer, that the present mesosphere could have been produced within a geologically reasonable time span. As a corollary it follows that global tectonic processes will cease when the asthenosphere has all been used up—sometime, according to Dickinson and Luth, within the next 1000 m yr. Their proposal is worth serious consideration; it is difficult to imagine any process by which the refractory lithosphere could be assimilated into the undepleted lower mantle and, as Ringwood (1969) has concluded, the refractory part of the descending lithosphere probably does not return to the shallower asthenosphere.

Wilson (1963a, b; 1965b) suggested that linear chains of islands and sea-mounts in the Pacific have been formed by the passage of moving lithospheric plates over plumes rising from (relatively) fixed sources deep in the mantle below the moving asthenosphere. Wilson (1965c) also used the concept of crustal plate motion over mantle hot spots to explain the origin of the Walvis, Iceland-Faeroe and other aseismic ridges. Morgan (1972) later showed that the Hawaiian-Emperor, Tuamotu-Line and Austral-Gilbert-Marshall island chains could be generated by the motion of a rigid Pacific plate rotating over three fixed hot spots. Morgan also showed that the relative plate motions deduced from fault strikes and spreading rates agree with the concept of rigid plates moving over fixed hot spots. Figure 8.14 shows the absolute motion of the plates over the mantle and predicts the trends of the island chains and aseismic ridges away from hot spots. Morgan (1971) proposed that these hot spots are manifestations of convection in the lower mantle which provide the main motive force for plate movements—in his model there are about twenty deep mantle plumes bringing heat and relatively primordial material up to the asthenosphere and horizontal currents in the asthenosphere flowing radially away from each of these plumes. Such currents will produce stresses on the bottom of the lithospheric plates. Although there are a number of localized upwellings, Morgan believes that there are no corresponding specific points of downwelling, the return flow being uniformly distributed through-out the mantle. Also, although the pull of the sinking plate is needed to explain the gravity minimum and topographic deep locally associated with the trench system (Morgan 1965)), Morgan does not believe that sinking lithospheric plates can provide the main motive force. The above model is also consistent with the observation that there is a very real difference between oceanic island and oceanic ridge basalts—those lavas that are partly generated from primordial material from deep in the mantle would be expected to be chemically distinct from those generated at shallower depths in the asthenosphere.

Morgan also pointed out that there appears to be a paucity of continental hot spots and suggested that a search for continental activity be made, particularly

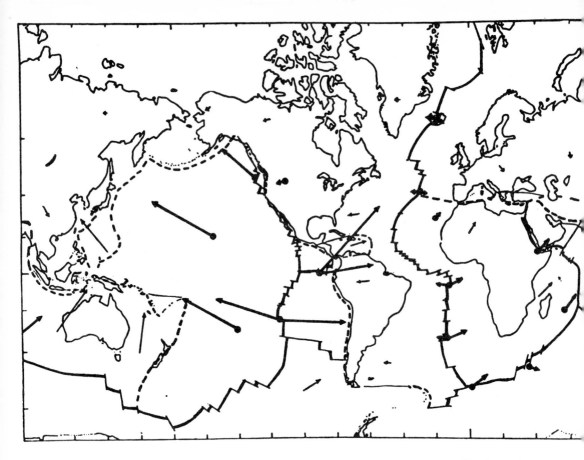

Fig. 8.14 Absolute motion of plates over the mantle. The arrows show the direction and speed of the plates—the heavier arrows show plate motion at hot spots. The synthesis was based on relative plate motion data (fault strikes and spreading rates) and predicts the directions of the aseismic ridges and island chains emanating from the hot spots (after Morgan (1971))

in Africa. Wilson (1972) followed this up (see Fig. 8.15). He suggested that the first visible stage would take the form of isolated uplifts capped by volcanoes (e.g. the Hoggar Massif in the Central Sahara). As a second stage, a series of plumes may become linked by rifts—many of the plumes forming triple points. The splitting of domes by widening rifts (as in the Red Sea and Gulf of Aden) marks the third stage: the rifts are formed by the lithospheric plates sliding off the uplifted domes. The Atlantic ocean is in the next stage—continental plates once joined have slid apart leaving halfdomes in western Africa and their counterparts in South America still linked by boundary ridges to islands over domes along the mid Atlantic ridge (see Fig. 8.15).

An interesting question is whether or not plumes are fixed to the mantle. If they are we have an absolute frame of reference which would enable true polar wandering, if any, to be distinguished from lithospheric plate motion. If polar wandering alone takes place by a rotation of the whole earth with respect to the rotational axis, a mantle fixed plume will also remain fixed to a lithospheric plate. If, however, lithospheric plate motion occurs, a plume trace will be produced as the plate moves over the plume. Assuming that plumes are fixed to the mantle and that igneous chains are plume traces on moving lithospheric plates, Duncan *et al.* (1972) investigated the history of two partly contemporaneous igneous chains in the European plate and concluded that polar wandering of about 23° has occurred over the past 50 m yr. On the other hand, McElhinny (1973) has estimated the vector sum of the horizontal displacements of all points on the earth's surface with respect to the axis of rotation over the past 50 m yr and

Fig. 8.15   The Atlantic Ocean and adjacent continents showing the mid-ocean ridge (dotted line), some lateral ridges (full lines in the oceans), domes and half-domes (hatched) on the continents connected by rift valleys (full lines on the continents). Note the isolated Hoggar Massif, and how many great rivers rise in domes often on the opposite side of the continent (after Wilson (1972))

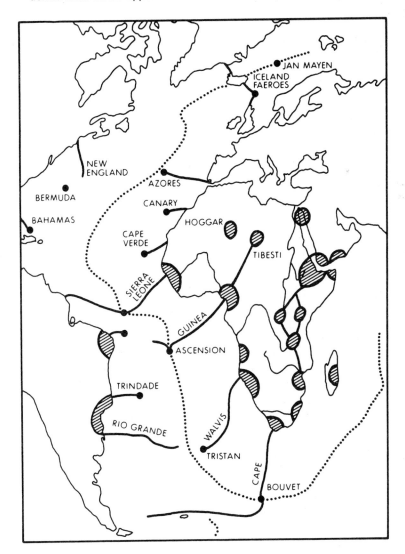

concluded that there has been no significant motion of the lithosphere as a whole with respect to the earth's rotational axis over this period of time. McElhinny thus questions the validity of the assumption of fixed plumes.

A further investigation of the fixity of plumes to the mantle has been carried out by Grommé and Vine (1972). If volcanic chains are plume traces and if the plumes are fixed to the mantle, then all the rocks in any given chain should have become magnetized at the latitude of the present active centre and should have the same palaeomagnetic inclination if no true polar wandering has occurred. Grommé and Vine have carried out a detailed palaeomagnetic study of two Midway Atoll drill cores, dated by the K–Ar method at 18 m yr. They obtained a palaeolatitude for the basaltic lavas of $14.7° \pm 4.2°$ which is not significantly different from the present latitude ($19°$) of the active volcanic island of Hawaii. It is significantly different, however, from the present latitude ($28°$) of Midway Atoll and Grommé and Vine suggest that the Pacific plate has drifted northward by about $13°$ during the past 18 m yr. It must be pointed out, however, that using European plate data, Duncan et al. (1972) found palaeomagnetic data to be consistent with fixed mantle plumes only if polar wandering has occurred; Grommé and Vine, on the other hand, using Pacific plate data can reconcile palaeomagnetic data with the assumption of fixed mantle plumes without invoking polar wandering.

Burke and Wilson (1972) have presented evidence which suggests that the African plate stopped moving relative to the mantle 25 m yr ago and has been at rest ever since. At the same time the pattern of spreading in the Indian Ocean altered and what had been the Chagos fracture zone became the crest of a new central Indian ridge (Fisher et al. (1971)). They further suggest that plumes have average lives of the order of 100 m yr and that it is the competition between vigorous new plumes and older ones that is the major cause of plate motion and of its changing patterns and rates. Kaula (1972) has discussed mantle convection pointing out the uncertainties in our knowledge of the mechanics of the flow system. The literature contains contradictory arguments on many questions such as the widths of up and down plumes, and the depth of the return flow.

Evidence for a possible deep mantle source for the Hawaiian basalts has recently been given by Kanasewich et al. (1972). Earthquakes from the Tonga and Samoa Islands recorded by the University of Alberta arrays and three Canadian network seismic stations arrive with phase velocities up to 15 per cent high. The results are not interpretable by anomalous conditions in the downgoing plate, in the source region, or at the receiver. The authors conclude that the observations are due to a heterogeneous region of high velocity in the mantle at the core-mantle boundary underneath the island of Hawaii. They point out that their results are consistent with Wilson's (1963) hypothesis for the origin of the Hawaiian Islands and also with Morgan's (1972) proposal for the existence of hot spots in the lower mantle. Davies and Sheppard (1972) in a study of the characteristics of seismic waves detected at the large aperture seismic array in Montana find three regions (including that beneath Hawaii) of anomalous conditions in the deep mantle in the vicinity of the downward extension of proposed hot spots by Morgan (1971).

It appears that plate tectonics has operated for at least the past 200 m yr. During this time virtually all the present oceans were created and others destroyed. Some 200 m yr ago the major continental masses were assembled into one super-continent Pangaea. It is more difficult to trace the geologic record back farther than this. Precambrian pole paths that have been constructed from palaeomagnetic data from Europe (west of the Urals), North America, and the East and Southern African Shields cannot be matched and the basic shape of these paths is different (Spall (1973)). The palaeomagnetic evidence thus suggests that these shield areas drifted relative to each other during the Precambrian. Also studies of mountain belts older than 200 m yr indicate that they probably owe their origin to processes operating at plate boundaries that are now extinct. The absence of well-defined zones of mountain building older than 2000 m yr suggests that some mechanisms other than plate tectonics were responsible for the evolution of the earth's crust in earlier times.

Many geologists have now accepted these newer ideas of plate tectonics. The literature has grown enormously in the last few years as different authors have attempted to reinterpret the geology of a particular area within the framework of plate motions. It is impossible to attempt in this book to summarize the 'new geology' of all the various regions of the earth. The review paper by Dewey and Byrd (1970) on 'Mountain belts and the new global tectonics' outlines in detail many of the geological and petrological aspects of plate tectonics with particular attention to compressive plate boundaries.

A number of books have very recently been published which describe in detail what has been called a 'revolution' in the Earth Sciences (see e.g. Vacquier (1972); Coulomb (1972); Hallam (1973); Le Pichon *et al.* (1973)). A special symposium on the mechanism of plate tectonics was held during the International Geological Congress in Montreal in 1972. The papers presented at this meeting have now been published as a special issue of Tectonophysics (Vol. **19**, No. 2, 1973). The American Geophysical Union (1972) has also issued a collection of some of the major papers published in the Journal of Geophysical Research between 1967 and 1972.

References

Anderson, D. L. (1967) Phase changes in the upper mantle, *Science*, **157**, 1165.
Anderson, D. L. (1971) The San Andreas fault, *Sci. Amer.* (Nov.).
Anderson, E. M. (1951) *The Dynamics of Faulting and Dyke Formation with Application to Britain*, (2nd edition) (Oliver and Boyd).
Banerjee, S. K. (1971) Decay of marine magnetic anomalies by ferrous ion diffusion, *Nature Phys. Sci.*, **229**, 181.
Barazangi, M., B. L. Isacks, J. Oliver, J. Dubois and G. Pascal (1973) Descent of lithosphere beneath New Hebrides, Tonga–Fiji and New Zealand: evidence for detached slabs, *Nature*, **242**, 98.
Briden, J. C. (1967) Recurrent continental drift of Gondwanaland, *Nature*, **215**, 1334.

Brune, J. N. (1968) Seismic moment, seismicity, and rate of slip along major fault zones, *J. geophys. Res.*, **73**, 777.

Brune, J. N. (1969) Surface waves and crustal structure, in: *The Earth's Crust and Upper Mantle*, p. 230 (Geophys. Monograph Series No. 13, Amer. Geophys. Union).

Bullard, E. C., J. E. Everett and A. G. Smith (1965), The fit of the continents around the Atlantic, in: A symposium on Continental Drift, *Phil. Trans. Roy. Soc.*, **A 258**, 41.

Burke, K. and J. T. Wilson (1972) Is the African plate stationary? *Nature*, **239**, 387.

Carmichael, C. (1970) The mid Atlantic Ridge near 45°N. VII. Magnetic properties and opaque mineralogy of dredge samples, *Can. J. Earth Sci.*, **7**, 239.

Chase, C. G. (1972) The N plate problem of plate tectonics, *Geophys. J.*, **29**, 117.

Coulomb, J. (1972) *Sea Floor Spreading and Continental Drift*, (D. Reidel Publ. Co., Holland).

Cox, C. S. (1971) The electrical conductivity of the oceanic lithosphere, in: *The Structure and Physical Properties of the Earth's Crust*, p. 227 (Geophys. Monograph Series No. 14, Amer. Geophys. Union).

Cox, C. S., J. H. Filloux and J. C. Larsen (1970) Electromagnetic studies of ocean currents and electrical conductivity below the ocean floor, in: *The Sea*, Vol. 4, Part I, 637 (ed. A. E. Maxwell) (Wiley-Interscience).

Davies, G. and J. N. Brune (1971) Regional and global fault slip rates from seismicity, *Nature Phys. Sci.*, **229**, 101.

Davies, D. and R. M. Sheppard (1972) Lateral heterogeneity in the earth's mantle, *Nature*, **239**, 318.

DeBoer, J. (1968) Palaeomagnetic differentiation and correlation of the late Triassic volcanic rocks in the Central Appalachians (with special reference to the Connecticut Valley), *Bull. geol. Soc. Amer.*, **79**, 609.

Dewey, J. F. and J. M. Bird (1970) Mountain belts and the new global tectonics, *J. geophys. Res.*, **75**, 2625.

Dickinson, W. R. and W. C. Luth (1971) A model for plate tectonic evolution of mantle layers, *Science*, **174**, 400.

Duncan, R. A., N. Petersen and R. B. Hargraves (1972) Mantle plumes, movement of the European plate, and polar wandering, *Nature*, **239**, 82.

du Toit, A. L. (1937) *Our Wandering Continents* (Oliver and Boyd).

Elsasser, W. M. (1969) Convection and stress propagation in the upper mantle, in: *The Application of Modern Physics to the Earth and Planetary Interiors* (ed. S. K. Runcorn) (Wiley–Interscience).

Engdahl, E. R. and E. A. Flinn (1969) Seismic waves reflected from discontinuities within the earth's upper mantle, *Science*, **163**, 177.

Evans, M. E. and M. L. Wayman (1972) The mid Atlantic Ridge near 45°N. XIX. An electron microscope investigation of the magnetic minerals in basalt samples, *Can. J. Earth Sci.*, **9**, 671.

Ewing, J. and M. Ewing (1967) Sediment distribution on the mid-ocean ridges with respect to spreading of the sea floor, *Science*, **156**, 1590.

Falconer, R. K. H. (1972) The Indian-Antarctic-Pacific triple junction, *Earth Planet. Sci. Letters*, **17**, 151.

Fisher, R. L., J. G. Sclater and D. P. McKenzie (1971) Evolution of the Central Indian ridge, Western Indian Ocean, *Bull. geol. Soc. Amer.*, **82**, 553.

Grommé, S. and F. J. Vine (1972) Palaeomagnetism of Midway Atoll lavas and northward movement of the Pacific plate, *Earth Planet. Sci. Letters*, **17**, 159.

Gutenburg, B. and C. F. Richter (1954) *Seismicity of the Earth* (2nd ed.) (Princeton Univ. Press).

Haggerty, S. E. and E. Irving (1970) On the origin of the natural remanence of the mid Atlantic Ridge at 45°N, *Trans. Amer. Geophys. Union*, **51**, 273.

Hallam, A. (1973) *A Revolution in the Earth Sciences* (Clarendon Press, Oxford).

Harrison, C. G. A. (1968) Formation of magnetic anomaly patterns by dyke injection, *J. geophys. Res.*, **73**, 2137.

Hayes, D. E. and W. C. Pitman III (1970) Magnetic lineations in the North Pacific, *Geol. Soc. Amer. Mem.*, **126**.

Heezen, B. C., E. T. Bunce, J. B. Hersey and M. Tharp (1964a) Chain and Romanche fracture zones, *Deep-Sea Res.*, **11**, 11.

Heezen, B. C., R. D. Gerard and M. Tharp (1964b) The Vema fracture zone in the equatorial Atlantic, *J. geophys. Res.*, **69**, 733.

Heirtzler, J. R., X. Le Pichon and J. G. Baron (1965) Magnetic anomalies over the Reykjanes Ridge, *Deep-Sea Res.*, **13**, 427.

Heirtzler, J. R. and D. E. Hayes (1967) Magnetic boundaries in the North Atlantic Ocean, *Science*, **157**, 185.

Heirtzler, J. R., G. O. Dickson, E. M. Herron, W. C. Pitman III and X. Le Pichon (1968) Marine magnetic anomalies, geomagnetic field reversals, and motions of the ocean floor and continents, *J. geophys. Res.*, **73**, 2119.

Herron, E. M. and J. R. Heirtzler (1967) Sea floor spreading near the Galapagos, *Science*, **158**, 775.

Hess, H. H. (1962) History of the ocean basins, in: *Petrological Studies: A Volume in Honor of A. F. Buddington* (ed. A. E. J. Engel, H. L. James and B. F. Leonard) (Geol. Soc. Amer, New York).

Hey, R. N., K. S. Deffeyes, G. L. Johnson and A. Lowrie (1972) The Galapagos triple junction and plate motions in the East Pacific, *Nature*, **237**, 20.

Irving, E. (1966) Palaeomagnetism of some carboniferous rocks from New South Wales and its relation to geological events, *J. geophys. Res.*, **71**, 6025.

Irving, E. (1970) The mid Atlantic ridge at 45°N. XIV. Oxidation and magnetic properties of basalt: review and discussion, *Can. J. Earth Sci.* **7**, 1528.

Isacks, B., J. Oliver and L. R. Sykes (1968) Seismology and the new global tectonics. *J. geophys. Res.*, **73**, 5855.

Isacks, B. and P. Molnar (1969) Mantle earthquake mechanisms and the sinking of the lithosphere, *Nature*, **223**, 1121.

Isacks, B. and P. Molnar (1971) Distribution of stresses in the descending lithosphere from a global survey of focal mechanism solutions of mantle earthquakes, *Rev. Geophys. Space Phys.*, **9**, 103.

James, D. E. (1971) Andean crustal and upper mantle structure, *J. geophys. Res.*, **76**, 3246.

Kanamori, H. and K. Abe (1968) Deep structure of island arcs as revealed by surface waves, *Bull. Earthquake Res. Inst., Tokyo*, **46**, 1001.

Kanasewich, E. R., R. M. Ellis, C. H. Chapman and P. R. Gutowski (1972) Seismic array evidence of a core boundary source for the Hawaiian linear volcanic chain, *Nature*, **231**, 99.

Karig, D. E. (1971a) Structural history of the Mariana Island arc system, *Bull. geol. Soc. Amer.*, **82**, 323.

Karig, D. E. (1971b) Origin and development of marginal basins in the western Pacific, *J. geophys. Res.*, **76**, 2542.

Kaula, W. M. (1972) Global gravity and mantle convection, *Tectonophysics*, **13**, 341.

Knopoff, L. (1964) The convection current hypothesis, *Rev. Geophys*, **2**, 89.

Lee, W. H. K. (1970) On the global variations of terrestrial heat flow, *Phys. Earth Planet Int.*, **2**, 332.

Le Pichon, X. (1968) Sea floor spreading and continental drift, *J. geophys. Res.*, **73**, 3661.

Le Pichon, X., J. Francheteau and J. Bonnin (1973) *Plate Tectonics* (Elsevier Publ. Co.).

Le Pichon, X. and J. R. Heirtzler (1968) Magnetic anomalies in the Indian Ocean and sea-floor spreading, *J. geophys. Res.*, **73**, 2101.

Lubimova, E. A. (1967) Theory of thermal state of the earth's mantle, in: *The Earth's Mantle* (ed. T. F. Gaskell) (Acad. Press).

Mascle, J. and J. D. Phillips (1972) Magnetic smooth zones in the South Atlantic, *Nature*, **240**, 80.

Matthews, D. H. and J. Bath (1967) Formation of magnetic anomaly pattern of mid-Atlantic ridge, *Geophys. J.*, **13**, 349.

Maxwell, A. E., R. P. Von Herzen, K. J. Hsu, J. E. Andrews, T. Saito, S. F. Percival, E. D. Milow and R. E. Boyce (1970) Deep sea drilling in the South Atlantic, *Science*, **168**, 1047.

McBirney, A. R. (1971) Thoughts on some current concepts of orogeny and volcanism, *Comments on Earth Sciences: Geophysics*, **2**, 69.

McElhinny, M. W. (1970) Palaeomagnetism of the southern continents, in: *Symposium on Continental Drift U.N.E.S.C.O., Paris, 1970* (ed. J. T. Wilson).

McElhinny, M. W. (1973) Mantle plumes, palaeomagnetism and polar wandering, *Nature*, **241**, 523.

McElhinny, M. W., J. C. Briden, D. L. Jones and A. Brock (1968) Geological and geophysical implications of palaeomagnetic results from Africa, *Rev. Geophys.*, **6**, 201.

McKenzie, D. P. (1967) Some remarks on heat-flow and gravity anomalies, *J. geophys. Res.*, **72**, 6261.

McKenzie, D. P. (1968) The influence of the boundary conditions and rotation on convec- in the earth's mantle, *Geophys. J.*, **15**, 457.

McKenzie, D. P. (1969a) Speculations on the consequences and causes of plate motions, *Geophys. J.*, **18**, 1.

McKenzie, D. P. (1969b) The relation between fault-plane solutions for earthquakes and the directions of the principal stresses, *Bull. seism. Soc. Amer.*, **59**, 591.

McKenzie, D. P. and R. L. Parker (1967) The North Pacific—an example of tectonics on a sphere, *Nature*, **216**, 1276.

McKenzie, D. P. and J. G. Sclater (1968) Heat flow inside the island arcs of the north-western Pacific, *J. geophys. Res.*, **73**, 3173.

McKenzie, D. P. and W. J. Morgan (1969) Evolution of triple junctions, *Nature*, **224**, 125.

Minear, J. W. and M. N. Toksöz (1970) Thermal regime of a downgoing slab and new global tectonics, *J. geophys. Res.*, **75**, 1397.

Morgan, W. J. (1965) Gravity anomalies and convection currents 2. The Puerto Rico trench and the mid-Atlantic rise, *J. geophys. Res.*, **70**, 6189.

Morgan, W. J. (1968) Rises, trenches, great faults and crustal blocks, *J. geophys. Res.*, **73**, 1959.

Morgan, W. J. (1971) Convection plumes in the lower mantle, *Nature*, **230**, 42.

Morgan, W. J., Plate motions and deep mantle convection, in: *Hess Memorial Volume* (ed. R. Shagam) (Geol. Soc. Amer. Mem. 132) (in press).

Oliver, J. and B. Isacks (1967) Deep earthquake zones, anomalous structures in the upper mantle, and the lithosphere, *J. geophys. Res.*, **72**, 4259.

Oxburgh, E. R. and D. L. Turcotte (1968) Mid-ocean ridges and geotherm distribution during mantle convection, *J. geophys. Res.*, **73**, 2643.

Oxburgh, E. R. and D. L. Turcotte (1970) Thermal structure of island arcs, *Bull. geol. Soc. Amer.*, **81**, 1665.

Oxburgh, E. R. and D. L. Turcotte (1971) Phase changes and mantle convection, *J. geophys. Res.*, **76**, 1424.

Ozima, M. (1971) Magnetic processes in oceanic ridge, *Earth Planet. Sci. Letters*, **13**, 1.

Pitman, W. C. III and J. R. Heirtzler (1966) Magnetic anomalies over the Pacific Antarctic Ridge, *Science*, **154**, 1164.

Pitman, W. C. III and D. E. Hayes (1968) Sea floor spreading in the Gulf of Alaska, *J. geophys. Res.*, **73**, 6571.

Plafker, G. P. (1965) Tectonic deformation associated with the 1964 Alaska earthquake, *Science*, **148**, 1675.

Raff, A. D. (1968) Sea floor spreading: another rift, *J. geophys. Res.*, **73**, 3699.

Ringwood, A. E. (1969) Composition and evolution of the upper mantle, in: *The Earth's Crust and Upper Mantle* (Geophysical Monograph 13, Amer. Geophys. Union).

Ringwood, A. E. (1972) Phase transformations and mantle dynamics, *Earth Planet. Sci. Letters*, **14**, 233.

Rona, P. A., J. Brakl and J. R. Heirtzler (1970) Magnetic anomalies in the northwest Atlantic between the Canary and Cape Verde islands, *J. geophys Res.*, **75**, 7412.

Schaeffer, R. M. and E. J. Schwarz (1970) The mid Atlantic Ridge near 45°N. IX. Thermomagnetics of dredged samples of igenous rocks, *Can. J. Earth Sci.*, **7**, 268.

Schubert, G., D. L. Turcotte, and E. R. Oxburgh (1970) Phase change instability in the mantle, *Science*, **169**, 1075.

Schubert, G. and D. L. Turcotte (1972) One-dimensional model of shallow-mantle convection, *J. geophys. Res.*, **77**, 945.

Sclater, J. G. (1972a) New perspectives in terrestrial heat flow, *Tectonophysics*, **13**, 257.

Sclater, J. G. (1972b) Heat flow and elevation of the marginal basins of the western Pacific, *J. geophys. Res.*, **77**, 5705.

Sleep, N. H. (1969) Sensitivity of heat flow and gravity to the mechanism of sea floor spreading, *J. geophys. Res.*, **74**, 542, 1969.

Spall, H. (1973) Review of Precambrian palaeomagnetic data for Europe, *Earth Planet Sci. Letters*, **18**, 1.

Stauder, W. (1968a) Mechanism of the Rat Island earthquake sequence of February 4, 1965 with relation to island arcs and sea floor spreading, *J. geophys. Res.*, **73**, 3847.

Stauder, W. (1968b) Tensional character of earthquake foci beneath the Aleutian Trench with relation to sea-floor spreading, *J. geophys. Res.*, **73**, 7693.

Stauder, W. and G. A. Bollinger (1966) The focal mechanism of the Alaska earthquake of March 28, 1964, and the aftershock sequence, *J. geophys. Res.*, **71**, 5283.

Sykes, L. R., (1966) The seismicity and deep structure of island arcs, *J. geophys. Res.*, **71**, 2981.

Sykes, L. R. (1967) Mechanism of earthquakes and nature of faulting in the mid-ocean ridges, *J. geophys. Res.*, **72**, 2131.

Sykes, L. R. (1968) Seismological evidence for transform faults, sea floor spreading and continental drift, in: *History of the Earth's Crust* (ed. R. A. Phinney) (Princeton Univ. Press).

Toksöz, M. N., J. W. Minear and B. R. Julian (1971) Temperature field and geophysical effects of a downgoing slab, *J. geophys. Res.*, **76**, 1113.

Tozer, D. C. (1967) Towards a theory of thermal convection in the mantle, in: *The Earth's Mantle* (ed. T. F. Gaskell) (Acad. Press).

Turcotte, D. L. and E. R. Oxburgh (1972) Mantle convection and the new global tectonics, *Ann. Rev. Fluid Mech.*, **4**, 33.

Utsu, T. (1967) Anomalies in seismic wave velocity and attenuation associated with a deep earthquake zone, 1. *J. Fac. Sci. Hokkaido Univ. Ser. 7, Geophys.*, **3**, 1.

Vacquier, V. (1972) *Geomagnetism in Marine Geology* (Elsevier Publ. Co.).

Vening Meinesz, F. A. (1962) Thermal convection in the earth's mantle, in: *Continental Drift* (ed. S. K. Runcorn) (Acad. Press).

Vine, F. J. (1966) Spreading of the ocean floor: new evidence, *Science*, **154**, 1405.

Vine, F. J. and D. H. Matthews (1963) Magnetic anomalies over oceanic ridges, *Nature*, **199**, 947.

Vine, F. J. and H. H. Hess (1970) Sea floor spreading, in: *The Sea*, Vol. 4, Part II, 587 (ed. A. E. Maxwell) (Wiley-Interscience).

Vogt, P. R., C. N. Anderson, D. R. Bracey and E. D. Schneider (1970) North Atlantic magnetic smooth zones, *J. geophys. Res.*, **75**, 3955.

Walcott, R. I. (1970) Flexural rigidity, thickness and viscosity of the lithosphere, *J. geophys. Res.*, **75**, 3941.

Wegener, A. (1966) *The origin of continents and oceans* (Dover Publ. Inc.).

Wilson, J. T. (1963a) Evidence from islands on the spreading of ocean floors, *Nature*, **197** 536.

Wilson, J. T. (1963b) Hypothesis of earth's behaviour, *Nature*, **198**, 925.

Wilson, J. T. (1963c) A possible origin of the Hawaiian Islands, *Can. J. Phys.*, **41**, 863.

Wilson, J. T. (1965a) A new class of faults and their bearing on continental drift, *Nature*, **207**, 343.

Wilson, J. T. (1965b) Submarine fracture zones, aseismic ridges and the International Council of Scientific Unions line: proposed western margin of the East Pacific Rise, *Nature*, **207**, 907.

Wilson, J. T. (1965c) Convection currents and continental drift, in: A Symposium on Continental Drift, *Phil. Trans. Roy. Soc.*, **A 258**, 145.

Wilson, J. T. (1970) Continental drift, transcurrent and transform faulting, in: *The Sea*, Vol. 4, Part II, 623 (ed. A. E. Maxwell) (Wiley-Interscience).

Wilson, J. T. (1972) New insights into old shields, *Tectonophysics*, **13**, 73.

9

The Rheology of the Earth

9.1 Introduction

Rheology is the study of the flow and deformation of matter, and a fundamental problem in the physics of the earth is the determination of the rheological conditions that exist at any point, i.e. the fundamental relationship between the stress and strain tensors σ and ε. Since the behaviour of some materials depends on the rate of strain and the rate at which the stresses are produced, the rheological equation may also include the time derivatives of σ and ε, and is thus of the general form

$$f(\sigma, \dot{\sigma}, \varepsilon, \dot{\varepsilon}) = 0 \qquad (9.1)$$

The simplest rheological equation is $\varepsilon = 0$, which implies that there is no strain, whatever stresses may be applied. This is the case for a *rigid body*, and is the simplest 'ideal' body. All rheological equations describe the behaviour of ideal bodies which may form increasingly better approximations to reality, yet nevertheless do not exist in nature. The two next simplest ideal bodies are the *Hooke solid* and the *Newtonian liquid*. For a perfectly elastic body below the elastic limit (a Hooke solid), the relation between shear stress and shear strain is

$$\sigma_{ij} = 2\mu\varepsilon_{ij} \qquad i \neq j \qquad (9.2)$$

where μ is the rigidity, while for a viscous (Newtonian) liquid

$$\sigma_{ij} = 2\eta\dot{\varepsilon}_{ij} \qquad i \neq j \qquad (9.3)$$

where η is the viscosity. Viscous flow is usually thermally activated, η depending exponentially on temperature. The theory of these two ideal bodies has been very highly developed mathematically, but cannot fully explain the flow and deformation of actual bodies, especially when the stresses are applied over any length of time.

Some materials behave as a Hooke solid when the stresses are applied for a short time, but over longer time intervals flow like a liquid, exhibiting the phenomenon of *creep*. A possible rheological equation for such a body is a combination of the rheological equations of the Hooke solid and the Newtonian liquid. Omitting the indices i, j, i.e. considering an isotropic, incompressible body, equations (9.2) and (9.3) may be combined to give

$$\dot{\varepsilon} = \frac{\dot{\sigma}}{2\mu_M} + \frac{\sigma}{2\eta_M} \tag{9.4}$$

where μ_M and η_M are constants of the body, known as Maxwell constants. A body which obeys equation (9.4) is known as a *Maxwell liquid*, after Maxwell, who first postulated such a relation in 1868. A Maxwell liquid exhibits what is known as *stress relaxation*, i.e. if the deformation is kept constant ($\dot{\varepsilon} = 0$), the stress diminishes exponentially, reaching a value of $1/e$ of its initial value after a time η_M/μ_M, known as the relaxation time. On the other hand if a constant stress is applied ($\dot{\sigma} = 0$), deformation occurs at a constant rate, i.e. the material exhibits creep.

Equations (9.2) and (9.3) could have been combined in another way. A body whose rheological equation is

$$\sigma = 2\mu_K\varepsilon + 2\eta_K\dot{\varepsilon} \tag{9.5}$$

is called a *Kelvin solid*. If the strain is constant ($\dot{\varepsilon} = 0$), the stress is constant, but if the stress is removed, the strain does not vanish at once, i.e. a Kelvin solid exhibits a *delayed elastic effect*, or *elastic after effect*. A material showing both Maxwell and Kelvin behaviour has been called a *Burgers body*. The strain-time curve (for a given stress) for a Kelvin body is shown in Fig. 9.1. If elastic deformation is represented by a spring and viscous flow by a dash-pot, a Maxwell liquid may be

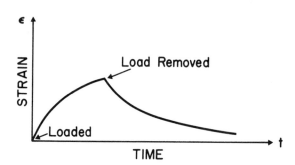

Fig. 9.1 Strain-time curve for a Kelvin body (after Reiner (1949))

represented by a dash-pot in series with a spring, and a Kelvin solid by a spring in parallel with a dash-pot.

More ideal bodies may be obtained by introducing a yield stress σ_Y. The simplest is a *perfectly plastic* or *Saint-Venant body*. For stresses below the yield stress, such a body behaves as a Hooke solid; when the yield stress is reached, the body becomes plastic, i.e. under the same stress, the body is continuously deformed at any strain rate. A *Bingham body* does not deform permanently until the yield stress is reached. Above the yield stress it flows like a viscous fluid, the strain rate being directly proportional to the stress. Its equation is

$$\sigma = \sigma_Y + 2\eta\dot{\varepsilon}, \qquad (\sigma > \sigma_Y) \tag{9.6}$$

Figure 9.2 shows the stress–strain relations for a Newtonian, perfectly plastic and Bingham body.

There are many possible physical mechanisms for anelasticity in the earth; these have been reviewed by Gordon and Nelson (1966) and Jackson and Anderson (1970). They conclude that the most likely mechanisms operative in the

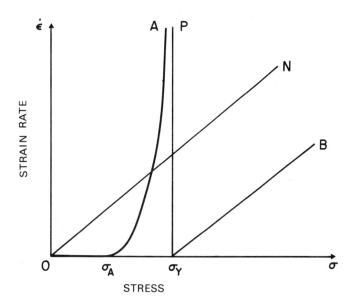

Fig. 9.2 Stress–strain relations for some rheological models and for high-temperature creep of polycrystalline aggregates. ON=Newtonian (viscous) body; Oσ_YB=Bingham body; Oσ_YP=perfectly plastic body; Oσ_AA=Andradean creep (characteristic of real materials) (after Ranalli (1971))

mantle are intergranular thermo-elastic relaxation, grain-boundary diffusion, and partial melting. Probably no single process is responsible for attenuation throughout the mantle. Smith (1972) has pointed out that the elastic properties of solids are controlled mainly by the interatomic forces of the crystalline lattice, whereas the anelastic properties are controlled by the diffusion of defect species, and thus may be affected by thermal diffusivity, grain size and the concentration and mobility of defects.

9.2 The response of the earth to stresses

Scheidegger (1957) has classified stresses according to whether their duration is short, intermediate, or long. Stresses are considered short if they have a typical duration of the order of 3 s, as intermediate if their typical duration is of the order of 3 yr, and as long if their typical duration is 100 m yr. The upper limit for stresses of short duration is about 4 h. In this time range the material of the crust and mantle behaves as an elastic solid with a rigidity of about 2×10^{12} dynes/cm^2

and a Young's modulus of about 5×10^{12} dynes/cm². If the elastic limit is exceeded, the material undergoes brittle fracture. This information is obtained from laboratory experiments and the propagation of seismic waves.

In a later paper, Scheidegger (1971) re-examined the rheology of the earth's crust and upper mantle for stresses in the short time range. The conclusion that he reached in his original paper (1957) namely that, to a first approximation, the earth behaves elastically for small stresses, still holds. New information has been obtained, however, from the damping of seismic waves and from free oscillation data which indicates that the earth does not respond perfectly elastically for stresses in this time range. The most significant result is that $Q*$ is essentially independent of frequency with values in the upper mantle around 100–500 (see Knopoff (1964); Jackson and Anderson (1970)). Neither a Kelvin solid nor a Maxwell liquid exhibits frequency independence of Q. In fact Knopoff (1964) showed that no combination of Maxwell and Kelvin properties can give a frequency independent Q. This can be achieved in linear models only if higher time derivatives are used in the stress-strain relationship and if the material is inhomogeneous. A constant Q demands a non-linear rheological equation, such as a logarithmic creep model.

For stresses acting over a longer period of time, the materials of the earth exhibit slow flowage or 'creep'. In order to determine the response of a material to stresses (i.e. to find the resulting displacements under given loading conditions), we need the relevant rheological equation of state, Newton's laws of motion, and the equation of continuity together with initial and boundary conditions. The rheological equations considered so far have all been linear; they lead to exponential creep. However, the creep behaviour of many materials, particularly rocks, is often not exponential, but logarithmic. Griggs (1939) found for compressional stresses that

$$\varepsilon = A + B \ln t + Ct \tag{9.7}$$

while for shear stresses Lomnitz (1956) found that

$$\varepsilon = \frac{\sigma}{\mu}\left\{1 + a \ln\left(1 + bt\right)\right\} \tag{9.8}$$

where A, B, C and a, b are constants. Much earlier Andrade (1910), as a result of laboratory measurements, proposed the following equation for the behaviour of materials under a constant stress σ_0,

$$\varepsilon = A + BE(t) + Ct \tag{9.9}$$

where $E(t)$ is an empirical function of time (see Fig. 9.3). Scheidegger (1970a) suggested the following approach to determining a rheological equation for logarithmic creep.

Assume that the strain-time relationship can be written as

$$\varepsilon = A + B \ln\left(a + bt\right) + Ct \tag{9.10}$$

---

* See § 4.1 for a definition of Q.

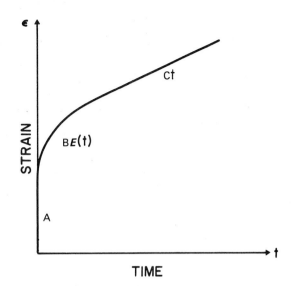

Fig. 9.3 Strain-time curve for materials under constant stress proposed by Andrade (1910)

Then

$$\dot{\varepsilon} = \frac{bB}{a+bt} + C$$

and

$$\ddot{\varepsilon} = \frac{-b^2B}{(a+bt)^2}$$

Eliminating t we obtain

$$\ddot{\varepsilon} + \frac{1}{B}(\dot{\varepsilon}-C)^2 = 0$$

Since the stresses σ are constant, Scheidegger proposed that

$$\dot{\sigma} = K\left\{\ddot{\varepsilon} + \frac{1}{B}(\dot{\varepsilon}-C)^2\right\} \tag{9.11}$$

Any homogeneous polynomial in $\dot{\sigma}$ could have been written for the left-hand side of equation (9.11) but the above value is the simplest. When $B\rightarrow\infty$, the equation reduces to that of a viscous fluid so that $K=2\eta$. Hence, writing $\beta=K/B$, Scheidegger arrived at the following non-linear rheological equation

$$\dot{\sigma} = 2\eta\ddot{\varepsilon} + \beta(\dot{\varepsilon}-C)^2 \tag{9.12}$$

It should be noted that it cannot hold under conditions of stress reversal.

The tectonically significant layers of the earth (the crust and upper mantle down to a depth of about 400 km) have been called the tectonosphere. Consider the rheological behaviour of the tectonosphere in the intermediate time range (4 hours to 1000 years). Information in this time range comes from a number of sources: direct observations of rock behaviour in the laboratory, seismic after-shock sequences, earth tides and the decay of the Chandler wobble (Scheidegger

265

(1970b)). It is difficult to reproduce in the laboratory the pressures and temperatures that exist deep within the earth and to maintain them for fairly long time intervals (weeks to years). All that we can hope to do is to model conditions prevailing in the upper crust. The results of such investigations indicate, as already mentioned, that most rocks exhibit logarithmic creep—a behaviour which cannot be described in terms of linear models. Originally Scheidegger (1957) tried to explain the response of the earth to seismic aftershock sequences and the damping of its free nutation—the Chandler wobble—in terms of a Kelvin model. In the Benioff strain rebound theory of aftershock sequences, the main shock relieves the build up of stress and subsequent aftershocks are believed to be the expression of the adjustment of a Kelvin material to stress redistribution. However, new information obtained over the last ten years shows that such a model cannot be valid. Ranalli and Scheidegger (1969) carried out a detailed analysis of a number of aftershock sequences and showed that they are a discontinuous manifestation of overall plastic creep, the earth adjusting itself to the redistribution of stresses. Explicitly they found that

$$\dot{\varepsilon} = At^{-1}$$

so that

$$\varepsilon = A \ln (Bt) \tag{9.13}$$

which equation is a form of logarithmic creep.

The gravitational attraction of the sun and moon on the earth produces deformation in the earth (earth tides). The time-constants of the orbital motions of the moon and the earth are such that the periods of these tidal deformations fall into the 'intermediate' time range. Unfortunately the mechanism of energy dissipation in earth tides and the damping of the Chandler wobble are not well understood (see § 4.1). However, a rheological model of the earth based on logarithmic creep is possible and does not contradict any of the observational evidence to date.

Finally, let us consider the response of the tectonosphere in the long time range, i.e. over periods longer than 1000 years. Some information in this time range can be obtained from the isostatic uplift and sinking of the earth's crust, the nonhydrostatic bulge of the earth, the theoretical behaviour of rocks at high temperatures and pressures and the stability of mountains (Scheidegger (1970c)). Most of our knowledge comes from vertical motions of the earth's crust; the most important data being that of postglacial isostatic recovery (Andrews (1968, 1970)). Attempts have also been made to deduce the rheological behaviour of the tectonosphere from the horizontal motion of plates (e.g. Jacoby (1970)). The mechanics of such motions are not sufficiently well understood, however, to be able to reach any reliable conclusions. Uplift curves for Lake Huron and Arctic Canada are shown in Fig. 4.10. Several empirical equations have been proposed to fit such data. An exponential decay was initially suggested but much better agreement was found using a logarithmic dependence (Andrews (1968, 1970))—an exponential dependence consistently gives present-day uplift rates that are too low (see also § 4.4).

266

The most common view of the rheological properties of the tectonosphere in the long time range has been that it behaves as a visco-elastic (Maxwell) material. However, such a model leads to estimates of the viscosity from the submergence of guyots and the postglacial rise of Fennoscandia and Lake Bonneville in Utah which differ by orders of magnitude. These values can be reconciled to some extent with a layered viscosity model; more realistic models still can be obtained by having a thin elastic layer on top (see § 9.3). In this way McConnell (1965) obtained a viscosity distribution for the upper mantle of the earth—similar results have been obtained by Anderson and O'Connell (1967) from an analysis of the non-hydrostatic bulge of the earth assumed to be due to its changing rate of rotation. Although a viscous model for the tectonosphere qualitatively fits some of the observed data if a low-viscosity layer is assumed in the upper mantle, viscosity always yields an exponential relaxation pattern which is not supported by the uplift curves (Fig. 4.10).

The rheological behaviour of the tectonosphere in the long time range is thus very similar to that in the intermediate time range, and there may not be any real distinction between its response over these two different time intervals. In fact the same deviation from a linear rheological equation is found for stresses over all time ranges, i.e. the same equation (the logarithmic creep law) can account for such diverse phenomena as seismic damping, seismic aftershock sequences and the uplift of Fennoscandia. However, creep alone cannot characterize the long term behaviour of the tectonosphere. The fact that mountains, the non-hydrostatic bulge of the earth and gravity anomalies persist indicates that the tectonosphere possesses some long term strength. Values of the shearing strength of the tectonosphere have been estimated to lie between 10^6–10^8 dynes/cm^2. This wide range of values could be reconciled if some of the theoretical models upon which the calculations were based were shown to be invalid. In this respect it is possible that some of the above features (e.g. mountains and the non-hydrostatic bulge of the earth) may be the expression of a dynamic steady state rather than of static equilibrium.

At temperatures and pressures of the order of those existing in the upper crust, the behaviour of polycrystalline aggregates is at first elastic. When the stress-difference reaches a given value σ_Y (the yield strength) the deformation becomes plastic. The long-term yield strength cannot be determined experimentally. At low temperatures, strain-hardening occurs, and after a stage of (usually) logarithmic transient creep, no additional deformation is (in general) observed. Steady-state creep occurs more often as the temperature increases, since strain-hardening is then counterbalanced by thermal relaxation processes. The relationship between stress and strain rate is

$$\dot{\varepsilon} = C\sigma^n \tag{9.14}$$

The effect of temperature T can be expressed by including a factor $\exp(-a/T)$ on the right-hand side of this equation. Laboratory experiments indicate a value of n between 3 and 5.

The simplest form of equation (9.14) is given by $n=1$, i.e. ordinary Newtonian viscous behaviour. Such linear creep is called Nabarro-Herring creep. It occurs when deformation is governed, not by the migration of line defects (dislocations), but by the mobility of point defects (atoms and vacancies) which tend to migrate from one grain boundary to another. Necessary conditions for Nabarro-Herring creep to occur are $T \geqslant \frac{1}{3}T_m$ (where T_m is the melting temperature), low stresses and small grain size. There have been a number of estimates of the upper limit for the stress above which Nabarro-Herring creep is not important; it is likely to be less than that in the upper mantle. For large stresses, creep becomes non-linear. Creep is also non-linear at low stresses if the grain size is not very small. Experimental results (reviewed by McKenzie (1968)) indicate that the average grain size for which Nabarro-Herring creep is significant is less than 1 mm. Since the coefficient of viscosity is proportional to the square of the mean grain size, the creep rate decreases very rapidly as the grain size increases and linear creep becomes insignificant. Grain sizes in the mantle cannot be estimated exactly, although all indications are that they should be fairly large. Plutonic rocks like gabbros are coarse-grained and high temperature annealing has probably reshaped the material in the mantle. Average grain sizes are likely to be at least a few cm. Thus for a number of reasons, a linear creep law for the mantle as a whole is very unlikely to hold.

Experimental results on polycrystalline aggregates at $T \geqslant \frac{1}{2}T_m$ indicate a strongly non-linear dependence of strain rate upon stress ('Andradean' creep)—Fig. 9.2 includes the stress-strain rate relationship for this type of deformation. The deformation of polycrystalline aggregates at high temperatures is much closer to ideal plasticity than to Newtonian viscosity (see e.g. Orowan (1965)). The 'ductility' (i.e. the facility with which a solid body can creep) of the mantle can be greatly enhanced by the presence of certain chemicals (e.g. water and the alkalis), even in very small concentrations. Most probably the absence of earthquakes at depth (except those near oceanic trenches) is due to a lowering of the fracture strength due in turn to an increase in the water pressure. Anderson and Sammis (1970) have considered in some detail conditions in the low velocity layer, and concluded that it probably contains small (~ 1 per cent) amounts of melt. The rheology of such a partially molten region is not known, although it would most probably have a low effective viscosity (Weertman (1970)).

There has been much controversy over the mechanics of the flow of ice in a glacier. Perutz (1947) pointed out that any theory of glacier flow must explain two outstanding features:
(1) glacial erosion accentuates the relief of the land surface
(2) the speed of movement of a glacier is very sensitive to changes in glacier thickness brought about by climatic variations.
Neither feature can be explained by treating a glacier as a Newtonian liquid of constant viscosity; in such a case the flow would be laminar and would smooth out the surface and its mean velocity would depend on the square of the thickness (experience shows that it is much more sensitive than this). Like all crystalline solids, ice is not viscous but plastic, and its law of deformation is given by equation (9.14). It undergoes practically no permanent deformation before the shear

stress reaches a fairly well-defined critical value—in a narrow stress interval the rate of deformation then increases from zero to very high values. The application of plastic flow to glaciers is due originally to the work of Perutz (1947), Orowan (1950) and particularly Nye (1951, 1952, 1957). The flow law of ice has been investigated experimentally by Glen (1955, 1958) who obtained a value of 4·2 for the exponent n in equation (9.14). Detailed measurements by Paterson and Savage (1963) of velocity and strain rates in the Athabasca glacier, Alberta, have shown Nye's theory of glacier flow to be substantially correct and confirmed Glen's value of 4·2 for the exponent n.

Berg (1969a) has pointed out that the most natural mechanism of permanent deformation which can occur in polycrystalline materials and which can lead to concentrated zones of deformation, as opposed to diffuse deformation which arises in Andradean or Newtonian viscous creep, is nonhardening plastic deformation. If convection in the mantle is the result of such a process, then discontinuous slip across discrete surfaces (slip bands) may occur, and the material may deform so that whole blocks of material slide across each other as rigid bodies, while all the deformation is confined to the thin lamella of material between the two blocks. The scale of distance in which the velocity gradients must be measured is the width of the slip bands in the material, which need only be a few multiples of the typical grain diameter. Berg estimates that in order for the convective velocity field to attain a steady state within 10^8 yr, the slip band thickness should be of the order of 6 km. In a later paper (Berg (1969b)) he showed that if the upper mantle behaved as a viscous fluid, a whole series of trenches should be found on the ocean floor near a continental boundary, not just a single trench. On the other hand a single trench in the zone of highest compressive loading near the continental edge is just what would be expected if the mantle deforms plastically.

9.3 The viscosity of the mantle

Temperatures in the mantle most closely approach the melting point in the low velocity (LV) zone at a depth of about 100–150 km. At greater depths the melting-point curve and the actual temperature diverge again. This behaviour is of the greatest importance to such rheological properties as viscosity, plasticity, creep and strength. At the very low strain rates, which are encountered in tectonic deformation, these properties depend in an approximately exponential manner upon the difference between the actual temperature and the melting temperature. Thus the viscosity and strength in the LV zone may be lower than in the regions above and below by many orders of magnitude, and the LV zone is likely to be the site of flow and deformation processes of tectonic importance. Between depths of about 50 and 250 km the mantle beneath the Precambrian shield is, on

269

the average, some 200°C cooler than under the oceans; such horizontal temperature differences would result in a wide variation in rheological properties. Thus the mean viscosity and strength of the upper mantle beneath shields will be far higher, probably by orders of magnitude, than under oceans.

Gordon (1967) and McKenzie (1968) have investigated a number of earth models assuming Nabarro-Herring creep as the controlling mechanism in the mantle. This implies a linear stress-strain rate relationship, and is unlikely to be true (see § 9.2). The resulting model of the mantle consists of a very high viscosity lithosphere ($\eta \gtrsim 10^{27}$ poise) about 100 km thick, a low viscosity ($\eta \sim 10^{21}$ poise) upper mantle approximately 500 km thick, and a rapid rise in viscosity throughout the lower mantle ($\eta \gtrsim 10^{26}$ poise below about 1000 km) reaching approximately 10^{27} poise at the base of the mantle. The low viscosity region is a result of the rapid increase of temperature with depth in the upper mantle, whereas in the lower mantle the effect of pressure predominates and causes an increase in the viscosity.

As already mentioned, direct estimates of the rheology of the mantle have been made from an analysis of the isostatic rebound of the earth's surface following the removal of a surface load such as an ice sheet. Takeuchi (1963) showed that much of the discrepancy between values of the viscosity of the upper mantle derived from the post-glacial uplift of Fennoscandia and the limits imposed by the persistence of long wavelength gravity anomalies could be resolved by assuming that flow takes place within a relatively thin viscous channel in the upper mantle. This was later confirmed by Takeuchi and Hasegawa (1965) using additional data for the uplift of Lake Bonneville, Utah. McConnell (1963) agreed in principle with Takeuchi but found that the model which best fits the data has an elastic surface layer and a low viscosity channel in the upper mantle below which there is a considerable, but not infinite, increase in viscosity.

The damping of seismic waves with distance and the decay with time of the earth's free oscillations can also be used to determine how much the earth departs from a perfectly elastic body. This seismic measure of anelasticity Q (see § 4.1) is roughly independent of frequency for homogeneous materials, and the observed frequency dependence of Q for long period surface waves and the earth's free oscillations may be attributed to the variation of Q with depth in the earth. Kovach and Anderson (1964) estimated that Q in the crust is about 450, dropping to about 60–130 in the LV zone depending on whether S or P waves are involved, and beginning to rise rapidly at a depth of about 400 km. The mean Q below 600 km is about 2200. For comparison, crustal rocks have Q in the range 100–200 at room temperature and pressure. Q thus varies by orders of magnitude throughout the mantle, the main features being the existence of a low Q (i.e. extremely dissipative zone) in the upper mantle (in the general region of the LV layer) and a rapid increase in Q below about 400 km. A similar trend for the variation of viscosity with depth in the upper mantle was found by McConnell (1963).

Anderson (1966) pointed out that the ratio of viscosity η to Q is roughly constant in the upper mantle. Assuming such a relation to be valid for the rest of the earth, Anderson estimated viscosity at greater depths, using average values of

270

Q in shear for the mantle obtained by Kovach and himself (1964).* His viscosities are all much lower than the 10^{26}–10^{27} poise estimated for the lower mantle by MacDonald (1963) and McKenzie (1966, 1967) on the assumption that the earth's non-hydrostatic bulge is due to the delayed readjustment of the earth's shape from a faster spin rate in the past. Goldreich and Toomre (1969) have challenged the 'fossil bulge' interpretation and shown that the non-hydrostatic part of the earth's inertia ellipsoid is definitely triaxial; they suggested that the bulge is merely a consequence of the earth rotating about its axis of greatest moment of inertia. In addition they showed that the rate and pattern of polar wandering inferred from palaeomagnetic studies suggests that the average effective viscosity of the mantle is less than 10^{24} poise—a value more in keeping with that of Anderson (1966b). An even lower value of 10^{22} poise for the viscosity of the deep mantle has been given by Dicke (1969). Dicke obtained this estimate from a re-examination of the average acceleration of the earth over the past 3000 yr based on a new analysis of ancient eclipses.

McConnell (1968a, b) carried out a harmonic analysis of the present level of former shore line features in south-east Fennoscandia and deduced a relaxation time spectrum for the uplift that gives a characteristic relaxation time for each Fourier component of the surface deformation. Comparing this spectrum with that calculated for a layered Newtonian viscous half-space overlain by an elastic lithosphere, he found a low viscosity region ($\eta \sim 10^{21}$ poise) between depths of about 200–400 km with the viscosity increasing to 10^{23}–10^{24} at about 1200 km. Such an interpretation is in agreement with Crittenden's (1967) value of 10^{20}–10^{21} poise under Lake Bonneville obtained for a uniform half-space model. It must be stressed that because of the limited areal extent of regions like Fennoscandia the uplift in these regions is determined primarily by the properties of the upper mantle. Thus it does not follow that values of the viscosity deduced from such studies apply to the major parts of the mantle.

Walcott (1973) has recently reviewed the rheological structure of the earth based on the isostatic rebound of the Laurentide, Fennoscandian and Innuitian uplifts. His preferred model is a 110 km thick lithosphere which, although requiring a viscosity of 10^{25} poise to explain some long-term behaviour, behaves elastically for time-scales of the order of a few thousand years; a thin low-viscosity channel about 100–500 km thick with a viscosity, dependent upon thickness, between 10^{19}–10^{21} poise; and a lower mantle with a viscosity greater than 10^{23} poise. The relative changes of viscosity in this model are similar to those obtained by Weertman (1970) from studies of the creep strength of the mantle from the standpoint of solid-state physics. The actual values of effective viscosity (i.e. the viscosity defined for a specific strain-rate) obtained by Weertman are, however, generally smaller.

---

* However, as Smith (1972) has pointed out, to attempt a direct comparison of a mechanism of viscous flow on a time-scale of a hundred thousand years with seismic attenuation at periods of a few seconds, involves extrapolation over more than 10 decades in frequency! No special significance should thus be attached to any apparent correlation of attenuation and viscosity.

9.4 Shock wave data

Studies of wave propagation measurements in rocks show that the velocity of compressional waves depends principally upon density and mean atomic weight. However, most common rocks have mean atomic weights close to 21 or 22 regardless of composition, and rocks or minerals of very different composition may have the same densities and seismic velocities. Thus it is not easy to infer chemical composition from seismic data, and laboratory experiments at the conditions of pressure and temperature that exist deep within the earth are highly desirable. Shock wave data are at present the only source of information on the compressibility and polymorphism of silicates and oxides at pressures in excess of 300 kbar.

The pioneering experimental work of Bridgman up to pressures of 100 kbar corresponds to a depth of only 300 km within the earth. However, in the last few years, dynamic determinations of the compressibility of minerals and rocks have been made by a number of workers up to pressures in excess of those at the centre of the earth. These high pressures are created for very short time intervals behind the front of a strong shock wave set up by an explosive charge, and are an order of magnitude greater than those which can be obtained by static methods. Although the pressures are maintained only for very short time intervals (usually less than a millisecond), they are long enough to allow measurements to be made of some geophysically important parameters. However, Ringwood (1970) has questioned whether this time interval is sufficient for the attainment of chemical equilibrium, and pointed out that the high pressure phases for which data are obtained may not be the same as those that occur in the deep mantle.

An equation of state obtained from shock wave techniques usually takes the form of a relationship between shock pressure p, shock-induced density ρ, and internal energy E, along a curve called the Hugoniot. The Hugoniot curves, upon reduction, yield pressure-density-temperature states for different materials. One form of the equation of state consists of pressure-density isentropes (i.e. constant entropy curves). Upon differentiation the isentropes provide knowledge of the seismic parameter $\phi = (\partial p/\partial \rho)_S$. Direct comparison is thus possible between values of ϕ obtained from the seismic velocities V_P and V_S (see equation 3.13) and values of ϕ measured for rocks and minerals in the laboratory under similar conditions of temperature and pressure.

The thermodynamic states which are produced behind shock waves are determined from the Rankine-Hugoniot equations which express the conservation of mass, momentum, and internal energy or enthalpy across the pressure discontinuity or shock front, and in 1955 Walsh and Christian carried out such calculations for the states produced by 500 kbar shock waves in metals. Since then a large number of papers have been published giving shock wave equation of state data for many metallic elements to pressures, in some cases, up to 9 Mbar (see e.g. Walsh et al. (1957); McQueen and Marsh (1960); Al'tshuler et al. (1958a, b, 1962)).

McQueen et al. (1964, 1967) have now determined the Hugoniots to pressures greater than 1 Mbar for a series of crustal and possible mantle type rocks (granite, diabase, dunite, eclogite bronzitite), monomineralic feldspar rocks, and a series of oxides and mixed oxides. Hugoniot data for feldspars, granite, basalt, sandstone, limestone and rutile have also been reported by Ahrens *et al.* (1969a, b) and other workers. All but the data for four oxides Al_2O_3, MgO, SiO_2 and MnO_2 showed convincing evidence of shock-induced transitions to one or more denser phases or phase assemblages. From the high pressure data for the Twin Sisters and Hortonolite dunites McQueen *et al.* (1967) constructed metastable Hugoniots that bracketed the seismologically determined density-depth profiles in the lower mantle. A considerable body of data for compounds, as well as for rocks and minerals, has been collected in the *Compendium of Shock Wave Data* (Van Thiel *et al.* (1967)).

It must be stressed that in order to interpret the results, the equation of state as determined from the shock wave data, which is neither adiabatic nor isothermal, must be reduced to a reference temperature. Temperatures in the shock front are not generally known and additional measurements or assumptions must be made to reduce the pressure-density data to those at absolute zero. Moreover, the shock-produced states are characterized by pressure and internal energy—thus an $E(p, T)$ equation of state is required before the data can effectively be used.

It is customary to express the thermal dependence of most equations of state in terms of Grüneisen's parameter γ, where γ represents some average of the volume derivatives of the normal modes. The approximation $\gamma = $ constant is not sufficiently accurate and, because a knowledge of γ at high temperatures and pressures is essential in the reduction of shock-wave data to an adiabat or isotherm, considerable effort has been expended in trying to predict the temperature and volume dependence of γ. Knopoff and Shapiro (1969) compared the various conventional methods of computing γ and showed that they lead to different isotherms when used in shock reduction, and ignore both the contribution of shear modes, which may be ten times as important as the compressional modes, as well as the dispersion of high frequency elastic waves. They further pointed out that lattice models may be of little use in shock wave reduction because the high temperatures associated with the shock probably cause the solid to melt or exceed its elastic limit. The pseudo-Grüneisen parameter for liquids has also been investigated by Knopoff and Shapiro (1970).

9.5 Equations of state

By an equation of state we mean a relationship between the pressure, specific volume and temperature of a material. An equation of state cannot involve the history of the material and thus non-hydrostatic stresses are excluded—large non-hydrostatic stresses in general lead to irreversible (plastic) deformations. The pressures and temperatures in the deep interior of the earth are, on the one hand sufficiently high that it is difficult to reproduce them in the laboratory, and on the other hand sufficiently low that the quantum-statistical models of Thomas-Fermi-Dirac are not applicable. Fortunately temperatures are generally above the characteristic Debye temperatures of the materials involved so that classical analysis can be used to describe their vibrational properties.

One approach which has been developed extensively in the last few years is empirical in nature. The objective has been to formulate some relationship between elastic wave velocities and density using mean atomic weight or some other property of the solid as a parameter. It is then hoped that the observed trends in the laboratory data will allow the prediction of the properties of unmeasured mantle-candidate minerals. Following Birch's (1961) empirical relation between compressional velocity, mean atomic weight and density (see § 3.9), Anderson (1967) suggested a seismic equation of state relating the parameter $\phi = V_P^2 - \frac{4}{3}V_S^2 = k_S/\rho$ (see equation 3.13) to the density ρ through the equation $\rho = A\bar{M} \phi^n$, where $\bar{M}$ is the mean atomic weight and n is a constant of the order $\frac{1}{4}$ to $\frac{1}{3}$ related to Grüneisen's parameter γ. Using the relation $\phi = (\partial p/\partial \rho)_S$, Anderson was able to show that his seismic equation of state is consistent with the functional form of a wide variety of theoretical equations of state.

A different approach has been to use finite strain theory. This involves the expansion of the internal energy of a solid as a series in some tensor measure of the strain. Since the choice of the strain tensor is quite arbitrary, and since all such expansions must be truncated to allow the coefficients to be evaluated in terms of the available data, a critical factor is the convergence of the series, i.e. at what compression do the higher order terms neglected in the truncation introduce appreciable errors? The best known isotropic finite strain theories are those of Murnaghan (1944) and Birch-Murnaghan (Murnaghan (1937); Birch (1938, 1947, 1952)). Anderson and Kanamori (1968) and Chung et al. (1970) have compared these two approaches and shown that the Murnaghan equation, which is based on the assumption that $d(k/k_0)/dp$ is constant, and the Birch-Murnaghan equation, which follows from a second order truncation of the free energy expansion in terms of the Eulerian strain invariants, give different curves for the seismic parameter ϕ as a function of pressure. The good agreement between the two equations of state for pressure as a function of density observed for some materials does not imply the same good agreement in the relationship between ϕ and pressure—the value of ϕ is extremely sensitive to the choice of equation of state, since ϕ is the derivative of pressure with respect to density. Thomsen and Anderson (1969) showed that the equations are strictly isothermal (or adiabatic, depending on the constants used) and are not consistent with either the Mie-Grüneisen or Hildebrand equations, which are usually used for calculations along thermodynamic paths that are neither adiabatic nor isothermal. Thomsen (1970, 1972) later developed what he called a fourth order anharmonic theory. Although this theory is in many ways better than the isothermal/adiabatic isotropic formulations, the usual problems of convergence still remain.

Finally there is the atomistic approach in which the free energy is written in terms of the inter-atomic potential which is then differentiated to give the pressure and elastic constants as a function of volume. Because all expressions are in closed form there are no convergence problems as in finite strain theories. However, there are problems arising from incomplete knowledge of the inter-atomic potential. It is usually assumed that the free energy can be expressed as the sum of two body interactions consisting of a coulomb attraction and a two-parameter empirical repulsion. The two repulsive parameters are fixed by laboratory data, usually the bulk modulus and density of the static lattice. This procedure was pioneered by Born and his co-workers between 1918 and 1950 and is called the Born model.

Gaffney and Ahrens (1970) used the Born model to investigate the relative stability of several high pressure polymorphs of possible mantle minerals by comparing their theoretical internal energies. Sammis (1970) has shown how the Born model can be applied to more complex structures. He computed the pressure dependence of the elastic constants of spinel under the assumption of a Born-Mayer two body potential. He also stressed the importance of retaining the identity of the various bonds and suggested that, to the first order, the repulsive parameters of a bond may depend only on the interacting species and are independent of the particular crystal. The atomistic approach has one advantage over finite strain methods in that it can be tested, since only a very small amount of ultrasonic data is needed for this purpose. However, much additional theoretical work is necessary before the elastic properties of high pressure polymorphs can be predicted with confidence. The primary hope in the atomistic approach is that details of the inter-atomic potential play a less important role than the crystal structure in the calculation of elastic properties. Finally it must be pointed out that, with the exception of the remarks on isotropic finite strain, the above theories have been concerned with single crystals—any application to the earth requires the calculation of isotropic elastic constants from single crystal values. Kumazawa (1969) has given an account of the various averaging techniques—any discussion of this problem is beyond the scope of this book. Zharkov et al. (1972) have recently reviewed the present semi-empirical methods of obtaining the equations of state of rocks and minerals.

References

Ahrens, T. J., D. L. Anderson and A. E. Ringwood (1969a) Equation of state and crystal structure of high pressure phases of shocked silicates and oxides, *Rev. Geophys.*, **7**, 667.

Ahrens, T. J., C. F. Petersen and J. T. Rosenberg (1969b) Shock compression of feldspars, *J. geophys. Res.*, **74**, 2727.

Al'tshuler, L. V., K. K. Krupnikov, B. N. Ledenev, V. I. Zhuckikhin and M. I. Brazhnik (1958a) Dynamic compressibility and equation of state of iron under high pressure, *Sov. Phys. JETP*, **34**, 606.

Al'tshuler, L. V., K. K. Krupnikov and M. I. Brazhnik (1958b) Dynamic compressibility of metals under pressures from 400,000 to 4,000,000 atmospheres, *Sov. Phys. JETP*, **34**, 614.

Al'tshuler, L. V., A. Bakanova and R. F. Trunin (1962) Shock adiabats and zero isotherms of seven metals at high pressure, *Sov. Phys. JETP*, **15**, 65.

Anderson, D. L. (1966) Earth's viscosity, *Science*, **151**, 321.

Anderson, D. L. (1967) A seismic equation of state, *Geophys. J.*, **13**, 9.

Anderson, D. L. and R. O'Connell (1967) Viscosity of the earth, *Geophys. J.*, **14**, 287.

Anderson, D. L. and H. Kanamori (1968) Shock wave equations of state for rocks and minerals, *J. geophys. Res.*, **73**, 6477.

Anderson, D. L. and C. G. Sammis (1970) Partial melting in the upper mantle, *Phys. Earth Planet. Int.*, **3**, 41.

Andrade, E. N. da C. (1910) On the viscous flow of metals and allied phenomena, *Proc. Roy. Soc.* **A 84**, 1.

Andrews, J. T. (1968) Postglacial rebound in Arctic Canada: similarity and prediction of uplift curves, *Can. J. Earth Sci.*, **5**, 39.

Andrews, J. T. (1970). Present and postglacial rates of uplift for glaciated northern and eastern North America derived from postglacial uplift curves, *Can. J. Earth Sci.*, **7**, 703.

Berg, C. A. (1969a) The diffusion of boundary disturbances through a non-Newtonian mantle, in: *The Application of Modern Physics to the Earth and Planetary Interiors* (ed. S. K. Runcorn) (Wiley-Interscience).

Berg, C. A. (1969b) The formation of oceanic trenches and the mechanism of permanent deformation in the mantle, in: *The Application of Modern Physics to the Earth and Planetary Interiors* (ed. S. K. Runcorn) (Wiley-Interscience).

Birch, F. (1938) The effect of pressure upon the elastic parameters of isotropic solids, *J. appl. Phys.*, **9**, 279.

Birch, F. (1947) Finite elastic strain of cubic crystals, *Phys. Rev.*, **71**, 809.

Birch, F. (1952) Elasticity and constitution of the earth's interior, *J. geophys. Res.*, **57**, 227.

Birch, F. (1961) The velocity of compressional waves in rocks to 10 kilobars, *J. geophys. Res.*, **66**, 2199.

Chung, D. H., H. Wang and G. Simmons (1970) On the calculation of the seismic parameter ϕ at high pressure and high temperature, *J. geophys. Res.*, **75**, 5113.

Crittenden, M. D. (1967) Viscosity and finite strength of the mantle as determined from water and ice loads, *Geophys. J.*, **14**, 261.

Dicke, R. H. (1969) Average acceleration of the earth's rotation and the viscosity of the deep mantle, *J. geophys. Res.*, **74**, 5895.

Gaffney, E. S. and T. J. Ahrens (1970) Stability of mantle minerals from lattice calculations and shock wave data, *Phys. Earth Planet. Int.*, **3**, 205.

Glen, J. W. (1955) The creep of polycrystalline ice, *Proc. Roy. Soc.*, **A 228**, 519.

Glen, J. W. (1958) The mechanical properties of ice, 1, The plastic properties of ice, *Advan. Phys.*, **7**, 254.

Goldreich, P. and A. Toomre (1969) Some remarks on polar wandering, *J. geophys. Res.*, **74**, 2555.

Gordon, R. B. and C. Nelson (1966) Anelastic properties of the earth, *Rev. Geophys.*, **4**, 457.

Gordon, R. B. (1967) Thermally activated processes in the earth: creep and seismic attenuation, *Geophys. J.*, **14**, 33.

Griggs, D. (1939) Creep of rocks, *J. Geol.*, **47**, 225.

Jackson, D. D. and D. L. Anderson (1970) Physical mechanisms of seismic wave attenuation, *Rev. Geophys.*, **8**, 1.

Jacoby, W. R. (1970) Instability in the upper mantle and global plate movements, *J. geophys. Res.*, **75**, 5671.

Knopoff, L. (1964) Q, *Rev. Geophys.*, **2**, 625.

Knopoff, L. and J. N. Shapiro (1969) Comments on the interrelations between Grüneisen's parameter and shock and isothermal equations of state, *J. geophys. Res.*, **74**, 1439.

Knopoff, L. and J. N. Shapiro (1970) Pseudo-Grüneisen parameter for liquids, *Phys. Rev.*, **B1**, 3893.

Kovach, R. L. and D. L. Anderson (1964) Attenuation of shear waves in the upper and lower mantle, *Bull. seism. Soc. Amer.*, **54**, 1855.

Kumazawa, M. (1969) The elastic constant of polycrystalline rocks and non-elastic behaviour inherent to them, *J. geophys. Res.*, **74**, 5311.

Lomnitz, D. (1956) Creep measurements in igneous rock, *J. Geol.*, **64**, 473.

MacDonald, G. J. F. (1963) The deep structure of continents, *Rev. Geophys.*, **1**, 587.

McConnell, R. K. (1963) Comments on letter by H. Takeuchi, 'Time scales of isostatic compensations', *J. geophys. Res.*, **68**, 4397.

McConnell, R. K. (1965) Isostatic adjustment in a layered earth, *J. geophys. Res.*, **70**, 5171.

McConnell, R. K. (1968a) Viscosity of the earth's mantle, in: *The History of the Earth's Crust* (ed. R. A. Phinney) (Princeton Univ. Press).

McConnell, R. K. (1968b) Viscosity of mantle from relaxation time spectra of isostatic adjustment, *J. geophys. Res.*, **73**, 7089.

McKenzie, D. P. (1966) The viscosity of the lower mantle, *J. geophys. Res.*, **71**, 3995.

McKenzie, D. P. (1967) The viscosity of the mantle, *Geophys. J.*, **14**, 297.

McKenzie, D. P. (1968) The geophysical importance of high temperature creep, in: *The History of the Earth's Crust* (ed. R. L. Phinney) (Princeton Univ. Press).

McQueen, R. G. and S. P. Marsh (1960) Equation of state for nineteen metallic elements from shock-wave measurements to two megabars, *J. appl. Phys.*, **31**, 1253.

McQueen, R. G., J. N. Fritz and S. P. Marsh (1964) On the composition of the earth's interior, *J. geophys. Res.*, **69**, 2947.

McQueen, R. G., S. P. Marsh and J. N. Fritz (1967) Hugoniot equation of state of twelve rocks, *J. geophys. Res.*, **72**, 4999.

Murnaghan, F. D. (1937) Finite deformations of an elastic solid, *Amer. J. Math.*, **59**, 235.

Murnaghan, F. D. (1944) Compressibility of media under extreme pressure, *Proc. Nat. Acad. Sci. U.S.*, **30**, 244.

Nye, J. F. (1951) The flow of glaciers and ice sheets as a problem in plasticity, *Proc. Roy. Soc.*, **A 207**, 554.

Nye, J. F. (1952) The mechanics of glacier flow, *J. Glac.*, **2**, 82.

Nye, J. F. (1957) The distribution of stress and velocity in glaciers and ice sheets, *Proc. Roy. Soc.* **A 239**, 113.

Orowan, E. (1950) Discussion on 'The flow of ice and of other solids', *J. Glac.*, **1**, 231.

Orowan, E. (1965) Convection in a non-Newtonian mantle, continental drift and mountain building, *Phil. Trans. Roy. Soc.*, **A 258**, 284.

Paterson, W. S. B. and J. C. Savage (1963) Measurements on Athabasca glacier relating to the flow law of ice, *J. geophys. Res.*, **68**, 4537.

Perutz, M. F. (1947) Report on problems relating to the flow of glaciers, *J. Glac.*, **1**, 47.

Ranalli, G. (1971) The expansion–undation hypothesis for geotectonic evolution, *Tectonophysics*, **11**, 261.

Ranalli, G. and A. E. Scheidegger (1969) Rheology of the tectonosphere as inferred from seismic aftershock sequences, *Ann. Geofis*, **XXII**, 293.

Reiner, M. (1949) *Twelve Lectures on Theoretical Rheology* (North Holland Publ. Co.).

Ringwood, A. E. (1970) Phase transformations and the constitution of the mantle, *Phys. Earth Planet. Int.*, **3**, 109.

Sammis, C. G. (1970) The pressure dependence of the elastic constants of cubic crystals in the NaCl and spinel structures from a lattice model, *Geophys, J.*, **19**, 285.

Scheidegger, A. E. (1957) Rheology of the earth, the basic problem of geodynamics, *Can. J. Phys.*, **35**, 383.

Scheidegger, A. E. (1970a) On the rheology of rock creep, *Rock Mech.* **2**, 138.

Scheidegger, A. E. (1970b) The rheology of the earth in the intermediate time range, *Ann. Geofis.* **XXIII**, 27.

Scheidegger, A. E. (1970c) The rheology of the earth in the long time range, *Ann. Geofis*, **XXIII**, 325.

Scheidegger, A. E. (1971) Rheology of the tectonosphere in the short time range, *Ann. Geofis*, **XXIV**, 311.

Smith, S. W. (1972) Anelasticity of the mantle, *Tectonophysics*, **13**, 601.

Takeuchi, H. (1963) Time scales of isostatic compensations, *J. geophys. Res.*, **68**, 2357.

Takeuchi, H. and Y. Hasegawa (1965) Viscosity distribution within the earth. *Geophys. J.*, **9**, 503.

Thomsen, L. (1970) On the fourth-order anharmonic equation of state of solids, *J. Phys. Chem. Solids*, **31**, 2003.

Thomsen, L. (1972) The fourth order anharmonic theory—elasticity and stability, *J. Phys. Chem. Solids*, **33**, 363.

Thomsen, L. and O. L. Anderson (1969) On the high temperature equation of state of solids, *J. geophys. Res.*, **74**, 981.

Van Thiel, M. (1967) Compendium of shock wave data, *Univ. Calif. Radiation Lab. Rept. UCRL 50108*.

Walcott, R. I. (1973). Structure of the earth from glacio-isostatic rebound, *Ann. Rev. Earth Sci.*, **1**, 15.

Walsh, J. M. and R. H. Christian (1955) Equation of state of metals from shock wave measurements, *Phys. Rev.*, **97**, 1544.

Walsh, J. M., M. H. Rice, R. G. McQueen and F. L. Yarger (1957) Shock-wave compressions of twenty-seven metals. Equation of state of metals, *Phys. Rev.*, **108**, 196.

Weertman, J. (1970) The creep strength of the earth's mantle, *Rev. Geophys. Space Phys.*, **8**, 145.

Zharkov, V. N., V. A. Kalinin and V. L. Panjkov (1972) Equations of state and their geophysical applications, *Phys. Earth Planet. Int.*, **5**, 332.

10

The Structure and Composition of the Mantle and Core

10.1 Introduction

The earth's interior has been divided into a number of regions based on the velocity-depth curves. Such a division and the nomenclature introduced by Bullen have been discussed briefly in § 3.5. This book has been mainly concerned with the interior of the earth and no account will be given of the detailed structure of the crustal layers. Features of the crust, such as thickness, seismic velocity distribution, density, and chemical composition vary in a complex manner, being different in continental shields, oceanic abyssal plains, and mountainous and tectonic regions of the world. Special attention will be paid to the question of phase changes, lateral inhomogeneities and the major discontinuities between the crust and mantle and between the mantle and core.

10.2 The Mohorovičić discontinuity

As mentioned in § 3.5, there is a major discontinuity in the velocity-depth curves just below the surface of the earth. This boundary (the Mohorovičić discontinuity or 'Moho') separates the crust from the mantle. Beneath the sedimentary layers the velocities V_P and V_S of P and S waves in continental regions are about 6·2 and 3·6 km/s respectively. These velocities increase with depth to about 7·0 and 3·8 km/s just above the Moho where they jump to about 8·1 and 4·7 km/s, although there are large regional variations, especially in V_P. These velocities in the crust are typical for common rocks such as granite and gabbro; below the Moho they correspond to those found for denser rocks such as dunite, peridotite and eclogite. It is not certain how sharp the discontinuity is but the change is thought to take place within a few km at most. The depth to the Moho is of the order of 30 to 50 km beneath the continents and only about 5 km under the ocean floors.

It has generally been supposed that the Moho is a chemical boundary separating crustal rocks with a high feldspar content from underlying dunite or peridotite, essentially magnesium-iron olivines and pyroxenes. It has been suggested, however, that the Moho may be a phase change, and during the last decade a considerable amount of work has been carried out to investigate this possibility. Fermor, as long ago as 1912, suggested that the Moho represented a phase change from gabbro to eclogite. It will be shown that this transition is unlikely to be the explanation of the Moho under oceanic or stable continental areas although it may be important in tectonically active areas where the Moho cannot be clearly identified, e.g. in regions of recent orogenesis, continental margins and island arcs.

The gabbro–eclogite reaction is not the only phase change that has been proposed for the Moho. Hess in 1959 suggested that the oceanic crust may perhaps be serpentine not basalt. Serpentine may be formed by hydration of olivine and decomposes above a temperature of about 500°C. Hess (1962) suggested that the sub-oceanic Moho was originally the locus of the 500°C isotherm which became 'fossilized' at the present low temperature of 150–200°C. The two phases can only be in equilibrium at the Moho if the temperature gradient is about three times greater than the oceanic mean. Hess thus proposed that all the (non-sedimentary) oceanic crust originates along the mid-ocean ridges where this condition may be satisfied. The oceanic crust and upper mantle are produced by 'freezing' this equilibrium state as the sea floor moves away from the ridge. The hydration of olivine involves only one mineral (unlike the gabbro-eclogite phase change) and thus the phase boundary is sharp. However, the seismic velocities observed at the base of the crust are lower than those in serpentine. Hess' suggestion cannot be entirely discounted, although it could only apply to the oceanic crust, and there are additional problems concerning the dynamical effects supposed to follow subsequent hydration and dehydration.

Let us consider now the gabbro-eclogite reaction in more detail. Gabbro and eclogite have essentially identical compositions but sharply contrasting mineralogy. The mean density of eclogite is about 10 per cent greater than that of gabbro which is sufficient to account for the change in seismic velocities at the Moho. The phase change depends of course on temperature and pressure. Thus if the temperature gradient is different under continents, mountains and oceans, the transition would take place at different depths under these regions (see Fig. 10.1), thereby explaining the varying depths to the Moho. Kennedy (1959) suggested that such a phase change could explain a number of geological problems including the uplift of large continental areas long near sea level to form high plateaux, the long lifetime of continents and mountain ranges (far longer than would be inferred from the rates of erosion), the ability of low density sediments to subside into the higher density substratum in the troughs along continental margins, and the rough equality of heat flow through different regions of the earth's surface.

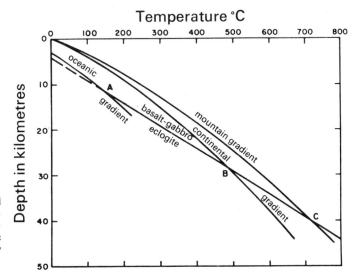

Fig. 10.1 Postulated temperature gradients under mountain ranges, continental areas and oceanic regions (after Kennedy (1959))

There are a number of difficulties in accepting the above hypothesis. It must be stressed, however, that as far as the consequences of a phase change for subsidence and uplift are concerned, it is immaterial whether or not the phase change is identified with the seismically defined Moho; all that is necessary is that significant volume changes accompany the change in phase. In this respect Wyllie (1963) has pointed out that the two alternatives of a chemical discontinuity and a phase change need not be mutually exclusive. He suggested that there may be two discontinuities beneath the crust: a chemical discontinuity of global extent between material of basaltic composition at the base of the crust, and material with the composition of feldspathic peridotite in the upper mantle. In addition there may exist (at the appropriate depth, depending on the geothermal gradient) a second discontinuity marking the gabbro-eclogite phase transition. Whichever discontinuity is nearer the earth's surface would be detected as the Moho by seismic measurements. It would probably be difficult by such methods to detect a deeper transition. In this respect it is interesting to note that magnetotelluric measurements in Massachusetts have indicated a rapid change in resisitivity at a depth of about 70 km, which is deeper than the Moho in that region.

A very thorough and detailed experimental investigation of the gabbro-eclogite transformation in several basalts has been carried out by Green and Ringwood (1967). They concluded (Ringwood and Green (1966)) that the hypothesis that the continental Moho is caused by an isochemical transformation from gabbro to eclogite must be rejected for a number of reasons. Before their important work is described in some detail, some general difficulties with the hypothesis will first be discussed. The principal minerals of gabbro are plagioclase feldspar and pyroxene, while those of eclogite are typically magnesium-rich garnet and soda-rich pyroxene. The relation between these rocks is not one of simple polymorphism; gabbro must recrystallize completely to form eclogite, no single mineral transforming without chemical change, even though the bulk composition remains fixed. Thus if the Moho represents a change in crystal structure it would not occur at a definite depth, but would extend over a range of pressures. Although the seismic evidence indicates that the discontinuity may possibly be spread over some km under the continents, it is extremely difficult to interpret the seismic data in this way under the oceans where the Moho is at a very shallow depth.

Bullard and Griggs (1961) re-examined the pressure-temperature relations in the earth and found that the oceanic and continental transitions could not both result in the high pressure form lying beneath the Moho—in fact the temperature-depth curve under the oceans as drawn in Fig. 10.1, is incorrect, the curve actually crossing the transition from below to above and not from above to below. They also showed that contrary to Kennedy's (1959) surmise, variations in the depth of the Moho from place to place are much less than would be expected from variations in the heat flow. An extremely large increase in crustal thickness under the oceans would be necessary if the surface heat flow is increased above average values. Similar results have been obtained by Wetherill (1961) as a result of much more detailed calculations.

Returning now to the experimental work of Ringwood and Green (1966), Fig. 10.2 shows, in a simplified form, the phase assemblages which they found

Fig. 10.2   Diagram illustrating the effect of chemical composition on the pressure required for the incoming of garnet (lower boundary) and the outgoing of plagioclase (upper boundary). All data are at 1100°C; compositions 1–6 are from Ringwood and Green (1966); 7, 8, 9, 10 are compositions from the references shown (after Green and Ringwood (1972))

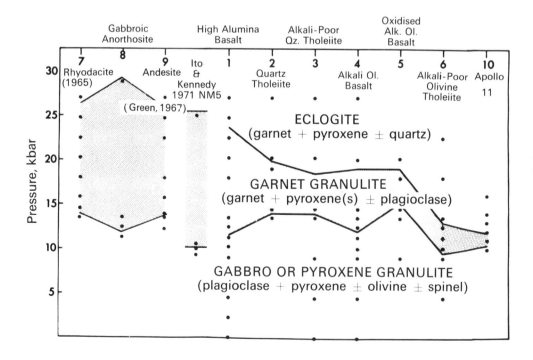

in a number of basalts at 1110°C as a function of pressure. For each basalt there are clearly three principal mineral stability fields corresponding closely with naturally observed mineral assemblages. The low pressure assemblage is that of gabbro or pyroxene granulite. In all the basalt compositions studied, the transformation from gabbro or pyroxene granulite to eclogite proceeded through an intermediate mineral assemblage characterized by co-existing garnet, pyroxene(s) and plagioclase. This intermediate assemblage possesses an extensive stability field varying from 3·5 to 12 kbar in width. Figure 10.2 also shows that rather modest changes in chemical composition cause large changes in the pressures and widths of the gabbro-eclogite transformation.

Figure 10.3 shows the extrapolation to lower temperatures of the phase boundaries between eclogite, garnet granulite and gabbroic mineral assemblages proposed by Ringwood and Green. It can be seen, for example, that for a basalt of quartz tholeiite composition the temperature on the garnet granulite/eclogite boundary at a pressure corresponding to that at the base of the normal continental crust (10 kbar) is 670°C. If the temperature at the base of the crust is lower than this, then eclogite would be the stable form of a basalt of this composition throughout the crust. In stable continental regions of normal crustal thickness and characterized by heat flows between 0·8 and 1·5 × 10^{-6} cal/cm^2 s the temperature at the base of the crust is usually less than 670°C on any reasonable assumption of the distribution of radioactivity. This conclusion is practically certain for Precambrian shields characterized by mean heat flows of 1·0 × 10^{-6} cal/cm^2 s; the temperature at the base of the crust in these regions is probably less than 450°C. It thus appears that eclogite is the stable modification of quartz tholeiite throughout very large regions of normal continental crust. This conclusion of Ringwood and Green was unexpected and has profound tectonic consequences.

Ito and Kennedy have carried out a number of experiments on an olivine tholeiite (NM5). Their most recent results (1970, 1971) differ significantly from their earlier data ((1968); also Cohen, Ito and Kennedy (1967)). In their 1971 paper, Ito and Kennedy state that their new data are 'remarkably in contrast to those published by Green and Ringwood (1967)'. They do not state where the supposed conflict in the experimental data occurs but imply that the problem of melting which in part invalidated their own earlier results also invalidates the work of Green and Ringwood. Green and Ringwood (1972) have shown that such is not the case, and that in fact the latest work of Ito and Kennedy is in excellent agreement with their (Green and Ringwood's) earlier work. (Ito and Kennedy's data are also included in Figs. 10.2 and 10.3.) The latest experimental work of Ito and Kennedy thus confirms the two major conclusions of the previous work of Green and Ringwood namely:

(1) The gabbro to eclogite transformation takes place through a rather broad transition interval in which assemblages appropriate to high pressure granulite assemblages are formed. The positions of the boundaries and the width of the field of the transitional garnet granulite mineral assemblage are very sensitive to variations in the chemical compositions of basalts.

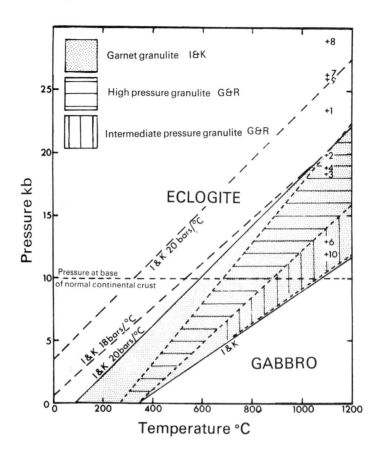

Fig. 10.3 Comparison of extrapolation to lower temperatures of phase boundaries between eclogite, garnet granulite and gabbroic mineral assemblages as advocated by Ringwood and Green (1966), Green and Ringwood (1967) (dashed lines and vertically and horizontally shaded fields labelled G & R) with the extrapolations advocated by Ito and Kennedy (1971) (stippled fields, lines labelled I & K). The numbered crosses at 1100°C mark the disappearance of plagioclase in the compositions 1–10 of Fig. 10.2. According to the arguments of Ito and Kennedy (1971) lines at 20 bar/°C could be drawn through each of these points to extrapolate the disappearance of plagioclase to lower temperatures; such a line for quartz tholeiite (composition 2) would be coincident with the garnet granulite → plagioclase eclogite boundary of Ito and Kennedy (1971) *if it is extrapolated at 20 bar/°C*. The dashed line labelled I & K 18 bar/°C is the line taken from Ito and Kennedy's own Fig. 4 marking the low pressure boundary of their plagioclase eclogite field. Ito and Kennedy (1971) drew this line at 18 bar/°C without justifying this by either experiment or argument. If it is argued that this boundary is more complex than the plagioclase-out boundary for NM5 and thus that Ito and Kennedy's arguments for a 20 bar/°C slope are not acceptable, then the only evidence on the slope of this boundary is that it lies within the experimental limits of Ringwood and Green's quartz tholeiite (composition 2) data, i.e. between 18 and 36 bar/°C (after Green and Ringwood (1972))

(2) The extrapolation of experimentally determined boundaries to lower temperatures suggests that eclogite is the stable mineralogy in dry basaltic rocks along normal geothermal gradients in the continental crust (stable or shield regions).

These two conclusions together with other arguments of Ringwood and Green (1966) make it extremely unlikely that the Moho in oceanic or stable continental regions is the result of the gabbro-eclogite transition.

10.3 The mantle

The upper mantle extends from the base of the crust (the Moho) to a depth of about 400 km. It consists of a thin (0–50 km thick) high velocity layer which together with the crust constitutes the lithosphere overlying the low velocity (LV) zone, a partly molten layer of the order of 100 km in thickness (the asthenosphere)—see also § 8.1. Between depths of about 150–400 km the mantle is relatively homogeneous, its properties being determined by self-compression in the presence of a moderate temperature gradient. The properties of the mantle are governed primarily by the response of the material to changes in pressure and temperature, variations in composition having but a minor effect. The distribution of pressure in the mantle is reasonably well known but temperatures are much less certain. At depths less than about 200 km the effect of temperature is comparable to that of pressure because of high thermal gradients, and, for certain properties, temperature may be the controlling factor. This is not the case at greater depths, and throughout most of the mantle pressure is the dominant parameter affecting the properties of the material.

The transition region (Bullen's region C, between depths of about 400–800 km) is characterized by a number of abrupt increases in seismic velocities which have been attributed to phase changes. This important region in the earth will be discussed in the next section. The lower mantle is relatively homogeneous with seismic velocities and densities increasing monotonically due to self-compression with no major phase changes.

The important minerals of the upper mantle are olivine, pyroxene and garnet. Comparison between seismic and ultrasonic data is made difficult by lateral heterogeneity, particularly in the upper 200 km of the mantle. Table 10.1 (after Anderson et al. (1971)) gives the velocities and densities of the mantle (at 200 km), mantle minerals and two mixtures of these minerals. The present uncertainty in seismic velocities in the upper mantle (below the LV layer) permits mineralogies ranging (in weight per cent) from 100 per cent olivine to 40 per cent olivine, 50 per cent pyroxene and 10 per cent garnet. On the basis of ultrasonic measurements of ultrabasic rocks Anderson (1970) concluded that the composition of the

Table 10.1 Velocities and densities of the mantle, of mantle minerals, and of various mixes

| Property | Mantle at 200 km | | Olivine | Pyroxene | Garnet | Mix 1 | Mix 2 |
	Ambient	STP					
V_P (km/s)	8·30–8·65	8·11–8·46	8·42	7·85	8·53	8·22	8·14
V_S (km/s)	4·52–4·60	4·80–4·88	4·89	4·76	4·76	4·83	4·81
ρ (g/cm^3)	3·50	3·42	3·31	3·34	4·16	3·38	3·39

The olivine is 93 mole per cent forsterite; the pyroxene is 85 mole per cent enstatite. The garnet is of almandite–pyrope composition. Mix 1 is pyrolite, with the following composition (in weight per cent): olivine, 56; pyroxene, 35; and garnet, 9. Mix 2 has the following composition (in weight per cent): olivine, 40; pyroxene, 50; and garnet, 10. STP, standard temperature and pressure. (After Anderson *et al.* (1971))

mantle at a depth of 200 km is consistent with a mineralogy of 45–75 per cent olivine, 20–50 per cent pyroxene and up to about 15 per cent garnet.

It is not known for certain what is the cause of the LV zone. It is possible, as suggested by Clark and Ringwood (1964), that it represents a different mineral assemblage to adjacent regions of the mantle or results from a thermal gradient so large that the effects of pressure are cancelled out. However, Anderson and Sammis (1970) have shown that the seismic velocities in the zone are too low to be attributed to these effects. The boundaries of the zone also appear to be quite sharp. Anderson and Sammis concluded that the low velocity, low Q and abrupt boundaries could best be explained by partial melting or dehydration of a mineral assemblage containing a small amount of water. Lambert and Wyllie (1970) made a similar suggestion, namely that the LV zone could be a layer of peridotite containing interstitial melt, overlain by a layer of peridotite containing interstitial water. In a later paper Anderson and Spetzler (1970) showed that if the lowest melting constituents were concentrated in narrow zones such as grain boundaries, quite a small amount of melt ($\simeq 1$ per cent) could account for both the V_P and V_S velocities in the LV zone.

Information on the composition and crystal structure of the lower mantle can be obtained from finite strain theory (see § 9.5). Birch (1952) pioneered research in this field, and many of his conclusions have since been verified. An excellent account of our present knowledge of the lower mantle has recently been given by Anderson *et al.* (1971). Figure 10.4, taken from their paper, shows how shock-wave data, seismic data and finite strain theory can be used to obtain information about the composition of the deep mantle. The mantle point for the particular model considered falls in the field defined by Fo_{90}, i.e. $\{(Mg_{0.90}Fe_{0.10})_2SiO_4\}$; Fo_{45}, i.e. $\{(Mg_{0.45}Fe_{0.55})_2SiO_4\}$, and SiO_2 (stishovite). The inferred composition of the lower mantle for this model is $0.32MgO + 0.18FeO + 0.50SiO_2$ (molar). Other models (also based on free oscillation data) require less SiO_2 in the lower mantle, but all are well above the high pressure olivine line. Anderson *et al.* obtained further information on the composition of the lower mantle using Birch's (1961) linear relationship between density and V_P at constant mean atomic weight (which, for the lower mantle, lies between

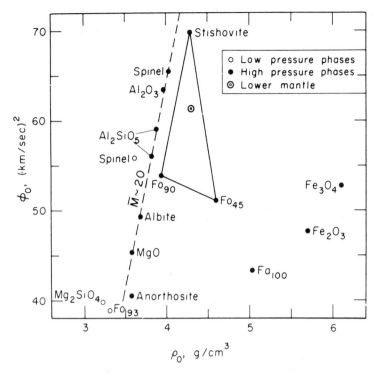

Fig. 10.4 A plot of ϕ_0 versus ρ_0 from shock wave experiments for various rocks and minerals (Anderson (1970)), and one solution for the lower mantle (Anderson and Jordan (1970)). Iron-free minerals have a mean atomic weight $\overline{M}$ near 20. Iron-bearing minerals fall to the right of the dashed line (after Anderson *et al.* (1971))

21·0–21·7). The compressional velocity is well above the spinel line as might be expected for the closer packed phases. Anderson *et al.* concluded that the density and velocity in the lower mantle are consistent with a denser ionic packing than occurs in such close packed oxides as MgO, $MgAl_2O_4$, and garnets. In these compounds the packing ratio (defined as the fraction of the cell volume that is occupied by ions), ranges from 0·63–0·66. The lower mantle has a packing fraction close to that of Al_2O_3, namely 0·72. For comparison stishovite, the high pressure form of SiO_2, with the Si ions in sixfold coordination with the oxygen ions, has a packing fraction of 0·84.

It might seem that phase transitions in the deep mantle in the pressure range 300–1300 kbar are not unlikely considering the number of transformations that occur at pressures of less than 300 kbar (see § 10.4). However, constraints arising from the density distribution and elasticity of the deep mantle severely limit the magnitude of allowable density changes arising from possible further transformations.

It is generally accepted that the lower mantle is significantly denser (by about 5 per cent) than an oxide mixture (SiO_2 as stishovite) isochemical with pyrolite.* In particular Ringwood (1972a) has suggested that further transformations in the mantle may have resulted in the formation of a mineral assemblage intrinsically denser than the isochemical mixed oxides; alternatively the lower mantle may be characterized by a higher Fe/Mg ratio than the upper mantle

Table 10.2 Model pyrolite composition

SiO_2	45·16
MgO	37·47
FeO	8·04
Fe_2O_3	0·46
Al_2O_3	3·54
CaO	3·08
Na_2O	0·57
K_2O	0·13
Cr_2O_3	0·43
NiO	0·20
CoO	0·01
TiO_2	0·71
MnO	0·14
P_2O_5	0·06
	100·00

(After Ringwood (1966a))

and transformations to phases denser than the mixed oxides do not occur. A comprehensive review on the state of mantle minerals has been given by Ahrens (1972).

Chinnery and Toksöz (1967) and Toksöz et al. (1967) have determined P wave velocities in the lower mantle from $dT/d\Delta$ measurements using the large aperture seismic array (LASA) in Montana and travel-times from the LONGSHOT nuclear explosion. Previous models for the P wave velocity profile were based primarily on observations of travel-times and amplitudes from earthquakes, with the inherent lack of accurate origin times and epicentres and the variability of measured amplitudes. They found that the velocity structure showed anomalous gradients or discontinuities at depths of about 1200 and 1900 km. It is not possible to say whether these discontinuities are global since the data used came from one small part of the earth, and there is evidence that lateral inhomogeneities persist to considerable depth. The discovery of such discontinuities indicates regions of the earth where there are either phase or composition changes or both, and, contrary to earlier beliefs, points to departures from homogeneity in the lower mantle. Using the extended array at the Tonto Forest Seismological Observatory in Arizona to measure $dT/d\Delta$ of direct P waves in the distance range 30°–100°, Johnson (1969) estimated P velocities in the lower mantle. He

* Pyrolite is a mixture of about three parts peridotite to one part basalt (see Ringwood (1966a)). The Fe/Fe + Mg ratio in pyrolite is 0·11 molecular. The model composition of pyrolite is given in Table 10.2.

288

found no large discrepancies with the traditional models of Gutenberg and Jeffreys, but increased velocity gradients near depths of 830, 1000, 1230, 1540, 1910 and 2370 km were indicated. Vinnik *et al.* (1972), using $dT/d\Delta$ measurements from several arrays in the USSR, found regions of high P wave velocity gradients at depths of about 900, 1300, 1700, 2000 and 2500 km. There appear to be lateral variations of P wave velocity in the lower mantle, more pronounced in south-east Asia than elsewhere. It must be stressed, however, that these anomalies in the lower mantle are spread over depth intervals of at least 50 km, and are an order of magnitude smaller than those which have been found in the upper mantle.

In a detailed study of the characteristics of seismic waves detected at LASA in Montana, Davies and Sheppard (1972) found additional indications of lateral heterogeneity in the earth's mantle, particularly near the core-mantle boundary. Anomalies associated with rays bottoming at 700–1500 km were also found but are not well understood; they may arise through contrasts between the structure beneath continents and oceans. They also found an anomaly beneath the Bonin Islands which they attributed to the influence of the descending slab of lithosphere on the discontinuity at 630 km.

Julian and Sengupta (1973) have presented further evidence from travel-time data that significant lateral variations occur in the lowest few hundred km of the mantle, this region being much more heterogeneous than that which lies above it. The size of some of the anomalies is of the order of 1000 km. It is interesting to speculate whether such deep-mantle variations be related in some way to the convection plumes that Wilson (1965) and Morgan (1971) have suggested exist in the deep mantle (see § 8.6). There is some support for this hypothesis in the region of the Hawaiian Islands, but unfortunately no other proposed plumes were well sampled by the data analysed by Julian and Sengupta—no other geological or tectonic features showed an obvious correlation with the inferred deep mantle velocity anomalies.

10.4 The transition region

The transition region (approximately $400 \sim 1000$ km) is characterized by several abrupt increases in velocity. Surface wave studies (see e.g. Anderson (1966, 1967)) have shown that the abnormally high velocity gradients in the transition region are concentrated in two relatively narrow zones some 50–100 km thick instead of being spread out uniformly over about 600 km as was previously thought. Later studies, using travel times and the apparent velocities of body waves, have verified the presence of these two transition regions (see Fig. 10.5). In most recent models of the upper mantle there is a very rapid increase in velocity between about 100 and 150–175 km and major discontinuities starting at about 350 and 630 km. The velocity gradients of the adjacent sections of the mantle are appropriate for normal compression, including that section of the mantle between about 440–620 km which lies in the middle of what has been designated as the

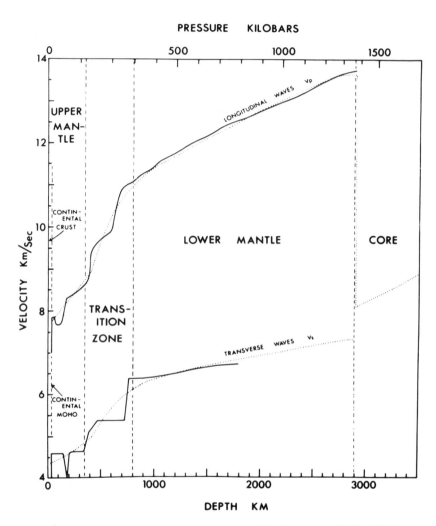

Fig. 10.5 Seismic velocity distribution in the mantle for P waves, solid line (Johnson (1967, 1969)); S waves, solid line (Nuttli (1969)); broken lines, P and S waves (Jeffreys (1937, 1939)) (after Ringwood (1972a))

transition region. It has been shown that the locations of these transition regions and their general shape and thickness are consistent with, first, the transformation of pyroxenes into a new type of garnet structure and of olivines to a spinel (or related) structure and, then, a further collapse to a material having approximately the properties of the component oxides, i.e. garnets and spinels transform to new phases possessing ilmenite, perovskite and strontium plumbate structures with densities and elastic properties resembling those of isochemical mixed oxides (see e.g. Ringwood (1969, 1970)).

The hypothesis that a series of major phase changes occurs between depths of about 300–1000 km was proposed by Birch in 1952. Although at the time Birch's arguments met with some opposition, experimental investigations have now confirmed them in all essential aspects. The difficulty about testing Birch's hypothesis experimentally was that, until 1963, available static high pressure-temperature (p, T) apparatus was incapable of reproducing the (p, T) conditions in the earth at depths greater than about 300 km. Thus prior to 1963 it was necessary to use *indirect* experimental methods based on thermodynamics, comparative crystal chemistry and, in particular, upon the study of germanate isotypes of silicates. The reason for this last method is the fortunate circumstance that germanates often display the same kinds of phase transformations as the corresponding silicates, but at much lower pressures. Ringwood (1969) has given an excellent summary with a very complete bibliography of the results of this indirect phase of investigations which covered the period 1956–1963 and in which he himself played a great part. All of the germanate olivines and pyroxenes which were studied in the pressure range 0–90 kbar were found to be unstable at high pressures and transformed to dense phases suggesting that the corresponding silicates would transform similarly at higher pressures. During this period Ringwood found that the silicate olivines Fe_2SiO_4, Ni_2SiO_4 and Co_2SiO_4 could be transformed to spinel structures at 20–70 kbar and that the spinels were about 10 per cent denser than the corresponding olivines. A high pressure form of silica (SiO_2), with the rutile structure in which silicon is in octahedral coordination with a density of about $4 \cdot 3$ g/cm^3, was synthesized by Stishov and Popova in 1961 at a pressure corresponding to a depth of about 100 km, and has since been identified in the crushed zone in the Barringer Meteorite Crater, Arizona (see § 1.6). This discovery suggested that the pyroxene family might transform to new structures such as ilmenite, characterized by octahedral coordination of Si^4.

In 1966 an apparatus capable of developing pressures above 200 kbar simultaneously with high temperatures was developed by Ringwood and Major (1966, 1968) in Australia. Other laboratories (in particular in Japan, see Akimoto and Ida (1966); Kawai (1966)) have now developed similar techniques. It is thus possible to synthesize most of the major mineral phases that probably occur in the mantle down to a depth of about 600 km. During 1966–1967 a number of important new transformations in natural mantle minerals were discovered. These included transformation of olivine into spinel and βMg_2SiO_4 structures, and of pyroxenes into new garnet forms containing some octahedrally coordinated structures. This work has been reviewed in detail by Ringwood (1969, 1970) and Ringwood and Major (1970).

It must be stressed, however, that although shock wave data are of great value in demonstrating the occurrence of phase transformations in silicates and oxides at pressures well above the range of static pressure apparatus, they do not identify the nature of the high pressure phases. In this respect indirect methods based upon crystal chemistry and germanate analogue studies complement the shock wave data. In many cases where shock transformations occur, the probable nature of the high pressure structure may be revealed by these indirect methods (see Ahrens *et al.* (1969)). The effect of the above transformations in the mantle on the movement of lithospheric plates has been discussed by Ringwood (1972b).

10.5 The core

There is no doubt about the sharpness of the mantle-core boundary (MCB) although there may be minor lateral variations in properties there. The most detailed calculations of the depth to the MCB are those of Taggart and Engdahl (1968) based on an analysis of P phases; they obtained for the radius r_c of the core the value 3477 ($\pm 2 \cdot 0$) km. There is also now a convincing body of evidence for a low velocity shell above the MCB. Both Ergin (1967) and Bolt (1970) found a negative velocity gradient in this region; in particular Bolt (1972) found a 2 per cent decrease in V_P through the lowest 150 km of the mantle. His results led to a value for $r_c = 3475$ ($\pm 2 \cdot 0$) km. Hales and Roberts (1970a) also suggested a decrease in V_S of about 3 per cent in a transition shell above the MCB. In a later paper (1970b) they obtained values of r_c of 3489·9 ($\pm 4 \cdot 7$) km and 3486·1 ($\pm 4 \cdot 6$) km for two possible mantle models, basing their calculations on the differences in travel times between ScS and S in the distance range $48° < \Delta < 70°$. The 10 km excess over the radius estimated from P phases is not significant at the 95 per cent confidence level.

Recent eigen vibration studies have confirmed that the outer core has very small rigidity. Earlier, Sato and Espinosa (1967) obtained an upper bound to the rigidity just below the MCB of $5 \cdot 5 \times 10^{10}$ dynes/cm^2 using spectral amplitude ratios of ScS with its multiple reflections. There is much uncertainty in this result however. Below the liquid outer core there is a slightly rigid ($\sim 0 \cdot 5 \times 10^{12}$ dynes/cm^2) transition shell (perhaps partially molten) surrounding a solid inner core ($r < 1216$ km) with a rigidity of the order of $1 \cdot 8 \times 10^{12}$ dynes/cm^2.

The first detailed study of the effect of the rigidity of the inner core on the earth's periods of free oscillations was undertaken by Derr (1969a). He found an average S velocity of 2·2 km/s in the inner core with a density jump of less than 2 g/cm^3 at the inner core boundary (see § 3.9). In a later paper Dziewonski and Gilbert (1971) have given convincing evidence for the solidity of the inner core based on their observations (1972) of normal modes from 84 recordings of the March 1964 Alaska earthquake. They also showed that, with the exception of $_2S_2$, the modes sensitive to the structure of the inner core were either mis-identified or remained unidentified by Derr (1969b). The existence of a solid inner core has now been definitely confirmed by the identification of the phase PKJKP on seismograms (Julian *et al.* (1972)).

The strong seismological evidence for a sharp inner core boundary, and one or more first order discontinuities in velocity at greater radii, raises the important question of possible liquid-solid or solid-solid phase transitions in either a homogeneous material or in a multi-component chemical system. If the transition from the inner to the outer core is a transition from the solid to the liquid form of a single material, then the boundary must be at the melting point and a constraint is put on the thermal regime of the earth's interior. This has been discussed in some detail in § 7.4

In 1949 Ramsey proposed that the lower mantle and core have the same chemical composition, the discontinuity at the MCB resulting from a change of mantle silicates to a high pressure liquid metallic form. However, the materials in the lower mantle are already tightly packed, and transformation to a metallic form is unlikely to increase the density. Also, as Birch (1968) pointed out, at 1 atm the mean atomic volumes of oxides and silicates are less than the mean atomic volumes of the pure metals of which they are made. The transformation to metallic form of mantle material, composed mainly of light elements, must result in a light metal, and metals lighter than chromium ($Z=24$) are all too light for the earth's core. Figure 10.6 is a plot of the hydrodynamical sound velocity $(\partial p/\partial \rho)_s^{1/2}$ against density along the Hugoniot compression curves for metals up to cobalt and for several rocks. The areas in which the corresponding quantities for the earth's mantle and core must lie are indicated by the pairs of dotted lines. On the diagram, Ramsey's hypothesis corresponds to a transition from a point such as A to a point such as B. The experimental evidence clearly shows that this corresponds to a change of atomic number, i.e. of chemical composition, and by a large amount (roughly from 12 to 23). For a given atomic number and given velocity, the rocks are both denser and less compressible than metals. Moreover, all experimental data have shown that the densities of rocks when compressed well beyond the pressure at the MCB fall on a smooth continuation of the curves for lower pressures and give no indication of transforming to core densities. Originally Ramsey put forward his theory to account for the differences in the observed densities of the terrestrial planets which he assumed to have a common primitive composition. More recent astronomical data have revised the older figures and also show that Ramsey's hypothesis is untenable. The sum total of all experimental data thus shows that Ramsey's hypothesis cannot be true.

Bullen found for his earth model A that there was no noticeable difference in the gradient of the incompressibility, dk/dp, between the base of the mantle and the top of the core. Moreover, there was only a 5 per cent difference in the value of k across the MCB. These features are in marked contrast to the large changes in density and rigidity at the boundary. The small change in k is a diminution from the mantle to the core. Because of the smallness in the change in k across the MCB, and because this change is opposite in sign to that predicted theoretically, Bullen (1949, 1950) proposed another earth model B in which he assumed that k and

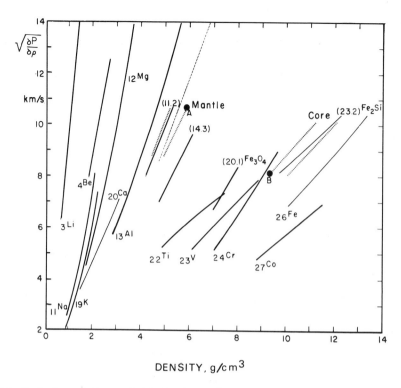

Fig. 10.6 Hydrodynamical sound velocity $(\partial \rho / \partial \rho)_S^{\frac{1}{2}}$ versus density along the Hugoniot compression curves for metals up to cobalt and several rocks. For a few materials the shock wave velocity has been plotted instead of the adiabatic sound velocity. Atomic numbers or representative atomic numbers (in parentheses) are attached to each curve. The line marked (11.2) shows values for Twin Sisters dunite, a rock composed mainly of olivine (92%) of composition about Fa_{10}; the line marked (14.3) shows values for hortonolite dunite, composed mainly of olivine (90%) of composition about Fa_{50}; the line marked (20.1) is for magnetite; the line marked (23.2) is for an iron–silicon alloy having nearly the composition Fe_2Si. Similar data for a large number of other rocks fall between the lines for the dunites, but have been omitted for the sake of clarity. The areas in which the corresponding quantities for the earth's mantle and core must lie are indicated by the pairs of dotted lines. Several oxides, silica, periclase, and corundum have been compressed to M bar pressures and also fall in the 'mantle' area (after Birch (1968))

dk/dp are smoothly varying functions throughout the earth below a depth of about 1000 km. This hypothesis, called the compressibility-pressure (k, p) hypothesis, implies that at high pressures the compressibility of a substance is independent of its chemical composition.

  On the basis of this hypothesis, Bullen found that there must be a concentration of more dense material near the base of the mantle (region D″). This material could be a mixture of metallic iron with silicates near the MCB, or an iron sulphide phase at the base of the mantle. If the entire core is liquid so that $V_S = 0$, then from equation (3.13), V_P is given by

$$V_P^2 = k/\rho \qquad (10.1)$$

Jeffreys' original velocity distribution showed a discontinuous jump across the boundary between regions F and G and, to accommodate this, k would have to increase by 32 per cent—excluding the highly improbable case that the density decreases with depth. Assuming Gutenberg's velocity-depth curve the effective increase in k would be 23 per cent. On the other hand, as Bullen first pointed out in 1946, if the inner core G is solid and thus capable of transmitting S waves, equation (10.1) is replaced by equation (3.13) in G, and the increase in V_P can be accounted for without violating his (k, p) hypothesis.

Bullen and Haddon (1967) and Haddon and Bullen (1969) have since revised Bullen's original model B and constructed a series of new earth models based on the (k, p) hypothesis. They found that the high densities of the original model B in the outermost several hundred km of the earth are no longer necessary, in fact one of the chief results of these revised models is that the differences between models of A and B type have tended to disappear (see also § 3.9).

The high pressure work of Ahrens et al. (1969) on magnesium oxide and aluminium oxide indicates that along an adiabat over the pressure range in the lower mantle, $\partial k/\partial p$ is not constant but decreases markedly. Lattice dynamical calculations also predict that $\partial k/\partial p$ should decrease as the pressure increases. Such a prediction is also given by quantum mechanical calculations of equations of state (such as the Thomas-Fermi-Dirac equation) and general theories of finite deformation (such as the Birch-Murnaghan equation). The results of shock wave experiments and theoretical considerations also indicate quite clearly that k and $\partial k/\partial p$ do not depend on pressure in the same way for all materials. The behaviour of solids in general is thus inconsistent with Bullen's (k, p) hypothesis, although it is a fairly good approximation in the case of the earth. This is in part fortuitous: the compressibilities of iron oxide, stishovite, aluminium oxide and magnesium oxide are similar at pressures comparable to that at the MCB. The gradients of the bulk modulus of such substances are not constant; the average values are, however, not so very different from the mean terrestrial value. If the mantle extended over a greater range of pressures, the agreement would become less close.

Bullen (1963) has also investigated the chemical inhomogeneity of the earth. The term chemical inhomogeneity is used here to include also inhomogeneity arising from phase changes. Assuming hydrostatic pressure (equation (3.29)) and an adiabatic temperature gradient, we have (by definition of ϕ),

$$k = \rho\phi$$

so that

$$\frac{dk}{dp} = \phi\frac{d\rho}{dp} + \rho\frac{d\phi}{dp} = \phi\frac{d\rho}{dp} - \frac{1}{g}\frac{d\phi}{dr}$$

For a chemically homogeneous region, $d\rho/dp = \rho/k$ (by definition), and thus

$$\frac{dk}{dp} = 1 - g^{-1}\frac{d\phi}{dr} \tag{10.2}$$

At any point of a region, whether chemically homogeneous or not, at which $d\rho/dr$ can be assumed to exist, we can write

$$\frac{dk}{dp} = \frac{-\phi}{g\rho}\frac{d\rho}{dr} - \frac{1}{g}\frac{d\phi}{dr}$$

i.e.

$$\frac{d\rho}{dr} = \frac{-\eta g\rho}{\phi} \qquad (10.3)$$

where

$$\eta = \frac{dk}{dp} + g^{-1}\frac{d\phi}{dr} \qquad (10.4)$$

When $\eta = 1$, equation (10.3) reduces to the Adams-Williamson equation (3.31), so that η is a measure of the departure from chemical homogeneity; it is the ratio of the actual density gradient to the gradient that would obtain if the composition were uniform. An excess temperature gradient normally reduces, while chemical inhomogeneity increases, the value of η.

From equation (10.4) it can be seen that η depends on dk/dp, g and $d\phi/dr$. On Bullen's compressibility-pressure hypothesis, dk/dp is slowly varying and lies between about 3 and 6 throughout most of the earth's deep interior. Uncertainties in estimates of g are not large, while values of $d\phi/dr$ are immediately derivable from the P and S wave velocity distributions. It follows then from equation (10.4) that η can be estimated in most parts of the earth's deep interior within limits that can be assigned. Thus it is possible to estimate the degree of departure from chemical homogeneity in any given region and to assess density gradients where the Adams-Williamson equation cannot be used. In a later paper, Bullen (1965a) refined equation (10.4); in particular he investigated the implications of the variation of k with composition and of the deviation of dk/dp from $(\partial k/\partial p)_{\text{constant composition}}$.

In the lower 200 km of the mantle (region D″), the seismic velocity distributions of both Jeffreys and Gutenberg indicate that $d\phi/dr \simeq 0$, so that $\eta \simeq dk/dp$. Bullen's model A gives $dk/dp \simeq 3$ in D″ indicating that the lower 100–200 km of the mantle is inhomogeneous. The inhomogeneity is not too severe, however; with $\eta = 3$ it contributes only an increase of 0·2 g/cm³ in density through D″. Using more recent values of the seismic velocities (including a decrease in D″), Bolt (1972) obtained a value of 4·9 for η leading to an increase in density of 0·33 g/cm³ through D″.

In the transition region F between the outer and inner core, Jeffreys' velocity distribution (characterized by a large negative P velocity gradient, i.e. $d\phi/dr \gg 0$) leads to a value of η of 38, entailing a density increase of the order of 3 g/cm³ through F. On the other hand Gutenberg's velocity distribution gives large negative values of $d\phi/dr$ so that η is significantly less than unity (actually negative), implying an unstable distribution of mass. It would seem that seismic velocity gradients much in excess of those in regions D and E cannot exist in the earth's deep interior. An infinite gradient (i.e. a velocity discontinuity) on the other hand is not impossible since the range of depth of any instability would then be zero.

In Bolt's (1962, 1964) revision of Jeffreys' distribution of the velocity of P waves in the deep interior of the earth, the core is divided into four distinct regions E′, E″, F and G. The velocity distribution down to the bottom of E′ is the same as that of Jeffreys for corresponding depths inside E. There are discontinuous jumps in V_P at the E″–F and F–G boundaries and V_P is constant in E″, F and G (see Fig. 10.7). Birch (1963), using shock wave data at pressures of the order of 10^6 bar, inferred that the density ρ_0 at the centre of the earth does

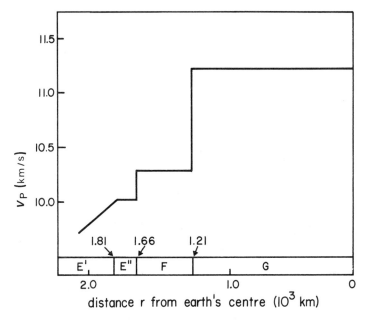

Fig. 10.7 P velocities and the regions E′, E″, F and G of the lower core (after Bullen (1965b))

not exceed 13 g/cm³. Bullen (1965b) investigated the consequences of this limiting value using Bolt's distribution of V_P in the core (Fig. 10.7). Since ρ is not likely to decrease with depth in the core, Bullen's (k, p) hypothesis implies, through equation (3.12), that departures from smooth variations of V_P with r are accompanied by similar departures in the variation of μ rather than of k. Bullen showed that it is impossible for ρ_0 to be as low as 13 unless there is substantial rigidity in both regions F and G. In addition it is essential that $d\mu/dr > 0$ over a significant range of depth in the lower core, i.e., the rigidity must *decrease* with increase in depth. Bullen found that ρ_0 could be as low as 12·6 g/cm³ if suitable assumptions (compatible with the seismic data) are made on the variations of k and μ in the lower core. The value of ρ_0 could actually be reduced to

12·3 g/cm^3 if E″ were chemically homogeneous—which is, however, rather improbable. If F and G are both fluid (i.e. complete absence of rigidity), ρ_0 must be at least 14·7 g/cm^3. This value is sufficiently in excess of 13 g/cm^3 to give additional support to the conclusion that the inner core is solid. Using a more recent P velocity distribution in the central core (due to Qamar (1971)), Bolt (1972) has shown that ρ_0 probably lies between 13·0 and 14·0 g/cm^3, the lower value being more compatible with shock wave data on iron. His preferred model is $\rho_0 = 13\cdot0$ g/cm^3, $k_0 = 15\cdot05 \times 10^{12}$ dyne/cm^2, and $\mu_0 = 1\cdot25 \times 10^{12}$ dyne/cm^2. The corresponding seismic velocities at the centre are $V_{P0} = 11\cdot35$ km/s and $V_{S0} = 3\cdot10$ km/s. None of Bolt's models showed a decrease of μ with depth in the inner core as was found by Bullen (1965b).

A pure iron-nickel core has too large a density and too small a bulk sound speed to be compatible with geophysical data (shock wave experiments by McQueen and Marsh (1966) on iron-nickel alloys indicate that pure iron is about 8 per cent denser than the outer core). A lighter alloying element that would increase the bulk sound speed is required. Two elements, silicon and sulphur have been proposed; both would have the desired properties of decreasing the density and increasing the seismic velocity of pure iron.

Silicon has long been the prime candidate for this element. Ringwood (1966) has summarized the geochemical arguments for silicon in the core. He favours a hot origin of the earth and believes that most of the sulphur escaped from the earth as volatiles. The inclusion of FeS in models of the early earth has been largely ignored in the past. However, Murthy and Hall (1970) have recently considered this possibility and come to a number of conclusions. They argue that the first melt formed during the early history of the earth was an FeS–Fe eutectic, which extracted chalcophilic elements from the lithosphere into the core. The amount of sulphur likely to be present in the core (they estimate it to be 15 per cent) should suffice to reduce the core density to well below that of any Fe–Ni alloy, bringing it into agreement with the estimates of density inferred from seismology and shock-wave experiments. Murthy and Hall also stress that the conditions under which sulphur and silicon go into the core are radically different. The presence of silicon in the core requires large-scale high temperature reduction, and the formation of a massive carbon monoxide atmosphere which would be blown off after accretion carrying with it other volatiles (Ringwood (1966)). On the other hand, even a relatively cold planet ($\sim$ 1000°C) could produce a sulphur-rich iron core.

Murthy and Hall compared the relative abundances of such volatile elements as the halogens, rare gases, sulphur and water in the crust and mantle with chondritic abundances. They found that sulphur was depleted in the crust and mantle by several orders of magnitude relative to other volatile elements. They concluded, on several geochemical grounds, that sulphur was more likely to be incorporated into the earth than not, and since it was deficient in the crust and mantle, it had probably been segregated into the core as a sulphur-rich iron melt.

As a direct consequence of the remarkable stability of alkali and alkaline metal sulphides, Lewis (1971) believes that one result of a differentiation history of the earth involving early FeS melting would be a concentration of K, Rb and Cs in the core. The degree of chemical reduction required to form a sulphide melt containing appreciable amounts of the alkali and alkaline metal sulphides is far less severe than that required to reduce silicon to the element. Thus it is far easier to make a case for sulphur (and hence potassium) in the core than to defend reduced silicon as the low density element in the core (Murthy and Hall (1970)). Hall and Murthy (1971) have also shown that during core formation it is possible to separate K from U and Th. Differential segregation of K into the core, leaving behind the initial complement of U and Th in the mantle and crust would also be consistent with the low K/U ratio of terrestrial rocks ($\simeq 1 \times 10^4$) relative to chondritic meteorites ($\simeq 7 \times 10^4$). From three-quarters to seven-eighths of the radio nuclide K^{40} on earth may be located in the core (Hurley (1968)), and thus a major source of heat, roughly 10^{20} erg/s may exist within the core. This is more than sufficient to sustain convection in the core and an abundant supply of energy should be available for core motions to maintain the earth's magnetic field, even if the efficiency of conversion is quite small (see § 5.6).

Oversby and Ringwood (1972) have examined the data on the distribution of potassium in meteorites and carried out experiments on the potassium distribution between silicate and metal-sulphide phases; they came to the conclusion that there can be no more than 1 or 2 per cent of the earth's potassium in an Fe–FeS core. Goettel and Lewis (1973) have disputed this and Oversby and Ringwood have replied to their criticisms. In view of the extremely important consequences of potassium in the core and the disagreements in the interpretation of existing data, the whole question warrants further investigation.

There has been much speculation on the origin and evolution of the earth's core, which is probably bound up with the origin of the earth itself. It is a common assumption in most theories of the origin of the earth that the proto-earth was homogeneous, and that the present differentiation into a core, mantle and crust occurred later. There are, however, as Ringwood (1966b) has pointed out, a number of difficulties with such a model, and there have been some qualitative considerations in recent years of a non-homogeneous accretion of the earth and planets.

Earlier views on the history of the solar system favoured a hot origin for the earth; these ideas gave way later to theories involving low temperature accretion of particles of roughly chondritic composition. Accretion was comparatively slow (~ 100 m yr) in order that the gravitational potential energy of the incoming particles could be largely radiated away and not raise the temperature of the earth significantly. Heating of the earth then takes place by long-lived radioactive isotopes giving rise to differentiation and formation of the core.

Elsasser (1963) has described in some detail this type of model of an original earth accreted cold with the material uniformly distributed. The main feature of such an earth model is that the melting point curve of the silicates rises much more steeply with depth than the actual temperature. This implies that the viscosity of the silicates should increase appreciably with increasing depth. As the original earth is heated by radioactivity the outer layers are then the first to become soft enough to permit iron to sink towards the centre; farther down the fall of iron is slowed by increased viscosity. It then forms a coherent layer which, however, is gravitationally unstable and results in the formation of quite large drops. The latter fall rapidly to the centre giving rise to a proto-core. Elsasser estimated that the formation of a proto-core slightly smaller than the present core probably took no more than several hundred m yr.

Tozer (1965) has reconsidered the question of the kinetics of core formation and concluded that the simple theory of falling iron masses in silicate material is untenable. He suggested as an alternative a mechanism based on the flow of iron along channels in the silicate phase. The acceptability of this theory depends quite critically on whether iron is able to flow over distances of the order of a km under such conditions. Tozer concluded that core formation, if still continuing, is at a much slower rate than in the past and was virtually complete very early in the earth's history. Levin (1972) has pointed out two further difficulties with such theories of core formation. It is difficult to explain from where the earth accreted so much metallic iron, or to explain the nature of the reduction process which had to operate within the earth if it accreted only oxidized iron. Secondly, the potential gravitational energy liberated when the metallic iron reached the core would be sufficient to increase the temperature of the whole earth by about 2000°C. This, together with radiogenic heat, would cause the present-day mantle to be entirely molten, unless the excess heat could be removed by convection.

During the last few years, a number of workers have returned to the idea of a hot origin for the earth. Ringwood (1960) proposed that the earth formed directly by accretion from the primitive oxidized dust in the solar nebula, and that reduction to metal, loss of volatiles, melting and differentiation occurred simultaneously and as a direct result of the primary accretion process. During the later stages of accretion, when the melting point of the surface was exceeded, metallic iron segregated into masses large enough to flow directly into the core. Gravitational energy liberated during the formation of the core would also contribute heat sufficient to result in complete melting throughout. On this model segregation of the core occurred as a continuous process during primary accretion. A catastrophic version of core formation is also a possibility (see § 7.3). Hanks and Anderson (1969), in a study of the thermal history of the earth, also concluded that the earth had a hot origin and that large-scale differentiation within the earth leading to core formation took place during an accretion period lasting 0·5 m yr or less (see § 7.5).

Wetherill (1972) has pointed out that measurements of the concentration of Rb and Sr and the Sr isotopic composition of lunar samples indicate that a major fractionation of Rb relative to Sr took place on the moon 4600 m yr ago. Similar studies of the highly differentiated lunar breccias provide additional evidence for a very early lunar differentiation. Wetherill concluded that the most likely heat source for this initial differentiation is the gravitational energy of lunar accretion and that, for a sufficiently high temperature to be reached during the accretion period, the time-scale for the accretion must be of the order of 1000 yr.

300

By analogy with the moon, Wetherill suggests that the entire earth was initially melted and fractionated, and that the age of the oldest rocks ($\simeq 3500$ m yr) indicates that it was not until then that the earth cooled sufficiently to enable the formation of extensive areas of stable crust.

This time-scale for the accretion of the earth (~ 1000 yr) is very different from that proposed by Safronov (1965, 1969) and others, namely about 100 m yr (see Levin (1972)). However, the process of accretion of the earth is very complex and imperfectly understood. The rate of mass growth would initially increase, because of the increase of cross-section of the embryo earth, pass through a maximum, and then decrease because of the exhaustion of accretable material. This characteristic of the accumulation process determines the length of the accumulation period and the amount of heating of the earth through the release of the gravitational energy of accretion.

Oversby and Ringwood (1971) have also produced evidence for the very early formation of the earth's core. The age of the earth (4550 m yr) is based on the discovery that the isotopic composition of modern terrestrial leads falls on the meteoritic isochron, which records a differentiation of lead relative to uranium at that time. It is assumed that the Pb/U ratio of the upper mantle and crust has remained essentially constant since that time. Ringwood had argued in 1960 that this assumption is probably not valid since iron descending during core formation would take with it substantial amounts of lead but not uranium, i.e. core formation would change the Pb/U ratio in the upper crust and mantle. He thus concluded that the 4550 m yr terrestrial event recorded by Pb/U geochronology referred to the time of core formation which must have taken place very soon after the formation of the earth. The main argument for Ringwood's conclusion is that during core formation the Pb/U ratio of the metal phase would be higher than that in silicates; the opposite point of view was taken by Patterson and Tatsumoto (1964). Oversby and Ringwood (1971) thus carried out experimental measurements of the distribution coefficient of lead between relevant metal and silicate systems to settle this question. Their results indicated that lead always ends up preferentially in the metal phase, and they deduced that the time interval between accretion and core formation was about 20 m yr—core formation could well have occurred simultaneously with accretion. These results strongly favour models of the earth in which much of the accretion occurred under sustained high temperature conditions, as against models in which the earth accreted in a cool, unmelted state and was heated later by the decay of long lived radioactive isotopes.

Orowan (1969) was one of the first to suggest that the earth may have accreted inhomogeneously. He pointed out that iron is plastic ductile, even at low temperatures, provided that it does not contain far more carbon than is found in meteorites. As a result, metallic particles would be expected to stick together when they collide because they can absorb kinetic energy by plastic deformation. They can therefore combine by cold or hot welding. Silicates, on the other hand, are brittle and break up on collision except within a narrow temperature range near their melting point. The accretion of the planets may thus have started with metallic particles. Once sufficiently large, a body could easily collect non-metallic particles by embedding them in ductile metal, and later by gravitational attraction. Orowan thus suggested that planets may arise cold in this way with a metal core already partially differentiated, and that subsequent melting could produce a sharp boundary between core and mantle.

Turekian and Clark (1969) also proposed a model of the planets stratified initially due to inhomogeneous accumulation of the elements. As the primitive solar nebula cooled, elements and compounds would condense in the order of increasing vapour pressure. Larimer (1967) calculated that the order of condensation would be iron and nickel, magnesium and iron silicates, alkali silicates, metals such as Ag, Ga, Cu, iron sulphide and finally metals such as Hg, Tl, Pb, In and Bi. Such an order of condensation is that grossly inferred in the earth, and usually attributed to differentiation. Turekian and Clark thus suggested that the earth's core formed by accumulation of the condensed iron-nickel in the vicinity of its orbit which, as in Orowan's model, became the nucleus upon which the silicate mantle was deposited. It must be pointed out, however, that according to Blander and Katz (1967) and Blander and Abdel-Gawad (1969), at the pressures that existed in the solar nebula, the silicates would have condensed before iron. Larimer has also revised his earlier conclusions and believes that iron condensed slightly later than the silicates (Anders (1968); Larimer and Anders (1970)).

The iron concentration of the terrestrial planets decreases monotonically with distance from the sun. Al'tshuler (1972) takes this as evidence that large-scale differentiation of the elements existed in the protoplanetary cloud at the time of formation of the solar system. Such an enrichment of the central zone of the cloud could be due, as suggested by Harris and Tozer (1967), to a slower dissipation of iron-nickel particles as a result of their cohesion and growth in the course of magnetic interactions.

An important boundary condition for the formation of the earth is that the outer core at present (and probably 3000 m yr ago) be molten. With the Turekian-Clark model there does not appear to be enough energy available to melt the outer core either during or after accretion; it is even more difficult to do so for a cold accretion model (Hanks and Anderson (1969)). This difficulty could be overcome if the lighter component of the core is sulphur. The eutectic temperature for the system Fe–FeS is 990°C and appears to be remarkably insensitive to pressure up to at least 30 kbar (Brett and Bell (1969)). In an earth of meteoritic composition a sulphur-rich iron liquid is the first melt to be formed. Core formation could proceed under these conditions at a temperature some 600°C lower than would be required to initiate melting in pure iron. In the vicinity of the core, the eutectic temperature is some 1600°C lower than that for pure iron.

Anderson and Hanks (1972) have recently reconsidered the inhomogeneous accretion model and concluded that, after all, it could account for the early melting of the core. They examined the condensation sequence in more detail— assuming the earth is fully accreted and solid, the zonation is as follows: a uranium-thorium rich central nucleus, composed primarily of Ca, Al and Ti rich silicates;

a shell of iron-nickel containing some of the earlier condensates which had not fully condensed or accreted when the iron and nickel condensed; a shell of less refractory silicates, mainly pyroxene and olivine; a shell of potassium and sodium rich silicates; and finally a shell of hydrated minerals and the volatile rich condensates. The proto-core is composed of the refractory nucleus and the Fe–Ni shell. The nucleus heats up rapidly because of its high U content and starts to melt the metal shell. The core is gravitationally unstable and when melting has progressed sufficiently it will overturn, plastering the refractory nucleus to the base of the solid mantle. This is unlikely to be a symmetric process and the nucleus will be plastered preferentially at the base of one hemisphere of the mantle. Anderson and Hanks estimate that the density of the nucleus at the MCB would be less than that of 'normal' mantle material at the boundary. This, together with its high radioactive content, may cause the nucleus to melt its way through the mantle and initiate continental growth in one hemisphere some 3800 m yr ago (Hurley and Rand (1969)). The anorthosite event 1300 m yr ago (see § 2.4) may also be related to the rise of the nucleus. Both events are unique happenings in the history of the earth and have been attributed to core formation or violent lunar related events.

The model of Anderson and Hanks could also account for possible undulations on the core-mantle boundary which have been suggested as a possible explanation of certain features of the earth's magnetic and gravity fields (see § 5.4)— pieces of the original nucleus which have been plastered against the top of the core could possibly depress the interface. If their model of the core is correct, it is necessary to reconsider the question of the nature of the lighter alloying material in the core; neither Si nor S would then be favoured. Likely candidates would be Ca–Al rich oxides and silicates—the high radioactivity of such material may also provide part of the energy for driving the geomagnetic dynamo. However, this raises the further question as to what extent such materials are soluble in molten iron at high temperature and pressure. If the Ca–Al oxides and silicates are not sufficiently soluble, perhaps there could be an emulsion or suspension of iron and silicates or oxides. Again, the presence of Ca–Al–U rich material at the base of the mantle may be the source of deep mantle convection plumes and hot spots (see § 8.6). There are still many unanswered questions concerning the evolution of the core but the recent work of Anderson and Hanks has given more evidence in favour of non-homogeneous accretion of the earth, and in addition suggested possible solutions to a number of other geophysical problems.

References

Ahrens, T. J., D. L. Anderson and A. E. Ringwood (1969) Equations of state and crystal structures of high pressure phases of shocked silicates and oxides, *Rev. Geophys.*, **7**, 667.

Ahrens, T. J. (1972) The state of mantle minerals, *Tectonophysics*, **13**, 189.

Akimoto, S. and Y. Ida (1966) High pressure synthesis of Mg_2SiO_4 spinel, *Earth Planet. Sci. Letters*, **1**, 358.

Al'tshuler, L. V. (1972) Composition and state of matter in the deep interior of the earth, *Phys. Earth. Planet. Int.*, **5**, 295.

Anders, E. (1968) Chemical processes in the early solar system as inferred from iron meteorites, *Acc. chem. Res.*, **1**, 289.

Anderson, D. L. (1966) Recent evidence concerning the structure and composition of the earth's mantle, in: *Physics and Chemistry of the Earth*, VI (Pergamon Press).

Anderson, D. L. (1967) Latest information from seismic observations, in: *The Earth's Mantle* (ed. T. F. Gaskell) (Acad. Press).

Anderson, D. L. (1970) Petrology of the mantle, in: *The Mineralogy and Petrology of the Upper Mantle* (ed. B. A. Morgan) (Min. Soc. Amer. Spec. Paper 3).

Anderson, D. L. and C. G. Sammis (1970) Partial melting in the upper mantle, *Phys. Earth Planet. Int.*, **3**, 41.

Anderson, D. L. and T. Jordan (1970) The composition of the lower mantle, *Phys. Earth Planet. Int.*, **3**, 23.

Anderson, D. L. and H. Spetzler (1970) Partial melting and the low-velocity zone, *Phys. Earth Planet. Int.*, **4**, 62.

Anderson, D. L., C. G. Sammis and T. Jordan (1971) Composition and evolution of the mantle and core, *Science*, **171**, 1103.

Anderson, D. L. and T. C. Hanks (1972) Formation of the earth's core, *Nature*, **237**, 387.

Birch, F. (1952) Elasticity and constitution of the earth's interior, *J. geophys. Res.*, **57**, 227.

Birch, F. (1961) Composition of the earth's mantle, *Geophys. J.*, **4**, 295.

Birch, F. (1963) Some geophysical applications of high pressure research, in: *Solids Under Pressure* (ed. W. Paul and D. M. Warschauer) (McGraw-Hill).

Birch, F. (1968) On the possibility of large changes in the earth's volume, *Phys. Earth Planet. Int.*, **1**, 141.

Blander, M. and J. L. Katz (1967) Condensation of primordial dust, *Geochim. cosmochim. Acta*, **31**, 1025.

Blander, M. and M. Abdel-Gawad (1969) The origin of meteorites and the constrained equilibrium condensation theory, *Geochim. cosmochim. Acta*, **33**, 701.

Bolt, B. A. (1962) Gutenberg's early PKP observations, *Nature*, **196**, 122.

Bolt, B. A. (1964) The velocity of seismic waves near the earth's core, *Bull. seism. Soc. Amer.*, **54**, 191.

Bolt, B. A. (1970) PdP and PKiKP waves and diffracted PcP waves, *Geophys. J.*, **20**, 367.

Bolt, B. A. (1972) The density distribution near the base of the mantle and near the earth's centre, *Phys. Earth Planet. Int.*, **5**, 301.

Brett, R. and P. M. Bell (1969) Melting relations in the Fe-rich portion of the system Fe–FeS at 30 kb pressure, *Earth Planet. Sci. Letters*, **6**, 479.

Bullard, E. C. and D. T. Griggs (1961) The nature of the Mohorovičić discontinuity, *Geophys. J.*, **6**, 118.

Bullen, K. E. (1946) A hypothesis of compressibility at pressures of the order of a million atmospheres, *Nature*, **157**, 405.

Bullen, K. E. (1949) Compressibility–pressure hypothesis and the earth's interior, *Mon. Not. Roy. Astr. Soc. geophys. Suppl.*, **5**, 355.

Bullen, K. E. (1950) An earth model based on a compressibility–pressure hypothesis, *Mon. Not. Roy. Astr. Soc. geophys. Suppl.*, **6**, 50.

Bullen, K. E. (1963) An index of degree of chemical inhomogeneity in the earth, *Geophys. J.*, **7**, 584.

Bullen, K. E. (1965a) On compressibility and chemical inhomogeneity in the earth's core, *Geophys. J.*, **9**, 195.

Bullen, K. E. (1965b) Models for the density and elasticity of the earth's lower core, *Geophys. J.*, **9**, 233.

Bullen, K. E. and R. A. W. Haddon (1967) Earth models based on compressibility theory, *Phys. Earth Planet. Int.*, **1**, 1.

Chinnery, M. A. and M. N. Toksöz (1967) P wave velocities in the mantle below 700 km, *Bull. seism. Soc. Amer.*, **57**, 199.

Clark, S. P. Jr. and A. E. Ringwood (1964) Density distribution and constitution of the mantle, *Rev. Geophys.*, **2**, 35.

Cohen, L. H., K. Ito and G. C. Kennedy (1967) Melting and phase relations in an anhydrous basalt to 40 kilobars, *Amer. J. Sci.*, **265**, 475.

Davies, D. and R. M. Sheppard (1972) Lateral heterogeneity in the earth's mantle, *Nature*, **239**, 318.

Derr, J. S. (1969a) Internal structure of the earth inferred from free oscillations, *J. geophys. Res*, **74**, 5202.

Derr, J. S. (1969b) Free oscillation observations through 1968, *Bull. seism. Soc. Amer.*, **59**, 2079.

Dziewonski, A. M. and F. Gilbert (1971) Solidity of the inner core of the earth inferred from normal mode observations, *Nature*, **234**, 465.

Dziewonski, A. M. and F. Gilbert (1972) Observations of normal modes from 84 recordings of the Alaska earthquake of 1964 March 28, *Geophys. J.*, **27**, 393.

Elsasser, W. M. (1963) Early history of the earth, in: *Earth Science and Meteoritics* (ed. J. Geiss and E. D. Goldberg) (North Holland Publ. Co.).

Ergin, K. (1967) Seismic evidence for a new layered structure of the earth's core, *J. geophys. Res.*, **72**, 3669.

Goettel, K. A. and J. S. Lewis (1973) Comments on a paper by V. M. Oversby and A. E. Ringwood (1972) (*Earth Planet. Sci. Letters*, **14**, 345), *Earth Planet. Sci. Letters*, **18**, 148.

Green, D. H. and A. E. Ringwood (1967) An experimental investigation of the gabbro to eclogite transformation and its petrological applications, *Geochim. cosmochim. Acta* **31**, 767.

Green, D. H. and A. E. Ringwood (1972) A comparison of recent experimental data on the gabbro–garnet granulite–eclogite transition, *J. Geol.*, **80**, 277.

Haddon, R. A. W. and K. E. Bullen (1969) An earth incorporating free earth oscillation data, *Phys. Earth Planet. Int.*, **2**, 35.

Hales, A. L. and J. L. Roberts (1970a) The travel times of S and SKS, *Bull. seism. Soc. Amer.*, **60**, 461.

Hales, A. L. and J. L. Roberts (1970b) Shear velocities in the lower mantle and the radius of the core, *Bull. seism. Soc. Amer.*, **60**, 1427.

Hall, H. T. and V. Rama Murthy (1971) The early chemical history of the earth: some critical elemental fractionations, *Earth Planet. Sci. Letters*, **11**, 239.

Hanks, T. C. and D. L. Anderson (1969) The early thermal history of the earth, *Phys. Earth Planet. Int.*, **2**, 19.

Harris, P. G. and D. C. Tozer (1967) Fractionation of iron in the solar system, *Nature*, **215**, 1449.

Hess, H. H. (1955) The oceanic crust, *J. Marine Res.*, **14**, 423.

Hess, H. H. (1962) History of the ocean basins, in: *Petrologic Studies, a volume to honor A. F. Buddington* (ed. A. E. J. Engel *et al.*) (Geol. Soc. Amer.).

Hurley, P. M. (1968) Corrections to: Absolute abundance and distribution of Rb, K and Sr in the earth, *Geochim. cosmochim. Acta*, **32**, 1025.

Hurley, P. M. and J. R. Rand (1969) Pre-drift continental nuclei, *Science*, **164**, 1229.

Ito, K. and G. C. Kennedy (1968) Melting and phase relations in the plane tholeiite–lherzolite–nepheline basanite to 40 kilobars with geological implications, *Contr. Min. Petrol.*, **19**, 177.

Ito, K. and G. C. Kennedy (1970) The fine structure of the basalt–eclogite transition, *Min. Soc. Amer. Spec. Paper*, **3**, 77.

Ito, K. and G. C. Kennedy (1971) *An experimental study of the basalt–garnet granulite–eclogite transition*, (ONR–CIRES Symposium volume).

Jeffreys, H. (1937) On the materials and density of the earth's crust, *Mon. Not. Roy. Astr. Soc. geophys. Suppl.*, **4**, 50.

Jeffreys, H. (1939) The times of P, S, and SKS and the velocities of P and S, *Mon. Not. Roy. Astr. Soc. geophys. Suppl.*, **4**, 498.

Johnson, L. R. (1967) Array measurements of P velocities in the upper mantle, *J. geophys. Res.*, **72**, 6309.

Johnson, L. R. (1969) Array measurements of P velocities in the lower mantle, *Bull. seism. Soc. Amer.*, **59**, 973.

Julian, B. R., D. Davies and R. M. Sheppard (1972) PKJKP, *Nature*, **235**, 317.

Julian, B. R. and M. K. Sengupta (1973) Seismic travel time evidence for lateral inhomogeneity in the deep mantle, *Nature*, **242**, 443.

Kawai, N. (1966) A static high pressure apparatus with tapering multipistons forming a sphere—I, *Proc. Japan Acad.*, **42**, 385.

Kennedy, G. C. (1959) The origin of continents, mountain ranges and ocean basins, *Amer. Sci.*, **47**, 491.

Lambert, I. B. and P. J. Wyllie (1970) Melting in the deep crust and upper mantle and the nature of the low velocity layer, *Phys. Earth Planet. Int.*, **3**, 316.

Larimer, J. W. (1967) Chemical fractionations in meteorites I. Condensations of the elements, *Geochim. cosmochim. Acta*, **31**, 1215.

Larimer, J. W. and E. Anders (1970) Chemical fractionations in meteorites, III. Major element fractionation in chondrites, *Geochim. cosmochim. Acta*, **34**, 367.

Lewis, J. S. (1971) Consequences of the presence of sulfur in the core of the earth, *Earth Planet. Sci. Letters*, **11**, 130.

Lewis, J. S. (1972) Metal/silicate fractionation in the solar system, *Earth Planet. Sci. Letters*, **15**, 286.

Levin, B. J. (1972) Origin of the earth, *Tectonophysics*, **13**, 7.

McQueen, R. G. and S. P. Marsh (1966) Shock-wave compression of iron–nickel alloys and the earth's core, *J. geophys. Res.*, **71**, 1751.

Morgan, W. J. (1971) Convection plumes in the lower mantle, *Nature*, **230**, 42.

Murthy, V. Rama and H. T. Hall (1970) The chemical composition of the earth's core: possibility of sulphur in the core, *Phys. Earth Planet. Int.*, **2**, 276.

Nuttli, O. W. (1969) Travel times and amplitudes of S waves from nuclear explosions in Nevada, *Bull. seism. Soc. Amer.*, **59**, 385.

Orowan, E. (1969) Density of the moon and nucleation of planets, *Nature*, **222**, 867.

Oversby, V. M. and A. E. Ringwood (1971) Time of formation of the earth's core, *Nature*, **234**, 463.

Oversby, V. M. and A. E. Ringwood (1972) Potassium distribution between metal and silicate and its bearing on the occurrence of potassium in the earth's core, *Earth Planet. Sci. Letters*, **14**, 345.

Oversby, V. M. and A. E. Ringwood (1973) Reply to comment by K. A. Goettel and J. S. Lewis, *Earth Planet. Sci. Letters*, **18**, 151.

Patterson, C. and M. Tatsumoto (1964) The significance of lead isotopes in detrital feldspar with respect to chemical differentiation within the earth's mantle, *Geochim. cosmochim. Acta*, **28**, 1.

Qamar, A. (1971) *Sesimic wave velocity in the earth's core: a study of PKP and PKKP* (Ph.D. Thesis, Univ. Calif. Berkeley).

Ramsey, W. H. (1949) On the nature of the earth's core, *Mon. Not. Roy. Astr. Soc. geophys. Suppl.*, **5**, 409.

Ringwood, A. E. (1960) Some aspects of the thermal evolution of the earth, *Geochim. cosmochim. Acta.*, **20**, 241.

Ringwood, A. E. (1966a) The chemical composition and origin of the earth, in : *Advances in Earth Science* (ed. P. M. Hurley) (MIT Press)

Ringwood, A. E. (1966b) Chemical evolution of the terrestrial planets, *Geochim. cosmochim. Acta*, **30**, 41.

Ringwood, A. E. (1969) Phase transformations in the mantle, *Earth Planet. Sci. Letters*, **5**, 401.

306

Ringwood, A. E. (1970) Phase transformation and the constitution of the mantle, *Phys. Earth Planet. Int.*, **3**, 109.

Ringwood, A. E. (1972a) Mineralogy of the deep mantle: current status and future developments, in: *The Nature of the Solid Earth* (ed. E. C. Robertson) (McGraw-Hill).

Ringwood, A. E. (1972b) Phase transformations and mantle dynamics, *Earth Planet. Sci. Letters*, **14**, 233.

Ringwood, A. E. and D. H. Green (1966) An experimental investigation of the gabbro-eclogite transformation and some geophysical implications, *Tectonophysics*, **3**, 383.

Ringwood, A. E. and A. Major (1966) Synthesis of Mg_2SiO_4–Fe_2SiO_4 spinel solid solutions, *Earth Planet. Sci. Letters*, **1**, 241.

Ringwood, A. E. and A. Major (1968) Apparatus for phase transformation studies at high pressures and temperatures, *Phys. Earth Planet. Int.*, **1**, 164.

Ringwood, A. E. and A. Major (1970) The system Mg_2SiO_4–Fe_2SiO_4 at high pressures and temperatures, *Phys. Earth Planet. Int.*, **3**, 89.

Safranov, V. S. (1965) Sizes of the largest bodies fallen on planets in process of their formation, *Astron. Zh.*, **42**, 1270.

Safranov, V. S. (1969) *Evolution of the Preplanetary Cloud and the Formation of the Earth and Planets* (Nauka, Moscow).

Sato, R. and A. F. Espinosa (1967) Dissipation in the earth's mantle and rigidity and viscosity in the earth's core determined from waves multiply reflected from the mantle-core boundary, *Bull. seism Soc. Amer.*, **57**, 829.

Stishov, S. M. and S. V. Popova (1961) New dense polymorphic modification of silica, *Geokhimiya*, **10**, 837.

Taggart, J. and E. R. Engdahl (1968) Estimation of PcP travel times and depth to the core, *Bull. seism. Soc. Amer.*, **58**, 1293.

Toksöz, M. N., M. A. Chinnery and D. L. Anderson (1967) Inhomogeneities in the earth's mantle, *Geophys. J.*, **13**, 31.

Tozer, D. C. (1965) Thermal history of the earth I. The formation of the core, *Geophys. J.*, **9**, 95.

Turekian, K. K. and S. P. Clark Jr. (1969) Inhomogeneous accumulation of the earth from the primitive solar nebula, *Earth Planet. Sci. Letters*, **6**, 346.

Vinnik, L. P., A. A. Lukk and A. V. Nikolaev (1972) Inhomogeneities in the lower mantle, *Phys. Earth Planet. Int.*, **5**, 328.

Wetherill, G. W. (1961) Steady-state calculations bearing on geological implications of a phase transition Mohorovičić discontinuity, *J. geophys. Res.*, **66**, 2983.

Wetherill, G. W. (1972) The beginning of continental evolution, *Tectonophysics*, **13**, 31.

Wilson, J. T. (1965) Convection currents and continental drift, in: A Symposium on Continental Drift, *Phil. Trans. Roy. Soc.*, **A 258**, 145.

Wyllie, P. J. (1963) The nature of the Mohorovičić discontinuity, a compromise, *J. geophys. Res.*, **68**, 4611.

Appendix A

The Roche Limit

Consider two small spherical bodies, each of mass m_0 and radius r to be in contact in the tidal field of a large body of mass M, distant a from their point of contact (see Fig. A.1). The attraction between the two small bodies is $Gm_0^2/4r^2$. The

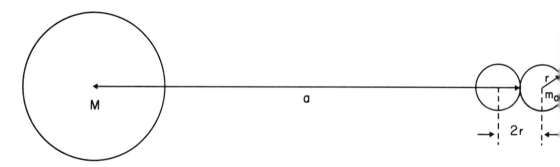

Fig. A.1 Tidal instability, as applied to the formation of planets

gravitational attraction of the large body upon the nearer body is $GMm_0/(a-r)^2$ and upon the more distant body $GMm_0/(a+r)^2$. The difference

$$GMm_0\left\{\frac{1}{(a-r)^2} - \frac{1}{(a+r)^2}\right\} \simeq GMm_0 \frac{4r}{a^3} \tag{A.1}$$

is the tidal disrupting force. The condensation will be stable if the mutual attraction of the two small bodies exceeds the disrupting force, i.e. if

$$\frac{Gm_0^2}{4r^2} > \frac{4GMm_0 r}{a^3}$$

i.e.

$$\frac{m_0}{r^3} > \frac{16M}{a^3}$$

If ρ is the density of the small bodies, $m_0 = \frac{4}{3}\pi\rho r^3$, and the above condition becomes

$$\rho > \left(\frac{3}{\pi}\right)\frac{4M}{a^3} \simeq \frac{4M}{a^3} \qquad (A.2)$$

If other forces are acting which tend to impede condensation (e.g. centrifugal forces due to rotation or magnetic forces), the numerical factor would be greater than 4.

The above tidal stability criterion is a necessary but not sufficient condition for the formation of a star—a hot gas will expand and not contract, unless the gravitating mass is very great. Let v and v_{escape} be the linear velocity and escape velocity of a mass m on a sphere of mass m_0 and radius R. The escape velocity can be found by equating the kinetic energy $\frac{1}{2}mv_{escape}^2$ to the work done against the gravitational field in going from the surface to infinity, i.e.

$$\frac{1}{2}mv_{escape}^2 = \int_R^\infty \frac{Gmm_0}{r^2}\,dr$$

$$= \frac{Gmm_0}{R}$$

or

$$v_{escape} = \sqrt{\left(\frac{2Gm_0}{R}\right)} \qquad (A.3)$$

In order that the condensation could retain the particle, $v < v_{escape}$; only a relatively cold gas would allow a condensation to grow.

Appendix B

Spherical Harmonic Analysis

B.1 Spherical harmonic functions

Fourier showed in 1807 that any function $f(t)$ which is defined in the interval $t=0$ to 2π and which satisfies certain conditions can be expressed as an infinite series of trigonometric functions, i.e.

$$f(t) = a_0 + \sum_{n=1}^{\infty} (a_n \cos nt + b_n \sin nt) \qquad (\text{B.1})$$

In geophysics a more practical problem is that of approximating a function $f(t)$ by a finite series, i.e. of finding a nearly equal function $f_k(t)$ of the form

$$f_k(t) = a_0 + \sum_{n=1}^{k} (a_n \cos nt + b_n \sin nt) \qquad (\text{B.2})$$

If the coefficients are chosen so that the square of the difference $f(t) - f_k(t)$ averaged over the interval $(0, 2\pi)$ is a minimum, it can be shown that

$$a_0 = \frac{1}{2\pi} \int_0^{2\pi} f(t) \, dt$$

$$a_n = \frac{1}{\pi} \int_0^{2\pi} f(t) \cos nt \, dt \qquad (\text{B.3})$$

$$b_n = \frac{1}{\pi} \int_0^{2\pi} f(t) \sin nt \, dt$$

It can also be shown that for each value of n ($n \leqslant k$), the coefficients a_n and b_n are independent of k. By adding additional terms to the series $f_k(t)$, the approximation is in general improved but it does not alter the values of the coefficients already obtained. This property is characteristic of the development of a function as a series of *orthogonal* functions.

310

A set of functions $f_0(t), f_1(t), f_2(t), \ldots$, which are defined in an interval $t_1 < t < t_2$, form an orthogonal system if

$$\int_{t_1}^{t_2} f_n(t) f_m(t) \, dt = 0 \qquad (n \neq m) \tag{B.4}$$

Further the function $f_n(t)$ is said to be normalized if

$$\int_{t_1}^{t_1} f_n^2(t) \, dt = 1$$

In deriving the expressions (B.3) use was made only of the orthogonality of the trigonometric functions.

An example of a non-orthogonal set of functions is the power series $1, x, x^2, \ldots$ $(-1 \leqslant x \leqslant 1)$. Thus if the function $f(x)$ is approximated by the series

$$f_k(x) = a_0 + \sum_{n=1}^{k} a_n x^n$$

and the coefficients so chosen as to minimize the square of the difference $f(x) - f_k(x)$ averaged over the interval $(-1, 1)$, different sets of coefficients will be obtained for different values of k. However, an orthogonal set of functions in the interval $(-1, 1)$ can be constructed by combining powers of x in the form of polynomials. Such a set is the Legendre function $P_n(x)$ defined by the equation

$$P_n(x) = \frac{1}{n! 2^n} \frac{d^n}{dx^n} (x^2 - 1)^n \tag{B.5}$$

Thus $P_0(x) = 1$, $P_1(x) = x$, $P_2(x) = \frac{1}{2}(3x^2 - 1), \ldots$. It is easy to show from equation (B.5), by repeated partial integration, that

$$\int_{-1}^{1} P_n(x) P_m(x) \, dx \begin{array}{l} = 0 \qquad \text{if } n \neq m \\[2ex] = \dfrac{2}{2n+1} \quad \text{if } n = m \end{array} \tag{B.6}$$

Thus the functions $P_n(x)$ are orthogonal although not normalized.

A spherical harmonic analysis may be regarded as a generalization of a Fourier analysis to three dimensions and gives an analytical representation of an arbitrary function of position on a sphere. By the substitutions (see Fig. B.1):

$$x = r \sin \theta \cos \phi$$
$$y = r \sin \theta \sin \phi \tag{B.7}$$
$$z = r \cos \theta$$

where $r > 0$, $0 \leqslant \theta \leqslant 180°$, $0 \leqslant \phi \leqslant 360°$, any function $f(x, y, z)$ can be expressed as a function $f(r, \theta, \phi)$. Thus any function of position on a sphere (radius a) is a function of θ and ϕ only.

Along any circle of latitude, defined by the polar distance θ_0, a function

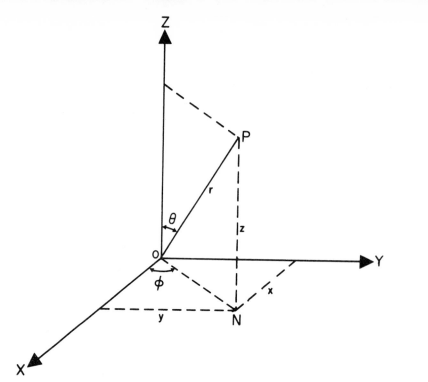

Fig. B.1

$f(\theta, \phi) = f(\theta_0, \phi)$ may be expressed by ordinary harmonic analysis as a trigono-metrical series of terms $a_m \cos m\phi$, $b_m \sin m\phi$. The coefficients (a_m, b_m) will be functions of θ_0. Dropping the suffix in θ_0, it follows that $f(\theta, \phi)$ can be represented by

$$f(\theta, \phi) = \sum_{m=0}^{\infty} \{a_m(\theta) \cos m\phi + b_m(\theta) \sin m\phi\} \tag{B.8}$$

The functions $a_m(\theta)$ and $b_m(\theta)$ may be expressed as orthogonal series of functions $P_n^m (\cos \theta)$ known as the associated Legendre functions defined by

$$P_n^m(\mu) = \frac{(1-\mu^2)^{1/2m}}{2^n n!} \frac{\mathrm{d}^{m+n}}{\mathrm{d}\mu^{m+n}}\{(\mu^2 - 1)^n\} \tag{B.9}$$

where $\mu = \cos \theta$. Since $P_n^m(\mu) = 0$ for $m > n$, we can write equation (B.8) for the expansion of $f(\theta, \phi)$ in the form

$$f(\theta, \phi) = \sum_{n=0}^{\infty} \sum_{m=0}^{n} S_n^m$$

$$= \sum_{n=0}^{\infty} \sum_{m=0}^{n} \{A_n^m \cos m\phi + B_n^m \sin m\phi\} P_n^m (\cos \theta) \tag{B.10}$$

S_n^m is called a surface spherical harmonic of order m and degree n. For $m=0$, f is independent of longitude and may be expressed as a series of zonal harmonics

$$S_n^0 = A_n^0 P_n^0 (\cos \theta)$$

312

where $P_n^0(\cos \theta) = P_n(\cos \theta)$, the ordinary Legendre function. For other values of m, $\sin m\phi$ or $\cos m\phi$ vanish along $2m$ meridians of longitude. The functions $P_n^m(\cos \theta)$ vanish along $(n-m)$ parallels of latitude, so that the surface of the sphere is divided by the nodal meridians and parallels into regions in which the solution is alternately positive and negative (see Fig. B.2). Larger values of m and n describe quantities of decreasing wavelength over the surface of the sphere.

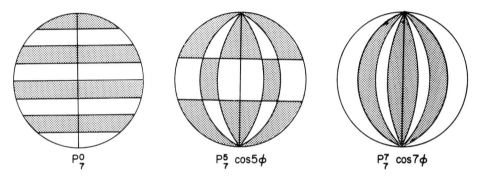

Fig. B.2   Equal area map of the zero lines over one hemisphere of a zonal harmonic P_7^0, a tesseral harmonic $P_7^5 \cos 5\phi$ and a sectorial harmonic $P_7^7 \cos 7\phi$. The central meridian corresponds to $\phi = 0$. The areas in which the sign of the function is positive are shaded

Further details of the properties of these functions may be found in any standard text on mathematical physics (see e.g. Jeffreys and Jeffreys (1962); Sneddon (1961)).

The numerical factors in the associated polynomials defined by equation (B.9) increase rapidly with m. In order that the coefficients in a harmonic analysis may have more physical significance with the terms they represent, various approximate normalizing factors have been introduced. A commonly favoured form is

$$ p_n^m = \left\{ \frac{(n-m)!}{(n+m)!} \right\}^{1/2} P_n^m \tag{B.11} $$

Other normalizing factors which have been used are $\{(n-m)!/n!\}$ (Jeffreys and Jeffreys (1962)) and $\{2(n-m)!/(n+m)!\}^{1/2}$ (Chapman and Bartels (1940)—an alternative which is widely used in geomagnetism).

B.2   The solution of Laplace's equation in spherical polar coordinates

The gravitational and magnetic fields of the earth as a whole can be analysed in terms of potentials of the fields which satisfy Laplace's equation at all external

313

points. In spherical polar coordinates (r, θ, ϕ), Laplace's equation for the potential V is

$$\nabla^2 V = \frac{1}{r^2} \frac{\partial}{\partial r}\left(r^2 \frac{\partial V}{\partial r}\right) + \frac{1}{r^2 \sin \theta} \frac{\partial}{\partial \theta}\left(\sin \theta \frac{\partial V}{\partial \theta}\right) + \frac{1}{r^2 \sin^2 \theta} \frac{\partial^2 V}{\partial \phi^2} = 0 \quad \text{(B.12)}$$

It may easily be verified that this equation has solutions of the form

$$V = (r^n, r^{-(n+1)})(\cos m\phi, \sin m\phi) P_n^m(\cos \theta) \quad \text{(B.13)}$$

where m and n are positive integers with $m \leqslant n$, and $P_n^m(\mu)$ satisfies the equation

$$(1-\mu^2) \frac{\mathrm{d}^2 P}{\mathrm{d}\mu^2} - 2\mu \frac{\mathrm{d}P}{\mathrm{d}\mu} + \left\{n(n+1) - \frac{m^2}{1-\mu^2}\right\} P = 0 \quad \text{(B.14)}$$

This is Legendre's associated equation, the solution of which, as can easily be verified, is given by equation (B.9). In geophysical problems it is more convenient to make the coefficients dimensionally uniform by relating r to the earth's radius a. The general expression for V as a sum of spherical harmonics is then

$$V = \frac{1}{a} \sum_{n=0}^{\infty} \sum_{m=0}^{n} P_n^m(\cos \theta) \left[\left\{c_n^m\left(\frac{r}{a}\right)^n + c_n'^m\left(\frac{a}{r}\right)^{n+1}\right\}\cos m\phi \right.$$

$$\left. + \left\{s_n^m\left(\frac{r}{a}\right)^n + s_n'^m\left(\frac{a}{r}\right)^{n+1}\right\}\sin m\phi\right] \quad \text{(B.15)}$$

where c_n^m and s_n^m are constant coefficients arising from sources external to the surface $r = a$, and $c_n'^m$ and $s_n'^m$ are coefficients arising from internal sources.

References

Chapman, S. and J. Bartels (1940) *Geomagnetism* (Clarendon Press, Oxford).
Jeffreys, H. and B. S. Jeffreys (1962) *Methods of Mathematical Physics* (Cambridge Univ. Press).
Sneddon, I. N. (1961) *Special Functions of Mathematical Physics and Chemistry* (Oliver and Boyd).

Appendix C

Derivation of Equation (3.35)

In a chemically homogeneous layer,

$$d\rho = \left(\frac{\partial \rho}{\partial p}\right)_T dp + \left(\frac{\partial \rho}{\partial T}\right)_p dT$$

$$= \frac{\rho}{k_T} dp - \rho\alpha \, dT \tag{C.1}$$

where k_T is the isothermal incompressibility and α the (volume) coefficient of thermal expansion. Using equation (3.29) this becomes

$$\frac{d\rho}{dr} = \frac{-g\rho^2}{k_T} - \rho\alpha \frac{dT}{dr} \tag{C.2}$$

The thermal gradient can be split into an adiabatic gradient and a non-adiabatic component τ (considered positive when the increase of temperature with depth exceeds the adiabatic). Since $(\partial T/\partial p)_S = T\alpha/\rho c_p$ where c_p is the specific heat at constant pressure, it follows that we can write

$$\frac{dT}{dr} = \frac{-Tg\alpha}{c_p} - \tau \tag{C.3}$$

Hence

$$\frac{d\rho}{dr} = \frac{-g\rho^2}{k_T} + \frac{T\rho\alpha^2 g}{c_p} + \alpha\rho\tau$$

Since

$$\frac{k_T}{k_S} = 1 - \frac{T\alpha^2 k_T}{\rho c_p}$$

this simplifies to

$$\frac{d\rho}{dr} = \frac{-g\rho^2}{k_S} + \alpha\rho\tau \tag{C.4}$$

$$= \frac{-g\rho}{\phi} + \alpha\rho\tau$$

Name Index

321

322

Subject Index

324

325